LUZERNE COUNTY
COMMUNITY COLLEGE
MAY 2000
LIBRARY

For a complete listing of *The Artech House Telecommunications Library* turn to the back of this book.

Introduction to Telecommunication Electronics

Second Edition

A. Michael Noll

Annenberg School for Communication
University of Southern California

Artech House
Boston • London

Library of Congress Cataloging-in-Publication Data
Noll, A. Michael
Introduction to telecommunication electronics / A. Michael Noll.–2nd ed.
p. cm.
Includes bibliographical references and index.
ISBN 0-89006-828-3 (alk. paper)
1. Telecommunication. 2. Electronics. I. Title.
TK5101.N78 1995 95-19091
621.382–dc20 CIP

British Library Cataloguing in Publication Data
Noll, A. Michael
Introduction to Telecommunication Electronics – 2Rev.ed
I. Title
621.382

ISBN 0-89006-828-3

685 Canton Street
Norwood, MA 02062

International Standard Book Number: 0-89006-828-3
Library of Congress Catalog Card Number: 95-19091

10 9 8 7 6 5

For the M's in my life: Mark, Martin, Marcella, and Marika

Contents

Preface to the First Edition

The material presented here is based on courses in understanding the fundamentals of communication technology taught by the author at the Interactive Telecommunications Program at New York University, and the Annenberg School for Communication at the University of Southern California.

These courses and this textbook are intended for persons without any technical background or experience. The book is intended as an introduction to the fundamental electrical and electronic concepts and principles that underlie modern communication systems and products. The objective of the material is to help nontechnical persons become literate in the terminology and concepts used by communication engineers in designing and developing communication systems and products. This literacy will result in improved communication between engineers and nontechnical persons.

The book starts with a variety of somewhat narrow topics which are necessary for the common background that is needed to explain and understand the broader principles and concepts of communication systems. Periodic and aperiodic signals, the time and frequency representation of a signal, spectrum and bandwidth, filters, electricity and Ohm's law, direct current and alternating current, and capacitance and inductance are some of the topics initially explained in the text. These initial topics are then broadened to include diodes and ac-to-dc conversion, amplification, vacuum tubes, transistors and integrated circuits, and decibels.

The book then moves on to communication systems concepts. The multiplexing together of a number of signals through frequency-division and time-division multiplexing is described. Also covered are amplitude modulation and frequency modulation with emphasis on their bandwidth and noise-immunity aspects. The process of converting an analog signal to a digital format is described along with such practical implementations as the use of analog-to-digital conversion in digital telephony and the digital audio compact disc.

The book winds up with digital computers and data communication. The interplay between computer hardware and software is described. Also discussed is the communication of data between computers including the use of devices called modems, which make data communication possible via analog transmission systems.

Simple mathematical formulations are developed and used where they help to explain fundamental concepts. Apprehension about technology is sometimes the result of a fear of mathematics, or "math anxiety," but the use of mathematics to explain physical phenomena can give a physical reality to mathematical concepts that otherwise might be too abstract to comprehend. So, it is hoped that some students will find the material in this book useful for becoming literate not only in technology but also in basic mathematics.

The material in this book is covered in a one-semester course at the graduate level for nontechnical students. Some material, such as power in reactive circuits, is optional, but the bulk of the material is treated in the course in the order in which it is explained in the book.

Preface to the Second Edition

Little has changed in the basic concepts central to modern telecommunication technologies and systems since the first edition of this book. However, improved ways to explain some of these concepts have been the natural result of the use of a book by the author in courses at the University of Southern California and at New York University. Some of the content in the first edition was outdated as well, particularly material dealing with computers and modems. Also, such topics as CD-ROM, error correction, quadrature amplitude modulation, and spread spectrum technologies now need adequate explanation. These reasons were the rationale for the second edition.

The creation of a second edition was an opportune time to improve and redraw all the many illustrations in the book. The drawings for the second edition were all drawn by the author on a Macintosh Quadra 650 computer using CA-Cricket® Draw III™ and printed on a Hewlett-Packard DeskWriter 520.

The material presented in this book has benefitted considerably from comments from the students taking the courses taught by the author both at the Annenberg School for Communication at the University of Southern California and at the Interactive Telecommunications Program at the Tisch School of the Arts at New York University. Students are a continued source of inspiration and enthusiasm, and their contributions have been considerable.

The first edition of the book was written at the Annenberg School, and the enthusiastic support of the School's then dean Dr. Peter Clarke is most appreciated. The second edition was written during a sabbatical from the University of Southern California, and the positive support of USC's Provost Dr. Lloyd Armstrong, Jr. is likewise most appreciated. Jeff Berg of NYU showed great patience and was invaluable in helping me learn the intricacies of Macintosh computers and programs.

Introduction

Modern communication technology and systems are essential to communication over distance, to the broadcasting of information on a mass or individual basis, and to innumerable forms of entertainment. Without communication technology and systems, the communication industry would lose most of its content and conduits.

Communication among people over distance is accomplished instantly in time by using modern telephone networks. These same telecommunications networks, when modified appropriately, facilitate the communication over distance between people and electronic computing machines that contain vast amounts of computerized data. Computers communicate with other computers over data networks that in some cases may span continents. New forms of communication systems enable groups of people located at distances from each other to conduct meetings. Forms of electronic mail enable people to send text-based messages instantly to each other.

Television networks broadcast their programs over satellite communication systems before the programs are finally broadcast locally over the air or over cable into our homes. Television and cinema material is also available on magnetic tape that is physically carried into our homes or recorded from our television sets. Computers and digital techniques are important tools in the production and processing of movies and other forms of visual entertainment.

The technology that has made possible all these new forms of communication contains components that communicate internally with each other. Electronic communication technology is very much at the root of the modern revolution in communications.

Electronic communication technology and systems are based on scientific principles and their engineering application. Advances in science, engineering, and technology have significantly influenced modern communication systems. Nearly everyone in business needs a basic understanding of the scientific and technological principles behind communication systems so that products and services can be designed with meaningful features in a timely fashion to meet the needs of the marketplace.

Marketeers, managers, and producers, to name only a few, should be conversant with technology so that they can better communicate their needs to engineers and better understand the ideas presented to them by engineers. Technology and an understanding of the principles of communication engineering can no longer be the exclusive province of engineers.

The required basic understanding of the principles behind modern communication technology and systems is not beyond the grasp of educated nontechnical people. These principles can be explained in an understandable fashion by using words and graphics devoid of the sophisticated mathematical analyses that are needed by engineers. The principles are basic to all communication systems so that an understanding of the basic principles is applicable to most segments of the communication industry. This book is intended to help develop this basic understanding of the scientific and technological principles behind modern electronic communication systems.

The following is a survey of the material presented in this book. It is easy to get lost in the details, and this survey is intended to present the grand view of our book's content and structure. We begin with the concept of a *signal.*

A signal is an event that varies with time. The concept of a signal is basic to communication and to modern communication technology and systems. Electrical signals travel over wires within digital computers and over long-distance telephone circuits. Acoustic signals propagate in the air in a room to create the sensation of sound. Light signals are created by turning on and off a light source, and are conveyed over thin transparent fibers. Signals can exist in many different forms and media, but all useful signals vary with time.

The electrical, acoustic, or other signal can be displayed as a function of time. The shape of this variation with time is called the *waveshape* (or waveform). There are an infinite variety of waveshapes, depending on the source of the signal and the characteristics of its associated wave. However, all waveshapes, regardless of how complex they may appear, can be decomposed into the combination of pure waves, called *sine waves.*

A sine wave is a smoothly varying wave, and if audible would sound like a pure tone. A sine wave consists of a fundamental shape that repeats itself, and the frequency of a sine wave is the rate at which the fundamental shape repeats itself. The different sine waves that comprise a complex waveshape can be plotted as a function of their frequency to produce a graph that is called the *frequency spectrum* of the waveshape or signal.

Most signals occupy a finite range of frequencies defined between a lower limit and an upper limit. This range or band of frequencies occupied by a signal is called the *bandwidth* of the signal. A communication channel or medium can pass only a range or band of frequencies, which is called the bandwidth of the channel. The bandwidth of a communications channel and the bandwidths of the various signals to be transmitted over it determine the numbers and types of signals that can be transmitted by a particular communication channel.

The concepts of spectrum and bandwidth are essential in characterizing signals and communication channels and media. The provision of bandwidth is costly, and so bandwidth is intimately related to the economics of communication systems.

Nearly all signals encountered in communication systems are electrical at one time or another. Two effects, *capacitance* and *inductance*, occur in electric circuits with time-varying signals. These two effects can occur on a stray basis and limit the bandwidth of electrical transmission circuits. The two effects can also be introduced deliberately into electrical circuits to shape and control the bandwidth of time-varying signals. Circuits that shape and control bandwidth are called *filters.*

Electricity is the flow of electrons in an electric circuit. The electrons are pumped around the circuit by an *electromotive force* (or electron moving force (EMF)) that is measured in a quantity called *volts.* The number of electrons flowing per second is the *electric current,* and it is measured in a quantity called *amperes.* Electric current can be constant with time, and such current is called *direct current,* or dc for short. Electric current can also vary with time in both amplitude and direction. Such varying current is called *alternating current,* or ac for short. The flow of electric current is opposed by *resistance* and the effects of capacitance and inductance. The opposition offered to the flow of ac by the effects of capacitance and inductance is called *reactance.* Reactance, unlike resistance, is dependent on frequency.

Most communication signals are processed as electrical signals, but, since many signals exist in forms other than electrical signals (for example, acoustic signals), we must find ways to convert signals from one form, or medium, into another. This conversion is accomplished by a class of devices called *transducers.* As examples, a microphone is a transducer that converts a sound wave into an electrical signal, and a loudspeaker is a transducer that converts an electrical signal into a sound wave.

Signals often are too small and need to be made larger through a process called *amplification.* The process of amplification is never perfect, and various distortions, noise, and bandwidth limitations are inadvertently introduced. The amount of amplification is measured in *decibels.*

Amplification is accomplished by such devices as the old-fashioned *electron tube* and the modern *transistor.* Hundreds and thousands of transistors are fabricated together on a single small device called an *integrated circuit,* which is known as a *chip* for short.

In many communication systems, a number of signals must all share the same medium. One way by which this sharing can be accomplished is to place each signal in its own band of frequencies within the total band of the medium. The combining of a number of signals to share a medium by dividing it into different frequency bands for each signal is called *frequency-division multiplexing.*

Frequency-division multiplexing requires the ability to move signals around within the frequency spectrum so that each multiplexed signal occupies a particular band. This frequency shifting is accomplished through a process called *modulation* in which a high-frequency sine wave "carries" the signal into the specified band. This sine wave is called the *carrier.* Either the amplitude or the frequency of the carrier wave can be varied, or modulated, in synchrony with the information-bearing signal. The respective methods are called *amplitude modulation* and *frequency modulation.*

The process of amplitude modulation is the technique used with AM radio. Amplitude modulation is a reasonably simple technology, and the bandwidth of the amplitude-modulated carrier is not more than twice the bandwidth of the modulating signal. However, an amplitude-modulated carrier signal is quite prone to the deleterious effects of additive noise.

The process of frequency modulation is the technique used with FM radio. Frequency modulation is a more complicated technology than amplitude modulation, and the bandwidth of the frequency-modulated carrier can be many times that of the modulating signal. However, the process of demodulating a frequency-modulated signal eliminates much of the deleterious effects of additive noise. This trade-off between bandwidth and noise immunity characterizes most communication systems.

Both amplitude modulation and frequency modulation are *analog* modulation schemes for multiplexing signals in the frequency spectrum. Another technique for combining together a number of signals is to sample each of them in time and then interweave all the various samples. This multiplexing technique is called *time-division multiplexing.*

Time-division multiplexing is normally used with *digital* signals. The process of digitizing a signal involves a number of steps and results in a binary digital signal that takes on one of two discrete values. This process results in considerable immunity to additive noise, but this benefit comes at the expense of a considerable increase in the bandwidth required for the binary digital signal.

The whole world appears to be going digital! Stereophonic audio has been forever revolutionized by the digital format of the compact disc. Telecommunication systems are increasingly using digital switching and such digital transmission media as optical fiber. In addition to noise immunity, the digital world offers the advantage of the readily available technology of digital processing by using digital computers.

A digital computer consists of electronic circuits that manipulate signals in the form of binary digits, called *bits.* The digital computer is a general-purpose machine that can be instructed to perform different functions in response to a set of instructions called a *program.* The circuits and physical devices of computers are known as *hardware,* while the term *software* refers to the program aspects. Any ready-to-use software applications built into the hardware of a particular computer are called *firmware.*

People need to communicate over distance with computers to gain access to the information stored within them. Also, computers frequently need to communicate over distance with other computers. All of this is accomplished by the communication of binary data over telephone lines by modulating a sine wave with the binary signal and then demodulating the sine wave at the receiving end. The two-way communication of data requires a modulator-demodulator, or *modem* for short, at each end.

This, then, is a summary of the basic electronic concepts central to the workings of modern communication systems and technology. The next step is the application of these basic principles to specific communication systems such as television, voice telephony, and data networks. This step should be much easier if the basic principles presented in our book are studied and understood.

The material presented in our book has been organized into a series of modular chapters, each covering a major area. Each chapter or module leads into another, gradually building in depth and complexity. The early modules help build familiarity with basic terminology and concepts. The later modules describe the operation of major systems and broader technologies.

The modules are shown by the following "map."

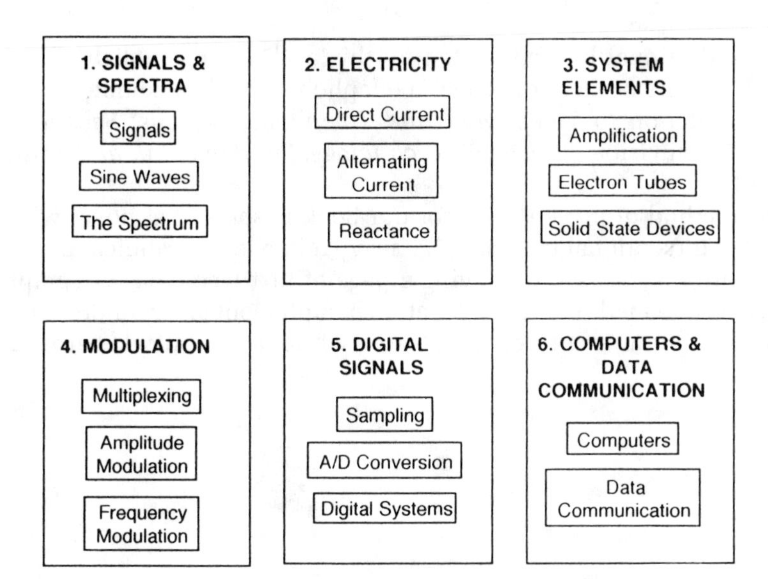

The map of the modules is reproduced on the title page of each chapter along with a detailed outline of the contents of the module. Each module begins with a short introductory summary of its overall contents.

The interrelationship among the topics presented in our book is truly two-dimensional. The following diagram depicts this interrelationship and the flow from one topic to another. The various topics presented in this guide are all relevant to understanding the technological workings of the systems shown in the right-hand portion of the diagram.

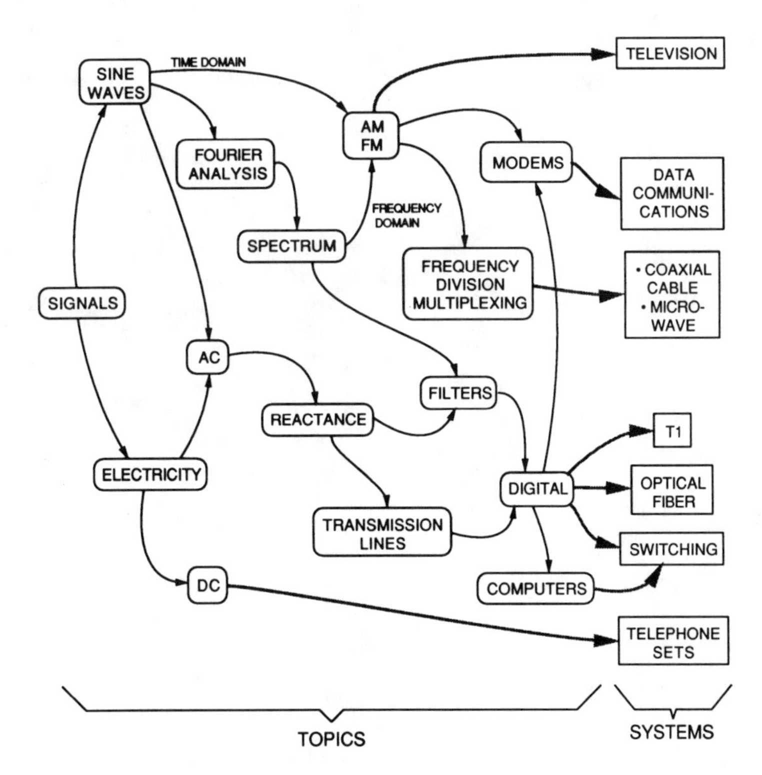

Two additional books, written by this author and published by Artech House, apply this basic knowledge to telephone and data communication systems (*Introduction to Telephones and Telephone Systems, Second Edition,* 1991) and to television (*Television Technology: Fundamentals and Future Prospects*, 1988).

The method of presentation for our book is somewhat novel when compared with conventional textbooks. A heavy reliance on graphical presentation coupled with text is used by having a page of text face a page of graphics. In this way, the two methods of presentation complement and reinforce each other in a manner similar to a teacher lecturing while using a blackboard.

Chapter 1: Signals and Spectra

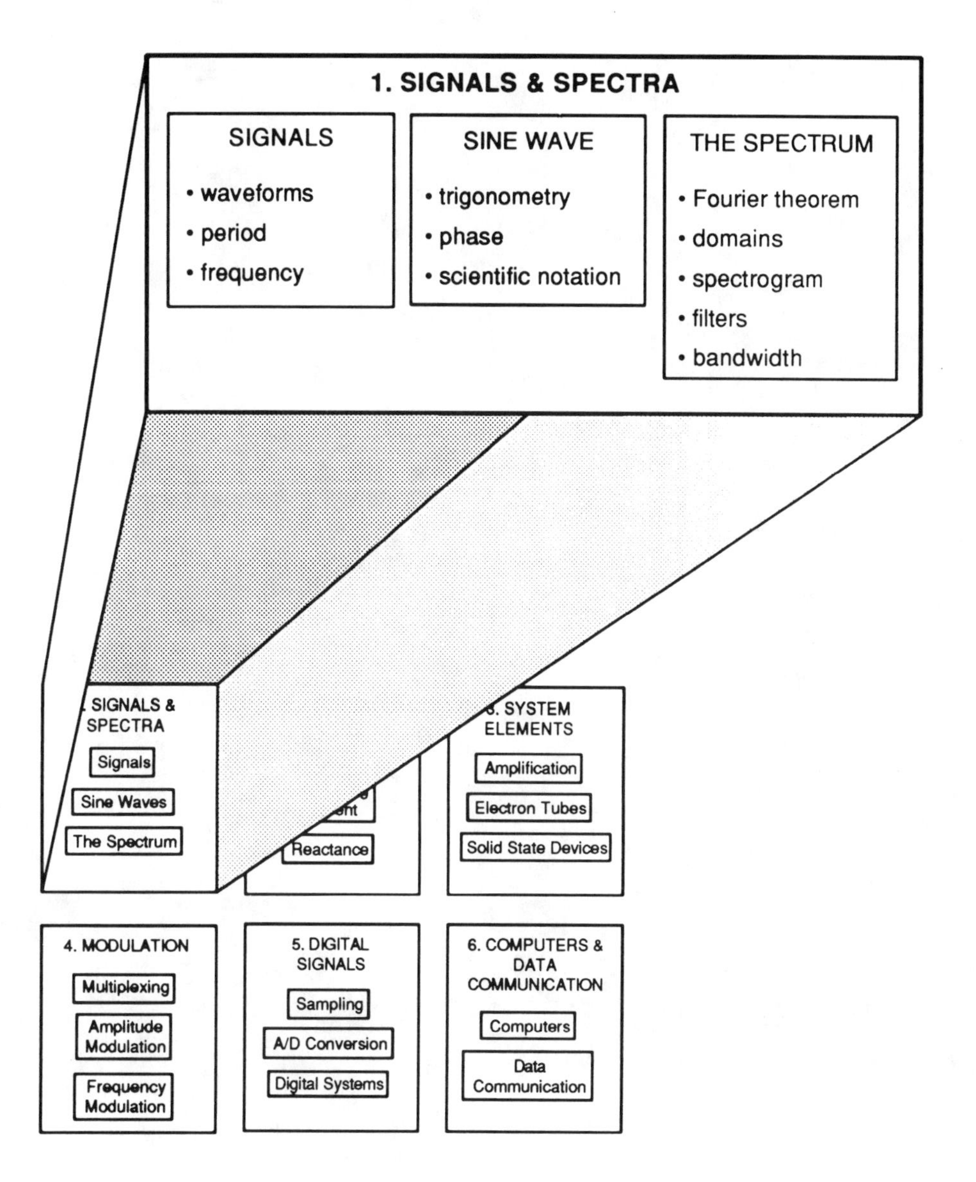

Introduction to Signals and Spectra

The concept of a *signal*—a time-varying event—is basic to communication. Signals can exist in many different types of media, but all signals vary with time. This module treats signals and various ways of characterizing them.

Time-varying signals have different shapes associated with their time variation. These *waveshapes* can be decomposed into combinations of pure tones, called *sine waves*, at different frequencies. *Frequency* is a measure of how frequently a wave repeats itself. The different sine waves that compose a signal can be plotted as a function of frequency to produce a graph that is called the frequency *spectrum* of the signal.

Most signals occupy a range of frequencies. This range of frequencies is called the *bandwidth* of the signal. A communication medium, or *channel*, can pass only a range or *band* of frequencies, which is called the *bandwidth* of the channel. The bandwidth of a communication channel and the bandwidths of the various signals determine the number and types of signals that can be transmitted over a particular communication channel.

The concepts of spectrum and bandwidth are very important in characterizing signals and communication channels and media. The provision of bandwidth is costly, and so bandwidth and spectra are closely related to the economics of communication systems.

The concept of bandwidth follows from the concept of the spectrum. The spectrum of a signal follows from the knowledge that a time-varying signal can be decomposed into a number of basic sine waves at particular frequencies. Hence, a knowledge of sine waves and frequency is essential to understanding the spectrum and bandwidth.

Signals

Communication manifests itself in the form of signals. Information from a source intended for a receiver is sent over a medium in the form of signals appropriate to that medium. There is a wide variety of signals and media.

Speech is used as a signal between human beings who are communicating with each other. The medium for interpersonal speech communication is air, since human speech is an acoustic signal. Broadcast television is an electromagnetic signal that is transmitted over the medium of space. For a newspaper, the characters printed on the page create visual signals that convey information from the publisher to the reader. Signals can exist in many different media and forms, for example, electrical, radio, acoustic, and visual signals.

An essential aspect of a signal is that it changes in some way. A sound wave or an acoustic signal utilizes changes in sound pressure. The broadcast television signal utilizes changes in an electromagnetic waveform. The printed symbols on the page of a newspaper utilize changes in reflected light intensity.

A signal that is constant and does not change in some way conveys little or no information. The more rapid the changes in the signal, the more information that the signal can convey. In the speech signal, changes in pitch, intensity, and resonant quality all combine to convey information. A signal with little change—for example, a whistle—would convey little information other than that the signal was there or not there—in this example, that the person whistling was still alive!

Signals

Signals:

Media:

- AIR (sound)
- SPACE (radio)
- PAPER (print, image)
- WIRE (telephone, CATV)

WAVEFORMS

A signal plotted as a function of time creates a shape called the *waveshape* or *waveform*. Time is plotted along the horizontal (or x) axis, and the corresponding values, or amplitudes, of the waveform for each instant in time are plotted along the vertical (or y) axis. The x-axis is sometimes called the abcissa, and the y-axis is called the ordinate.

Positive amplitudes are plotted above the x-axis, and negative amplitudes are plotted below the x-axis. The origin represents an amplitude of 0 along the y-axis, and the beginning of time (or $t = 0$) along the x-axis. A waveform usually has a peak, or maximum, positive amplitude over some specified interval of time and also a peak negative amplitude.

Periodic waveforms have a basic, or fundamental, shape that repeats over and over in time. Nonperiodic waveforms have no such repetitive pattern.

Waveforms

Graphical Representation:

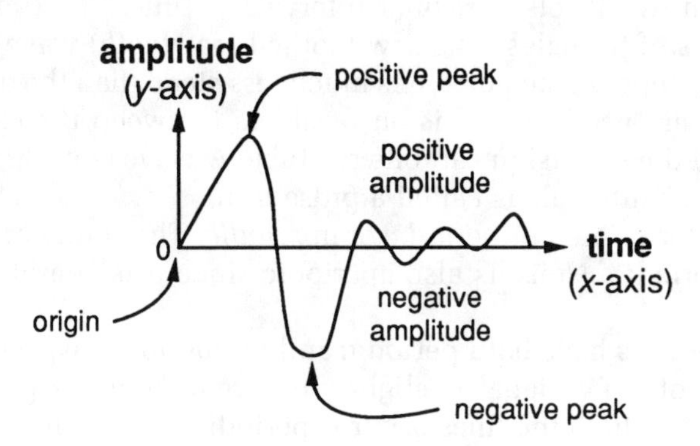

Periodic:

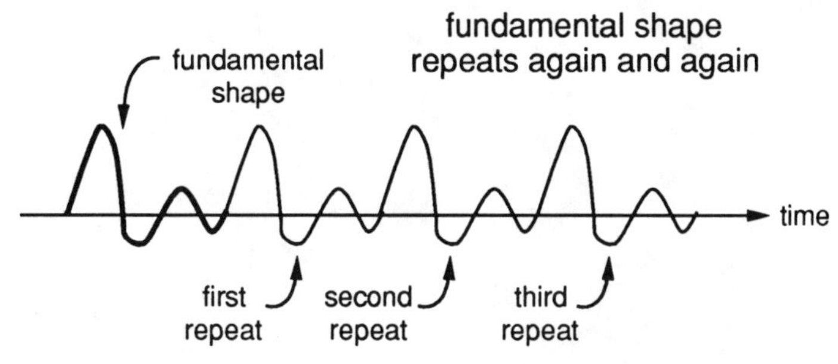

Nonperiodic:

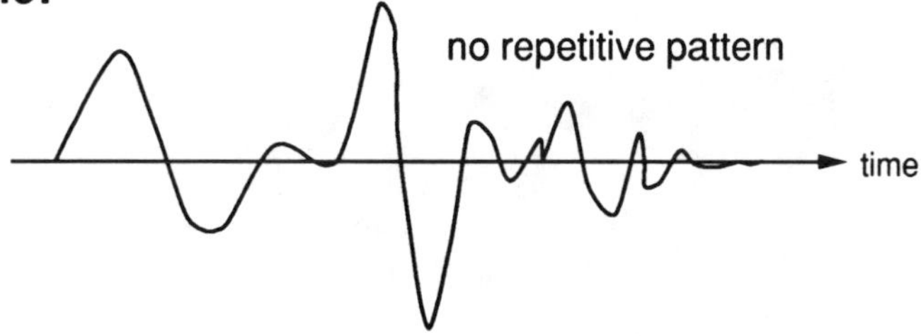

WAVEFORM EXAMPLES

Some periodic waveforms occur so frequently that they have been given names that are characteristic of their shape. The *square wave* instantaneously changes its value between two levels at regular intervals of time. The *triangular wave* looks like a series of triangles. The saw-toothed (*sawtooth*) wave is triangular in shape, but the slope of one portion is much less steep than the other portion. Sawtooth waves are used in television receivers to sweep the electron beam across the face of the television cathode ray tube. A wave consisting of a series of spikes at regular intervals is called a pulse train.

Other waveforms are *nonperiodic* or *aperiodic*. The sound of a door slamming shut is aperiodic. Noise is also aperiodic since it is completely random in shape.

Some waveforms have both periodic and nonperiodic aspects. Each individual scan line of a TV signal is slightly different from the preceding scan line, but the rate at which the lines occur is periodic. A speech signal is nearly periodic over short intervals of time, but changes in the waveshape are constantly occurring, although somewhat slowly. Speech is called a *quasiperiodic signal*.

Waveform Examples

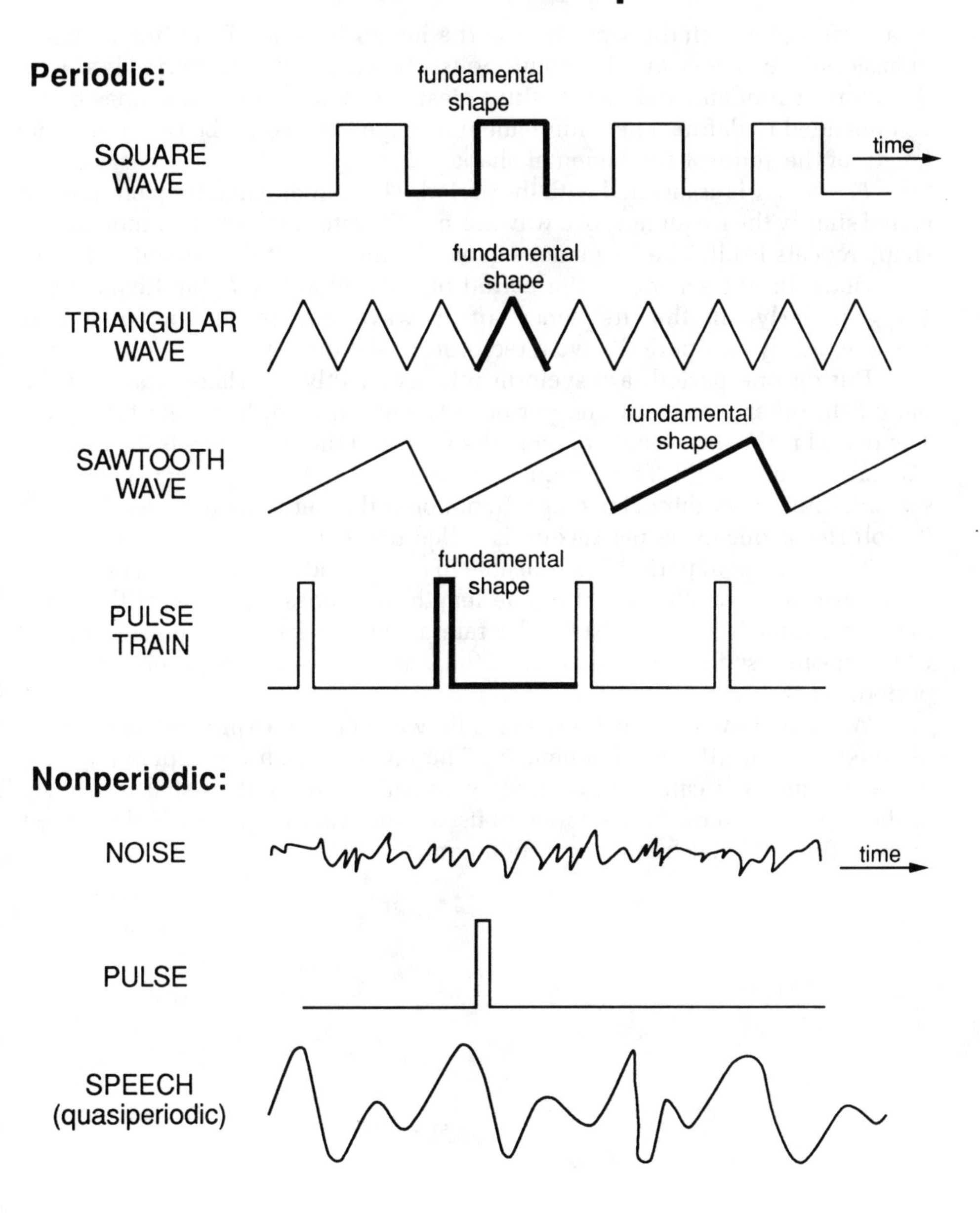

PERIOD AND FREQUENCY

The *period* of a periodic waveform is the length in time of the fundamental, or basic, shape of the wave that continuously repeats itself. The period is always the shortest fundamental shape, since clearly two fundamental shapes could be appended to define a new fundamental shape that would be twice the time length of the shortest fundamental shape.

Frequency is associated with the period. The fundamental frequency, often called simply the frequency, of a waveform is the rate at which the fundamental shape repeats itself. The frequency equals the number of fundamental shapes, or periods, in one second. If the period of a waveform is T, the frequency is $1/T$. Similarly, if the frequency of a waveform is F, the period is $1/F$. Frequency and period have a reciprocal relationship.

During one period, a waveform returns exactly to where it started the period. In other words, in one period a waveform completes one full cycle. The period is the time length of one full cycle, and the frequency is the number of cycles in one second. Thus, frequency is a measure of the number of cycles per second, sometimes abbreviated cps. In honor of the German physicist Heinrich Rudolf Hertz, one cycle per second is called one hertz, abbreviated Hz.

The concept of period is applicable to any periodic waveform. The period of any periodic waveform is the time length of the basic portion of the wave that continuously repeats itself. The rate at which that basic portion of the wave repeats itself is the frequency of the wave, and it is the reciprocal of the period.

We will show later that any periodic wave can be expressed as the sum of sine waves at different frequencies. The rate at which a complex periodic wave repeats itself can more accurately be called the fundamental frequency of the wave. The term fundamental is used, since that frequency is the lowest or most fundamental frequency which comprises the wave.

Period and Frequency

Period:

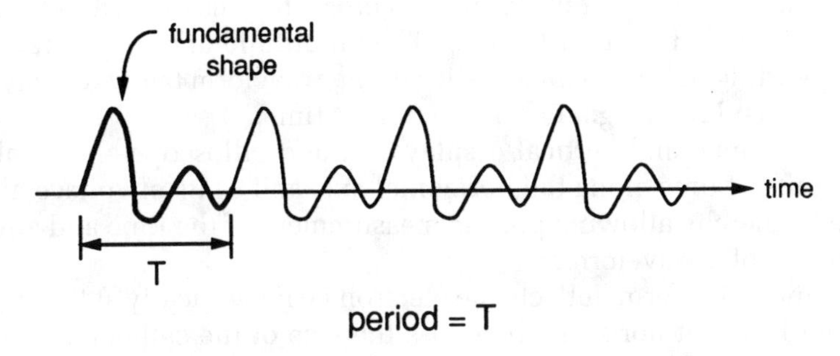

Fundamental Frequency:

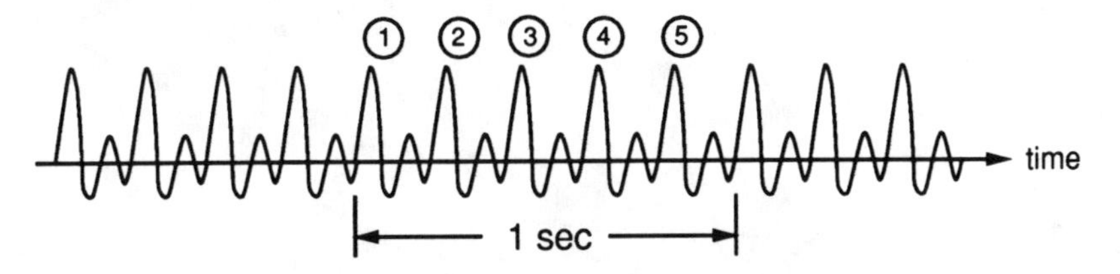

fundamental frequency equals
number of periods in 1 second

$$F = \frac{1}{T}$$

FUNDAMENTAL FREQUENCY (in hertz)

PERIOD (in sec)

OSCILLOSCOPE

Engineers use an electronic instrument to display electrical waveforms. The electrical waveforms are used to control the vertical deflection of a beam of electrons that are swept at a uniform rate across the face of a cathode ray tube. This device is called an *oscilloscope.* The amplitude and sweep rate can be varied, depending on the characteristics of the waveform being displayed. The oscilloscope displays a signal as a function of time.

The horizontal and vertical display on the oscilloscope are usually calibrated in units of time along the horizontal axis and units of voltage along the vertical axis, thereby allowing precise measurement of the time and amplitude characteristics of a waveform.

The input waveform deflects the electron beam vertically at the same time as the beam is swept horizontally across the face of the cathode ray tube. An internal sawtooth wave controls the horizontal sweep, and the frequency of the sweep signal determines the time resolution of the display.

Oscilloscope

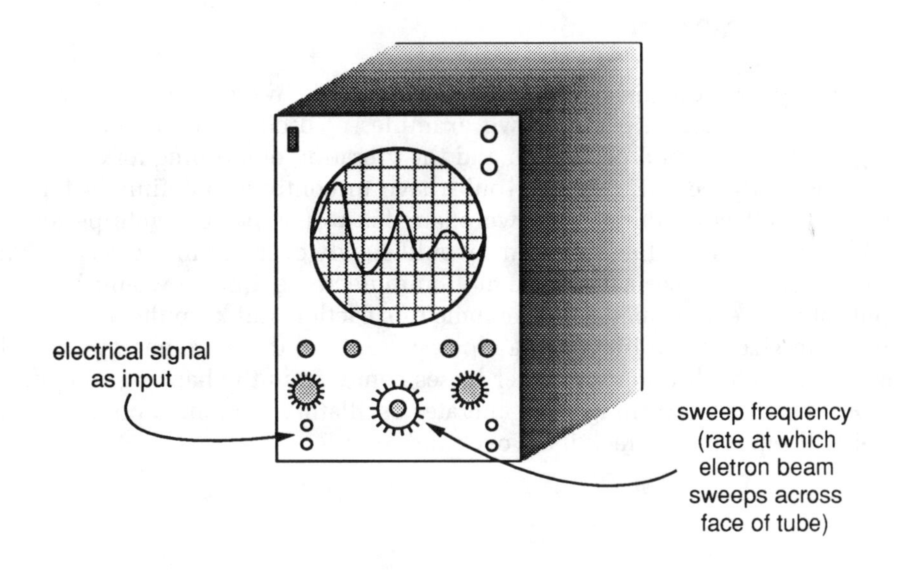

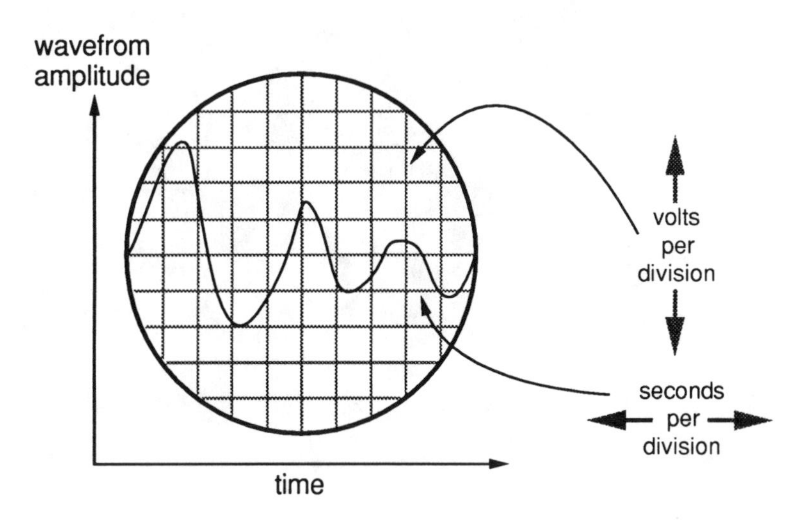

Sine Waves

OSCILLATORY MOTION

A perfectly periodic signal, and its associated periodic waveform, occurs whenever *oscillatory motion* occurs. Two examples of oscillatory motion are the swinging to and fro of a pendulum and the vibration of a tuning fork.

A perfectly periodic signal would have to continue indefinitely for all time, and so it is a fiction. The two examples would not generate perfectly periodic signals, since the waveforms would slowly decrease in size as friction and other energy losses caused the motion to decrease. Energy would have to be put into these movements to overcome this friction and keep the movement constant in size or amplitude. In a clock pendulum, the extra energy needed to compensate for friction and other losses comes from the hanging weights.

A wheel that is turning also generates oscillatory motion. A point on the wheel will repetitively track itself over and over.

Oscillatory Motion

Oscillatory Motion:

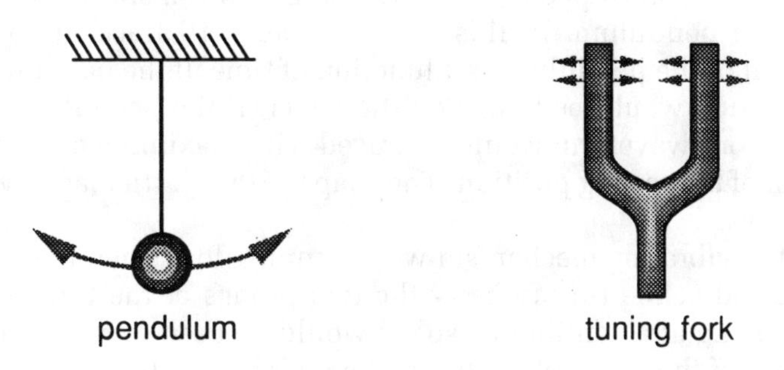

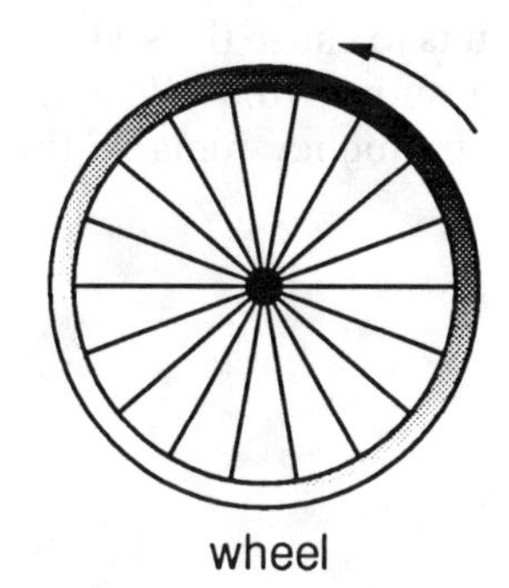

THE SINE WAVE

The shape of the movement of a pendulum can be displayed by placing a marking pen at the tip of the pendulum. It will be assumed that the pendulum is very long so that the swinging motion is essentially along a horizontal line. A continuous roll of paper that moves at a uniform speed can then be placed behind the pendulum. In this way, the pen will trace the displacement or movement of the pendulum as a function of time. If the pendulum were at rest, a straight line would be traced on the paper. If the pendulum were swinging, an oscillatory waveform would be traced with maximum excursions to the left and right of the resting position. The shape of this particular waveform is called a *sine wave*.

If the vibrating mechanism were a tuning fork, then the pen would need to be placed at the tip of one of the two prongs of the tuning fork. A roll of paper moving at a continuous speed would be used to record and display the movement of the prong of the tuning fork. As before, the shape of the waveform would be a sine wave, assuming that there were some way to overcome losses so that the vibration of the tuning fork never died down.

The effect of friction is to cause the swinging of the pendulum to gradually decrease. The waveform shows this effect in terms of gradually decreasing amplitude, although the frequency remains the same.

Sine Wave

Trace of Motion:

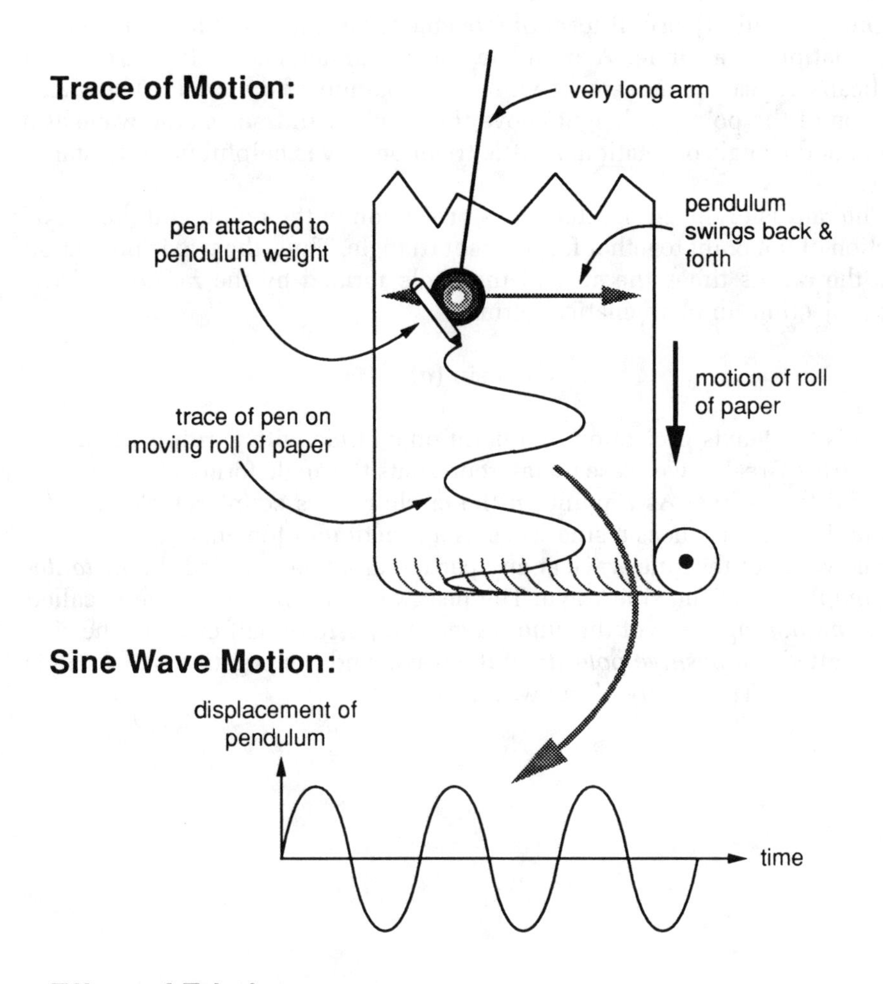

Effect of Friction:

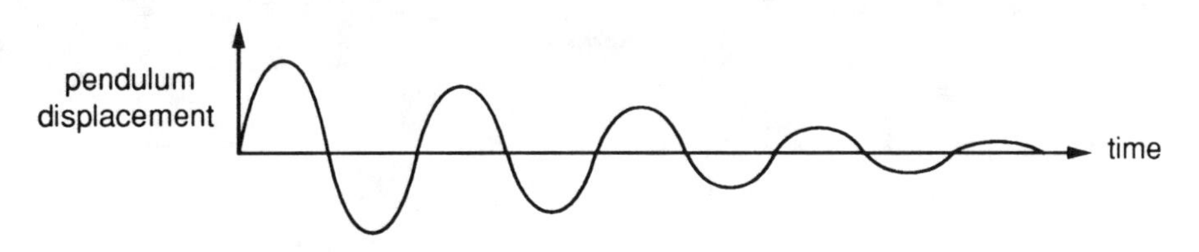

THE SINE WAVE (cont'd)

One particularly useful form of movement for understanding sine waves is the rotation of a circle. A point on the circumference of the circle will periodically repeat its movement with each rotation of the circle. The y-axis projection of the point, or height above the x-axis, will trace a sine wave as a function of the angle of rotation. A little trigonometry is helpful in understanding this.

The radius of the circle, the y-axis projection of the point, and the x-axis projection of the point together form a right triangle. Thus, the y-axis projection equals the radius times the sine of the angle formed by the radius and the x-axis projection. In mathematical terms,

$$a = A \sin(\theta)$$

where a is the y-axis projection of a point on a circle with a radius of A. The symbol θ (the Greek lower-case theta), represents the angle formed between the x-axis and the radius. As a point on the circle rotates counterclockwise, the y-axis projection produces a sine wave as a function of the angle θ.

The value of the sine wave at any particular angle is called the *instantaneous amplitude* of the sine wave. The peak value of the sine wave is called the *maximum amplitude* of the sine wave. The positive half-cycle of the sine wave is called the *positive polarity* of the wave, and the negative half-cycle is called the *negative polarity* of the wave.

Sine Wave (cont'd)

Trigonometry:

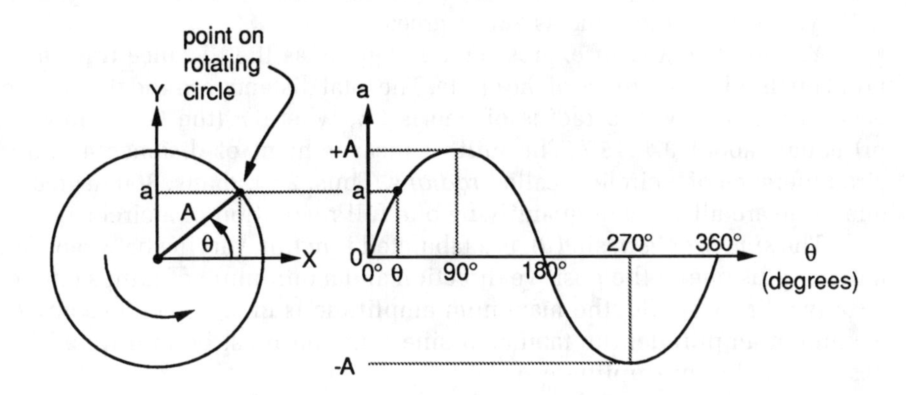

A = radius of circle (maximum amplitude)
a = height of point (instantaneous amplitude)

$$a = A\sin(\theta)$$

Waveform (one cycle):

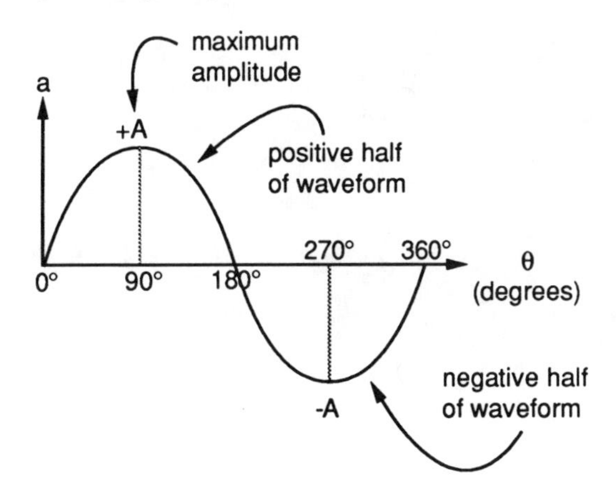

SINE FUNCTION

There are a number of ways to express an angle. We are all very familiar with the use of degrees, with 360 degrees being one complete revolution. Another way to express an angle is as a fraction of a complete revolution, or cycle. One-half cycle then is the same as 180 degrees.

Yet another way of expressing an angle is as the distance traveled by a point on the circumference of the circle. The total distance around the circumference of a circle with a radius of one is 2π, where π (the Greek lower-case pi) equals about 3.14159. The unit of measurement of distance around the circumference of a circle is called *radians*. Thus, 2π radians, 360 degrees, and one cycle are all ways of quantifying one full revolution of a circle.

The *sine function*, $\sin(\theta)$, is a tabulated function varying between a maximum of plus one in the positive direction and a minimum of minus one in the negative direction. So, the maximum amplitude is one. To achieve any other maximum amplitude, the tabulated sine function must be multiplied by the desired maximum amplitude A.

The table shows values of the sine function for a selection of angles between 0 degrees and 90 degrees. The sine function for angles greater than 90 degrees can be easily obtained from a table for values between 0 degrees and 90 degrees. That is because the sine function is symmetric about 90 degrees, or, in other words, the values for 90 degrees to 180 degrees are simply the values for 90 degrees to 0 degrees in descending order. The values of the sine function for 180 degrees to 360 degrees are simply the negative of the values for 0 degrees to 180 degrees.

Sine Function

Angles:

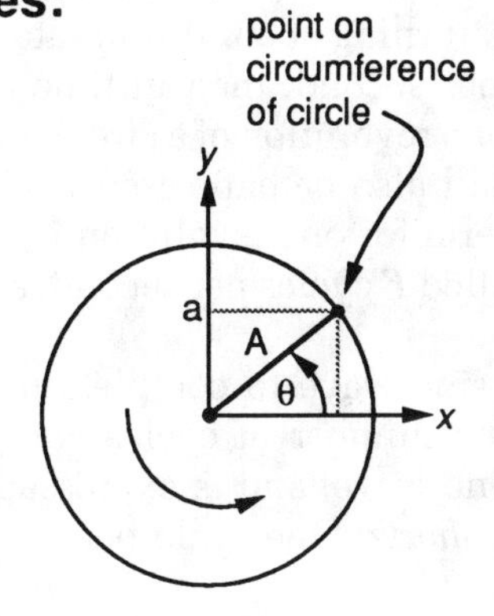

angle θ in:		
cycles	**degrees**	**radians**
0	0°	0
1/8	45°	π/4
1/4	90°	π/2
1/2	180°	π
3/4	270°	3π/2
1	360°	2π

Radians:

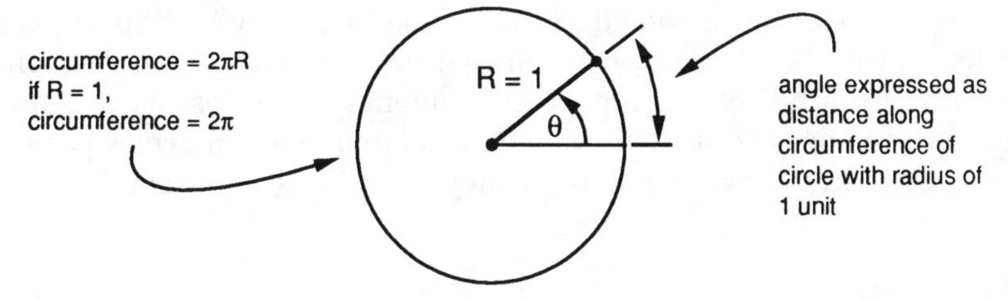

Sine Function:

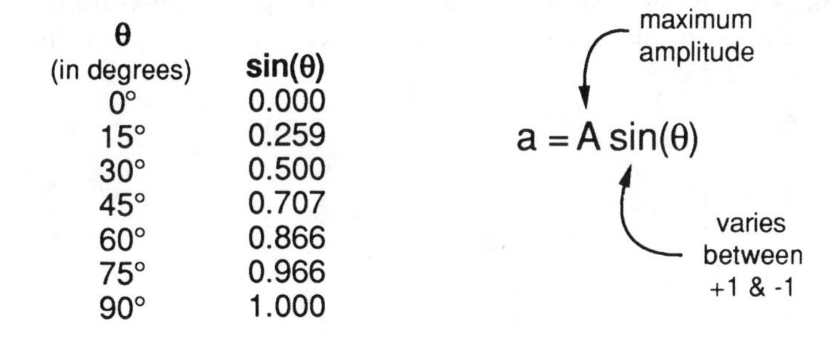

θ (in degrees)	sin(θ)
0°	0.000
15°	0.259
30°	0.500
45°	0.707
60°	0.866
75°	0.966
90°	1.000

SINE-WAVE FREQUENCY

The sine wave generated by the rotation of a circle can be plotted as a function of either time or the angle through which the circle has turned. We will always assume in our discussion that the circle is turning at a uniform rate. If the rate of rotation is specified as F revolutions per second, then in time t the circle would turn through Ft revolutions. Since one revolution of a circle is equivalent to 360 degrees, the rate of revolution could also be expressed in degrees per second and would equal 360 F. Another term for one revolution is a *cycle*. So, the rate at which the circle turns can be called F cycles per second, abbreviated cps.

The amount of time required for a sine wave to complete one cycle is called the *period* T of the sine wave. The number of cycles completed per second is called the *frequency* F of the sine wave, and is expressed in cycles per second. Another unit for frequency is *hertz*. One cycle per second is the same as one hertz, abbreviated Hz.

The frequency F of a sine wave is the reciprocal of the period T, or $F = 1/T$. For example, if a sine wave had a period of 0.1 seconds, its frequency would be 1/0.1 or 10 Hz.

Engineers are accustomed to expressing frequency in radians per second, and use ω (the Greek lower-case omega) to represent radian frequency. The letter F is usually used to represent frequency in cycles per second, or Hz. Since 2π radians is equivalent to one cycle, frequency in cycles per second or Hz can be easily converted to frequency in radians per second, according to the equation:

$$\omega = 2\pi F$$

where ω is frequency in radians per second and F is frequency in Hz.

Engineers are accustomed to dropping the term "maximum" when referring to the maximum amplitude of a sine wave and simply refer to A as the amplitude of the sine wave.

Sine-Wave Frequency

Frequency:

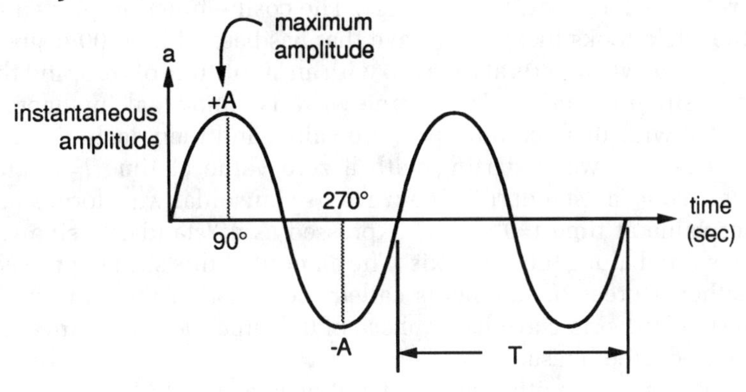

$$F = \frac{1}{T}$$

frequency in hertz (F); period in sec (T)

Equation:

F = frequency in hertz

$$\theta_{cycles} = F\,t$$

$$\theta_{degrees} = 360\,F\,t$$

$$\theta_{radians} = 2\pi\,F\,t$$

$$a = A\sin(\theta)$$

PHASE

Another trigonometric function is the *cosine function*. The radius times the cosine of the angle gives the *x*-axis projection of the radius. The cosine of 0 degrees is 1 for a unit-radius circle. The cosine function plotted as a function of the angle looks like a sine wave that has been shifted 90 degrees to the left.

A sine wave starts at time t=0 with an amplitude of zero, and then increases in a positive direction. The cosine wave is sinusoidal in shape, but starts at time t=0 with its maximum positive value, and then declines.

The sine wave starting with a zero value at time t=0 can perhaps be visualized as a "standard" sine wave. A sinusoidal waveform starting at some other value at time t=0 can be expressed as a "standard" sine wave that has been shifted along the time axis. The amount of this shift expressed as an angle in either degrees or radians is called the *phase* of the sinusoidal waveform. The standard sine wave has a phase of 0 degrees; a *cosine wave* is a sine wave with 90-degree phase.

A sine wave with a phase of 180 degrees would be a standard sine wave inverted in amplitude. If two sine waves with the same maximum amplitude but differing in phase by 180 degrees were added together, they would exactly cancel each other. The two sine waves are said to be 180 degrees *out of phase* with respect to each other.

If two sinusoidal waveforms have the same frequency, but differ in phase, the difference in phase between them is called the *phase shift* between the two waveforms.

Phase

Phase:

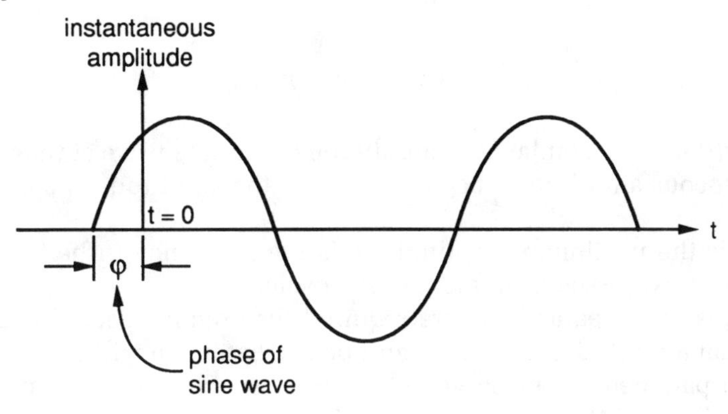

Cosine Wave:

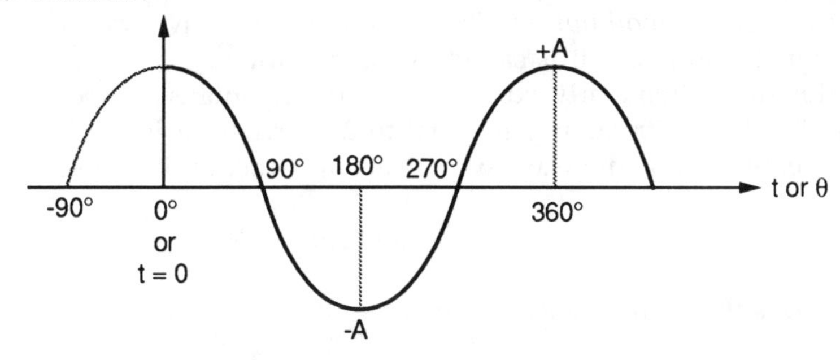

Cancellation:

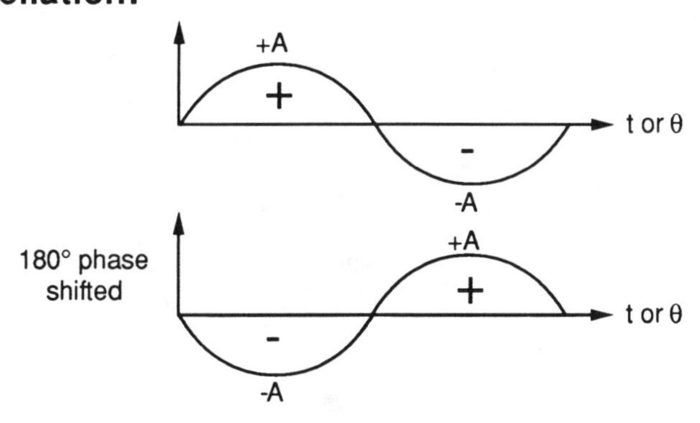

MATHEMATICAL EXPRESSION FOR A SINE WAVE

The general mathematical expression for a sine wave is important, since it will be helpful later in understanding various types of modulation. The equation is

$$a(t) = A \sin (Ft + \phi)$$

where $a(t)$ is the instantaneous amplitude of the sine wave at time t. It is called instantaneous amplitude since it represents the amplitude at a general instant of time.

A is the maximum amplitude, F is the frequency in hertz, and ϕ is the phase shift expressed as a fraction of a cycle.

Thus, three parameters are required for exactly specifying a sine wave: maximum amplitude, frequency, and phase. Later, we will show how any one of these parameters can be varied in exact synchrony with an information-bearing signal so that information can be transmitted over great distances and many signals combined together for transmission over a single medium. This process is called *modulation*. The modulated sine wave "carries" the information signal, and hence it is called the *carrier wave*.

Engineers frequently express frequency in radians per second, symbolized by ω. The radian frequency is equal to $2\pi F$, where F is the frequency in Hz. The equation for a sine wave when the sine function is in radians is

$$a(t) = A \sin(2\pi Ft + \phi)$$

where ϕ is the phase angle in radians.

Equation

$$a(t) = A \sin(Ft + \varphi)$$

SCIENTIFIC NOTATION

In scientific notation, numbers are expressed as powers of ten. For example, one thousand would be written as ten raised to the third power:

$$1{,}000 = 10^3 = 10 \times 10 \times 10$$

Fractional powers of ten would usually be expressed as negative powers of ten so that one one-hundredth would be written as ten raised to the negative two power:

$$1/100 = 0.01 = 10^{-2}$$

The great advantage of this system is that it is easy to multiply powers of ten because we only need to add the exponents, while, of course, taking into account their sign. Thus, for example,

$$10^6 \times 10^{-3} = 10^{6-3} = 10^3$$

Powers of ten appear so often that Greek words have been assigned as prefixes to denote various powers in steps of three. The following table gives the powers and terms most frequently encountered for representing both positive and negative powers of ten.

Number	*Term*	*Abbreviation*
10^{12}	tera	T
10^{9}	giga	G
10^{6}	mega	M
10^{3}	kilo	k
10^{-3}	milli	m
10^{-6}	micro	μ
10^{-9}	nano	n
10^{-12}	pico	p

By using these terms, a frequency of 5,000 Hz would be written 5 kHz and pronounced 5 kilohertz. A sine wave with a period of 0.2 nanosecond (ns) would have a frequency of 5 GHz.

Scientific Notation

Examples:

exponent

$1{,}000 = 10^3$

$340 = 3.4 \times 10^2$ [note: $10^0 = 1$]

$0.05 = 5.0 \times 10^{-2}$

Mathematics:

for multiplication:
ADD exponents

$10^3 \times 10^2 = 10^{3+2} = 10^5$

for division:
SUBTRACT exponents

$10^3 \div 10^2 = 10^3 \times 10^{-2} = 10^{3-2} = 10^1$

Terminology:

BIG	10^{12}	tera (T)
	10^9	giga (G)
	10^6	mega (M)
	10^3	kilo (k)
	10^0	1
SMALL	10^{-3}	milli (m)
	10^{-6}	micro (μ)
	10^{-9}	nano (n)
	10^{-12}	pico (p)

The Spectrum

FOURIER THEOREM

The sine wave is important because it is the fundamental waveform from which more complex waveforms can be created. The basic theorem for this principle was stated by the French mathematician, J. B. Fourier, during the early 19th century. The principle is called the *Fourier series expansion* of a complex periodic function.

The *Fourier theorem* states that any periodic function or waveform can be expressed as the sum of sine waves with frequencies at integer or harmonic multiples of the fundamental frequency of the waveform and with appropriate maximum amplitudes and phases. The Fourier theorem also specifies the procedure for analyzing a waveform to determine the amplitudes and phases of the sine waves that comprise it.

A sine wave has an average value of zero. Likewise, the average value of a sum of harmonically related sine waves, averaged over the time of the period of the fundamental frequency, is zero. Thus, a constant term may need to be added to the Fourier sum of sine waves if the original periodic waveform has a nonzero average value. The Fourier addition of sine waves is accomplished by algebraically adding the instantaneous amplitudes of all the sine waves at each particular instant in time.

The nth harmonic of a sine wave at a fundamental frequency of F_0 is a sine wave with a frequency of nF_0. So, the frequency of the second harmonic of a waveform with a fundamental frequency of 1 kHz is 2 kHz. The use of the term harmonic is somewhat different in music, where the first harmonic has a frequency which is twice that of the fundamental.

Fourier Theorem

Any periodic wavefrom can be expressed as a sum of sinusoidal components at harmonic multiples of the fundamental frequency. Each component will have its own appropriate maximum amplitude and phase.

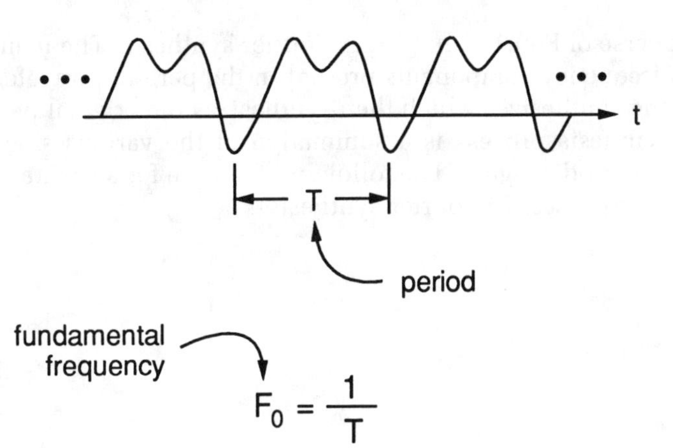

$$\text{HARMONICS} = n\,F_0$$

integer

FOURIER ANALYSIS AND SYNTHESIS

The determination of the specific sine waves and their corresponding frequencies, maximum amplitudes, and phases needed to represent a periodic signal is accomplished by Fourier analysis. Fourier analysis is accomplished, in effect, by performing a correlation of various sine waves with the original periodic waveform to determine how much of each sine wave is present in the signal. In mathematical terms, integration is performed. The result is an identification of each frequency F_n along with its corresponding maximum amplitude A_n and phase φ_n.

The inverse of Fourier analysis is Fourier synthesis. The identification of the various frequency components present in the periodic waveform are used as input to the synthesizer which then synthesizes the original periodic waveform. The synthesis process is a summation of the various sine waves that comprise the periodic signal. The following example for a square wave should help clarify the process of Fourier synthesis.

Analysis and Synthesis

Fourier Analysis:

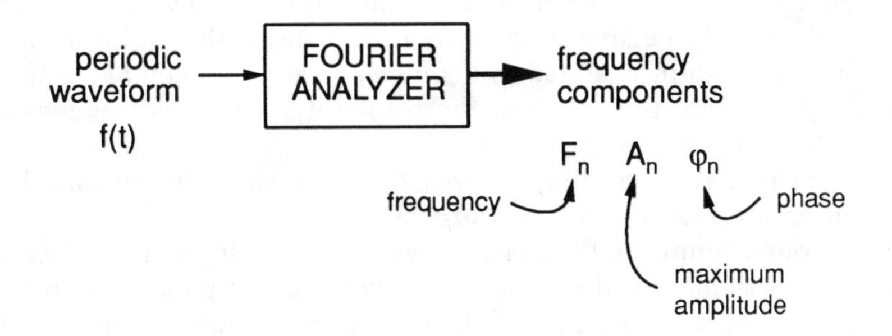

Fourier Synthesis:

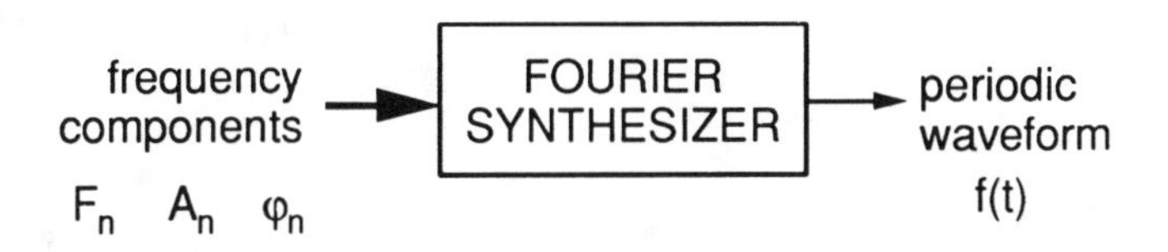

FOURIER SYNTHESIS

By using Fourier analysis, it is possible to show that a square wave is composed of sine waves at odd multiples or harmonics of the fundamental and with amplitudes inversely proportional to the harmonic number. Said somewhat differently, if a sine wave at the fundamental frequency F_0 with an amplitude of one is added to a sine wave at frequency $3F_0$ with amplitude 1/3, and this resultant waveform is added to a sine wave at frequency $5F_0$ with amplitude 1/5, and so forth, then each of the resultant waveforms becomes closer and closer to a square wave.

The process of adding together the sine waves to recreate the complex waveform is called *Fourier synthesis.*

In our example for the square wave, an interesting mathematical phenomenon occurs as more and more higher harmonics are added together. Ears form at the short vertical portions of the synthesized square wave. This is because the mathematics of the synthesis cannot handle the sharp discontinuity of the sudden vertical change from one amplitude to another. This phenomenon is named after the mathematician who first described it and is therefore called Gibbs' phenomenon.

Fourier Synthesis

Square Wave Synthesis:

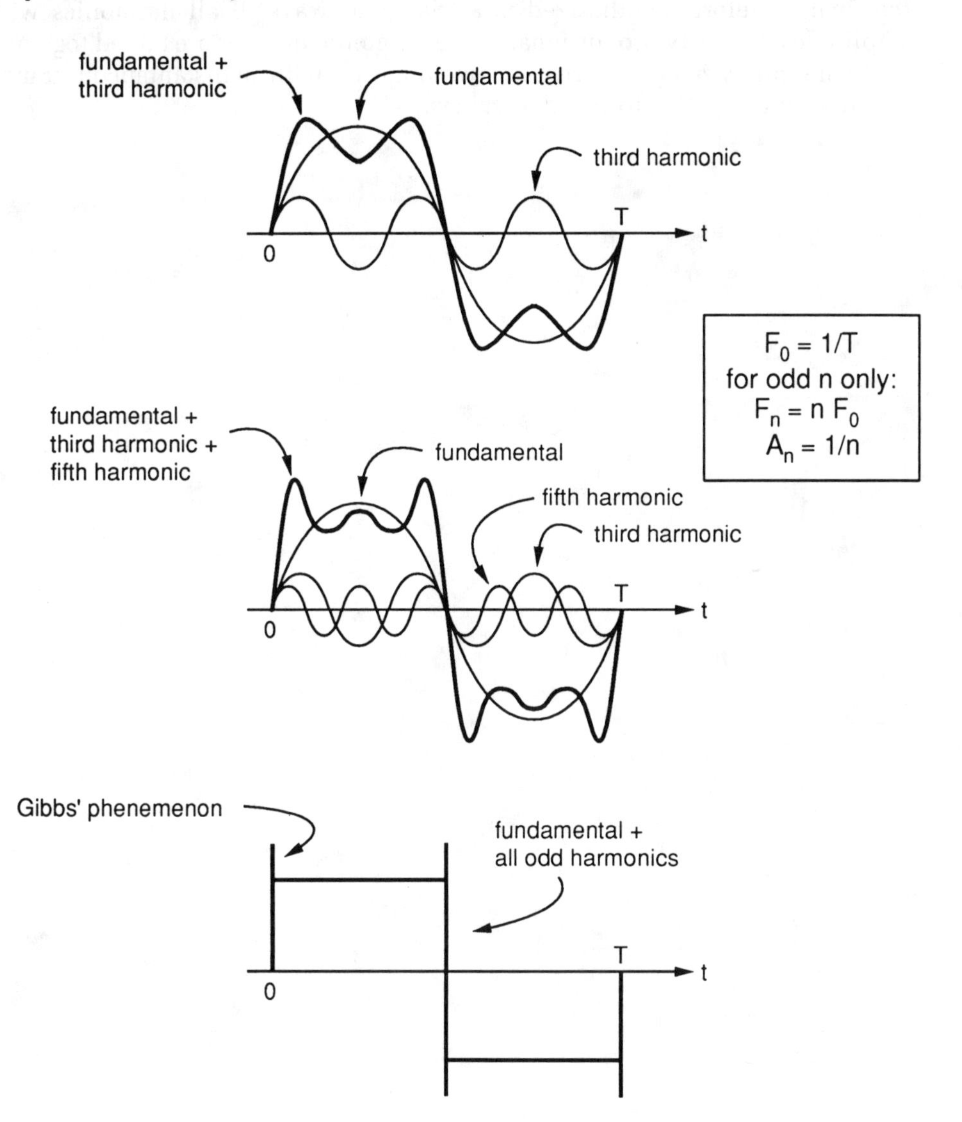

FOURIER SYNTHESIS (cont'd)

If all the odd harmonics of a cosine wave, with amplitudes inversely proportional to the square of the harmonic number, are added together, then the final waveform synthesized is a triangular wave. If all harmonics with amplitudes inversely proportional to the harmonic number are added together, the final synthesized waveform is a sawtooth wave with an instantaneous transition from the negative to positive values.

Fourier Synthesis (cont'd)

Triangular Wave:

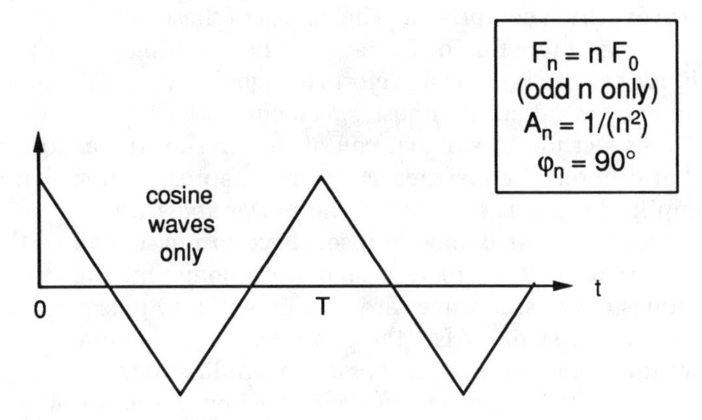

Sawtooth Wave:

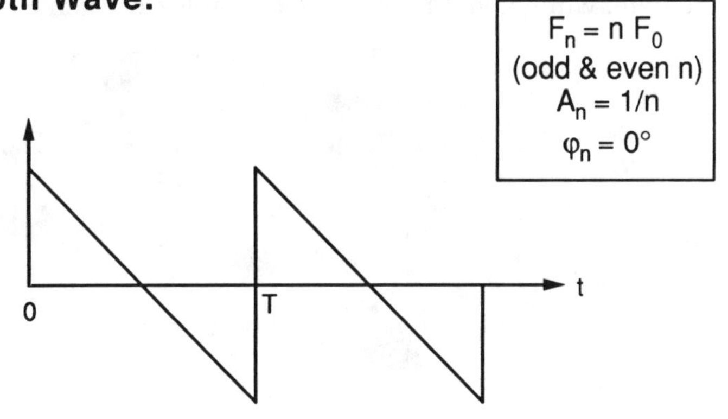

THE SPECTRUM

Suppose we have a complex periodic waveform or signal that has been Fourier analyzed to determine the frequencies, amplitudes, and phases of the various sine waves which comprise it. The values of these amplitudes could be plotted graphically as a function of frequency. The resulting plot or graph is called the *amplitude spectrum* of the waveform or signal. A plot of the phases as a function of frequency is called the *phase spectrum*.

The spectrum of a signal can also represent the energy or power in the signal at different frequencies. A spectral representation that is the square of the amplitude spectrum is called the *power spectrum*.

The human ear is mostly insensitive to phase, and so the phases of the sine waves used to recreate a signal are sometimes ignored. However, the amplitudes of the sine waves are very essential to preserving the tonal quality of the recreated signal. Also, the phase spectrum is not nearly as important as the amplitude spectrum in most practical applications. As a result, the tendency is to speak of the spectrum of a signal when we mean either the amplitude spectrum or the power spectrum.

The spectrum of a signal tells us much about the signal. Depending on its harmonic structure, a signal can be complex or pure. A whistle is an example of a pure tone, while a violin note has a complex harmonic structure.

The Spectrum

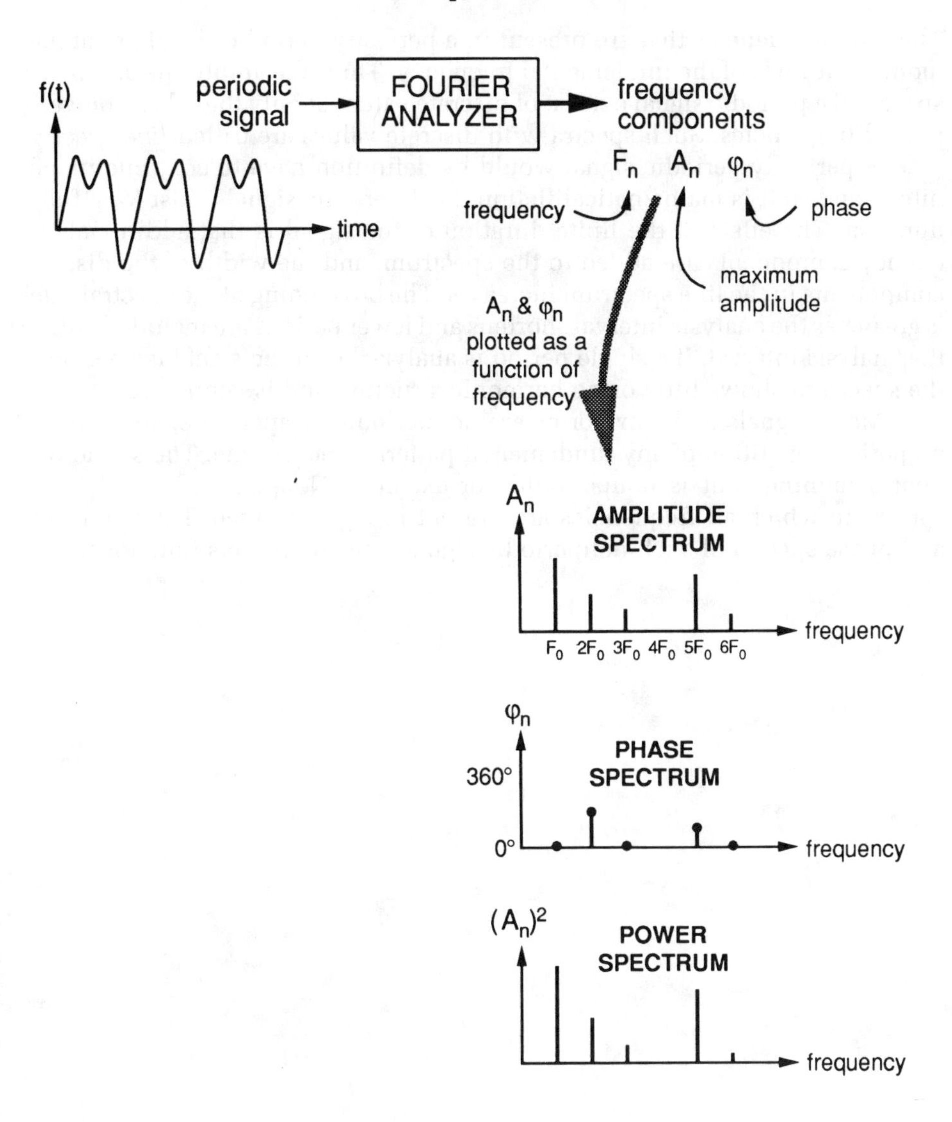

CONTINUOUS SPECTRA

The only frequencies that are present in a perfectly periodic signal are at harmonic multiples of the fundamental frequency. Thus, the amplitude and phase spectra of a periodic signal consist of discrete values at only these harmonically related frequencies. Such spectra with discrete values are called *line spectra*.

A perfectly periodic signal would by definition have to continue indefinitely, and so it is mathematical fiction. Real periodic signals exist for a finite duration. The effect of the finite duration of the signal is that additional frequency components are added to the spectrum, and the width of the discrete components in the line spectrum increases. The broadening of the spectral lines is greater as the analysis interval shortens and fewer periods are included within the analysis interval. If a single period is analyzed or if the signal is aperiodic, the spectrum shows little or no harmonic structure and is continuous.

Many signals and waveforms are nonperiodic or aperiodic, and there is no perfect repetition of any fundamental pattern over all time. The sound of a door slamming shut is nonperiodic, for example. Nonperiodic signals have spectra in which all frequencies are present in a given range. The amplitude and phase spectra of such nonperiodic signals are continuous functions.

Continuous Spectra

Line Spectrum:

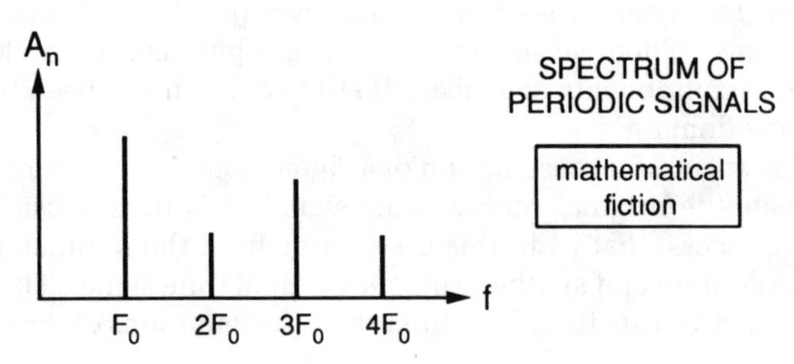

Continuous Spectrum:

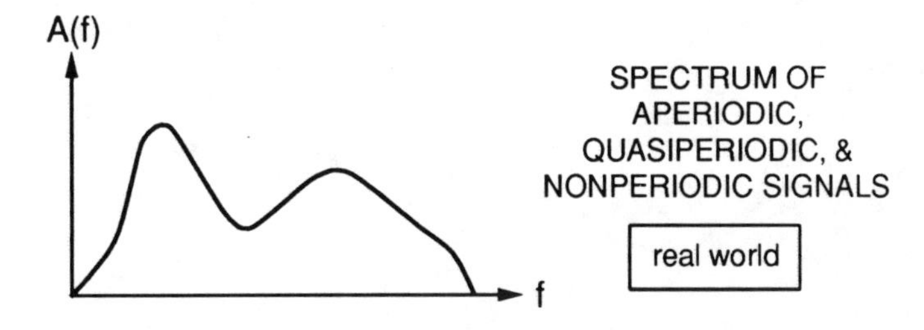

TIME AND FREQUENCY DOMAINS

The Fourier analysis of a signal is also called *spectrum analysis*. The device that performs such an analysis is called a *spectrum analyzer*. A spectrum analysis of a signal gives a representation of the signal in terms of its frequency components. When we say a time-varying signal is represented in terms of its frequency components, we mean the time domain has been translated into the frequency domain.

The spectral representation of a signal contains the information necessary to recreate the original time-varying signal. It is then possible to perform the inverse process, that is, in this case, going from the frequency domain back to the time domain and synthesizing the original time signal. The time signal itself and its spectrum are therefore different representations of the same information.

Domains

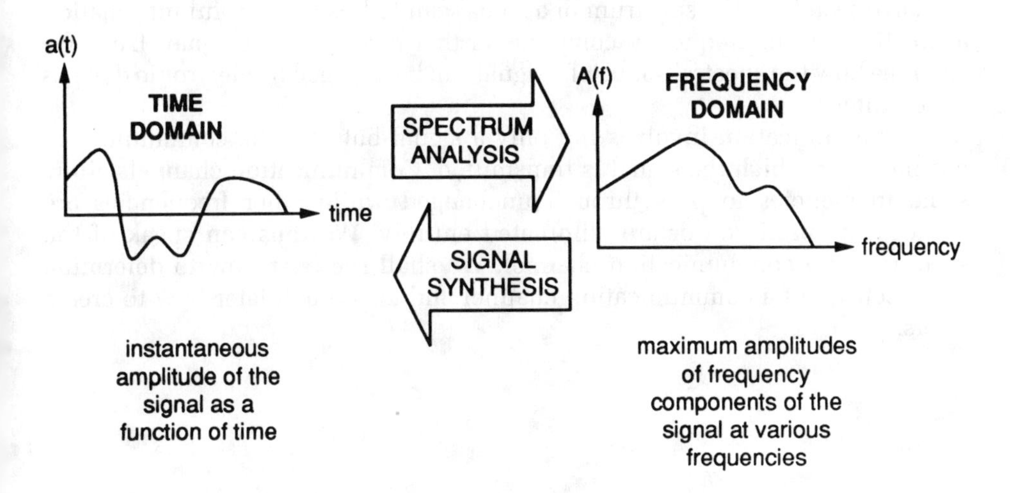
a(t)
TIME DOMAIN
time
SPECTRUM ANALYSIS
SIGNAL SYNTHESIS
A(f)
FREQUENCY DOMAIN
frequency
instantaneous amplitude of the signal as a function of time
maximum amplitudes of frequency components of the signal at various frequencies

USES OF SPECTRUM

We have seen how the spectrum of a signal can tell us very useful information about the various frequency components that comprise the signal. Later we shall see how the spectral shape of a signal can be changed by electronic devices called filters.

Communication involves not only a signal but also the communication channel over which the signal is transmitted. Communication channels allow some frequencies to pass through unchanged while other frequencies are reduced in amplitude or are eliminated entirely. We thus can speak of the spectrum of a communication channel. We shall see later how to determine the spectrum of a communication channel and also much later how to create filters.

Spectrum Uses

Signal Analysis:

Communication Channel:

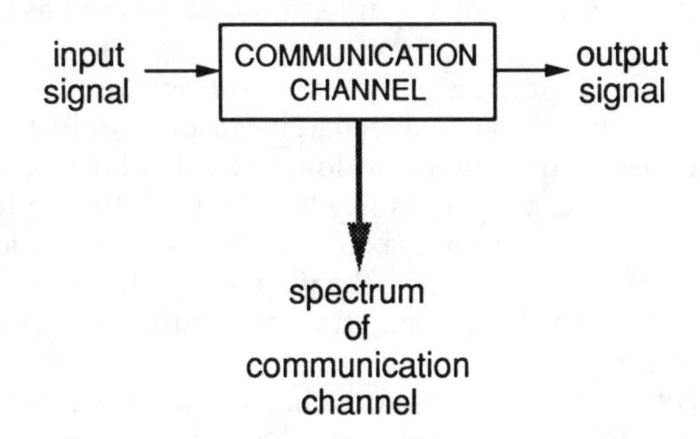

THE SPECTROGRAM

We now explore the concept of time-varying spectra.

Consider a signal consisting of three portions, each of equal length in time. The first portion is a low-frequency tone, the second portion is a high-frequency tone, and the third portion is silence. Rather than produce a single spectrum for the entire signal, we will produce a series of spectra over a number of time segments. Each analysis segment is chosen to equal the length of time of each portion of the signal. The analysis segments move along the signal in jumps of half the length of each portion, thereby creating segments that overlap, as shown. The analysis segments are labeled A, B, C, D, and E. A spectrum is plotted for each analysis segment. The spectrum for the first segment shows only the energy of the low-frequency tone. The spectrum for the second segment shows energy for both tones since the analysis interval overlaps portions of both tones. The series of spectra are shown plotted as a series along a vertical axis.

Suppose the analysis segments did not jump suddenly but slid slowly across the signal to be analyzed. The result would be a very closely spaced series of spectra that could no longer be displayed discretely. Some form of three-dimensional display is needed. The solution is to plot frequency along the *y*-axis, time along the *x*-axis, and spectral energy as a third dimension in terms of darkness. The darker the display, the higher the spectral energy. Such a three-dimensional representation of the time-varying spectra of a signal is called a spectrogram.

Spectrograms were used during World War II to analyze speech signals as a means to crack speech encryption techniques and also to display the underwater sounds of submarines. A crude example of a speech spectrogram is shown; the resolution and grey scale in a real spectrogram are considerably better.

Different speech sounds have different spectra. As the speech signal changes, the spectrum likewise changes, corresponding to the varying changes in the tonal and harmonic structure of the speech signal. Voiced segments of speech, such as vowels, typically have bands of energy at specific frequencies that are characteristic of the spoken sound. These bands of energy correspond to resonants in the vocal tract and are called formants. Fricative sounds, such as the sound "s," produce high frequency bursts of energy with no regular structure or pattern, since such sounds are like noise.

Spectrograms today are calculated by computer analysis of signals. Spectrograms display not only speech signals but the sounds of birds, whales, and machinery.

It is possible to collapse, or average out, the time variation in a spectrogram to create a single plot of the average energy in the speech signal as a function of frequency.

Spectrogram

Time-Varying Spectra:

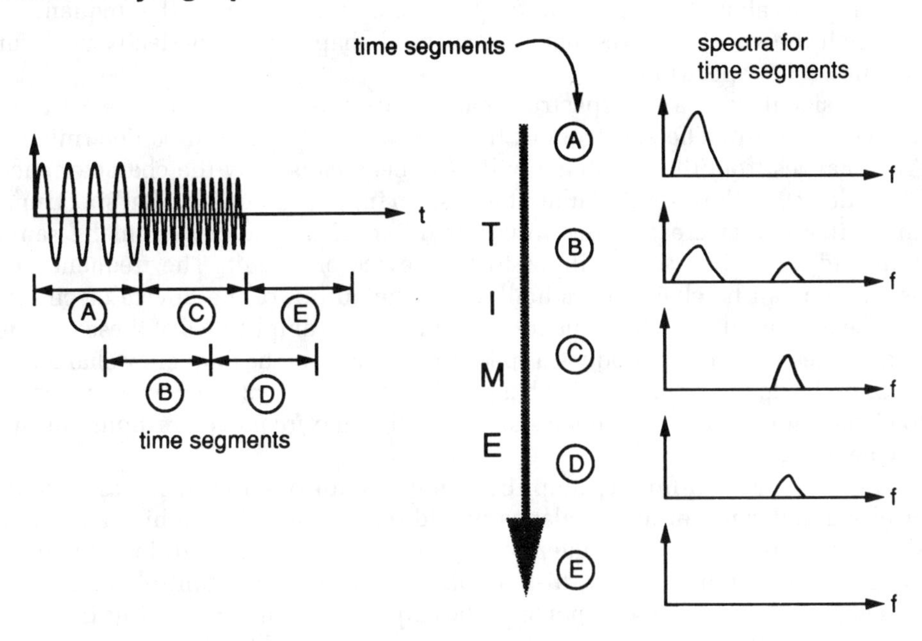

Spectrogram:

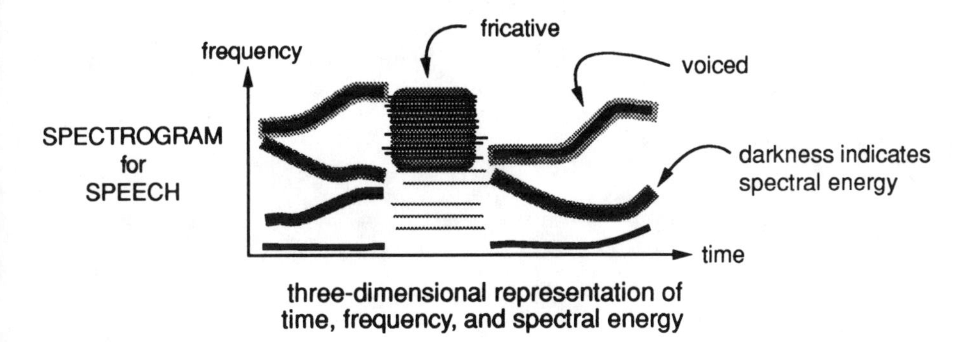

SPECTRUM SHAPES

The spectrum of a signal can be virtually any shape. Some spectra have peaks at certain frequencies. The spectrum of a voiced speech sound, like a vowel, typically has about three peaks at frequencies corresponding to the frequencies at which the vocal tract resonates. A spectral shape that is perfectly uniform is called a flat spectrum.

A signal with a flat spectrum can be used as input to some electronic device or circuit. The spectrum of the output signal can then be determined. Any changes from the flat shape will have been caused by the characteristics of the device or the circuit. Rather than use an input signal with a flat spectrum, an oscillator that creates sine waves at different frequencies with the same amplitudes can be used as input to the device or circuit. The frequency of oscillation can be changed gradually, and the output of the device or circuit may be measured at each frequency. The plot of the amplitudes of these output frequencies would also produce a spectrum that shows the frequency characteristics of the device or circuit. Such a plot is the response of the device or circuit to signals at different frequencies, and is called the *frequency response* of the device or circuit.

Clearly, a high-fidelity amplifier should amplify all frequencies equally or else a distortion of tonal balance would occur. Therefore, a hi-fi amplifier should ideally have a flat frequency response. However, since different people prefer different tonal quality, hi-fi amplifiers usually have controls to change the shape of the frequency response of the amplifier. Such tone controls increase and decrease the higher frequencies (treble control) and the lower frequencies (bass control).

Spectrum Shapes

Shapes:

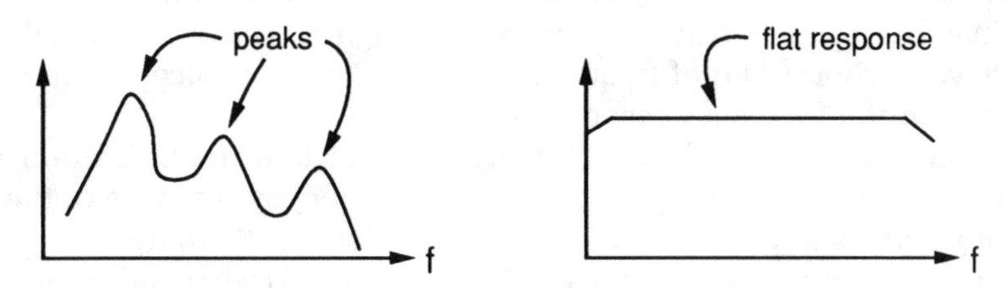

Frequency Response:

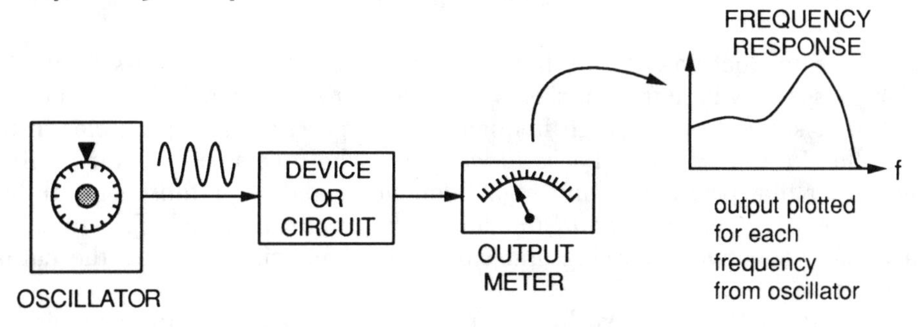

FILTERS

Some electronic devices are deliberately designed to have nonflat frequency responses. One group of frequencies, called a *frequency band*, might be amplified more than another. In the extreme case, a whole band of frequencies might be eliminated altogether. Any signal passed through such a device, called a *filter*, would have a whole band of frequencies eliminated or reduced in amplitude, depending on the frequency response of the filter.

The net effect of a filter on an input signal is a multiplication of the frequency spectrum of the input signal by the frequency response of the filter. If the filter passes only high frequencies, the effect on the output signal is to remove the low frequencies from the input signal. Algebraically, the output spectrum $O(f)$ equals the product of the input spectrum $I(f)$ by the frequency response of the filter $H(f)$:

$$O(f) = H(f)\, I(f)$$

A filter that passes only low frequencies is called a *low-pass filter*. One that passes only high frequencies is called a *high-pass filter*. A filter that eliminates, or stops, low and high frequencies and passes only frequencies in the middle of a range is called a *bandpass filter*. The inverse of a bandpass filter is a filter that passes low and high frequencies and stops those in a middle band, which is called a *bandstop filter*. The frequency at which a filter has a transition between passing and stopping frequencies is called the cut-off frequency.

Only in the ideal world can a filter change abruptly from passing to stopping frequencies in its frequency response. In the real world, such abrupt changes are not possible. The transition region between passing and stopping frequencies would be gradual and would have a slope associated with it. The steepness of this slope would determine the cost and complexity of the filter. Also, the passband would usually not be perfectly flat, but would have slight ripples in its frequency response. The size of these ripples would be less, depending on the specific design and complexity of the filter. The various trade-offs between the desired frequency characteristics of the filter and its cost and complexity are an engineering design decision.

Filters

Definition:

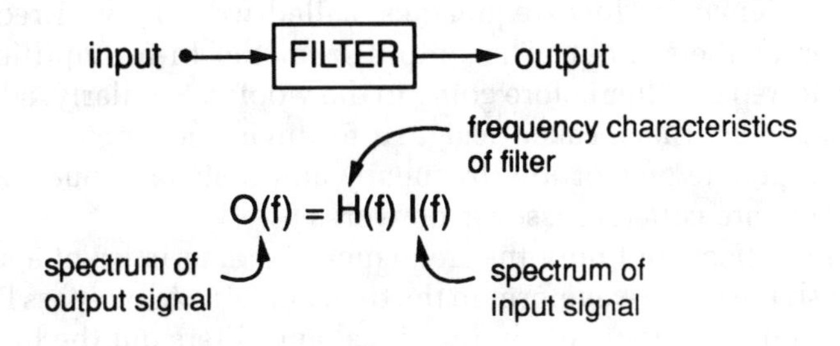

Types:

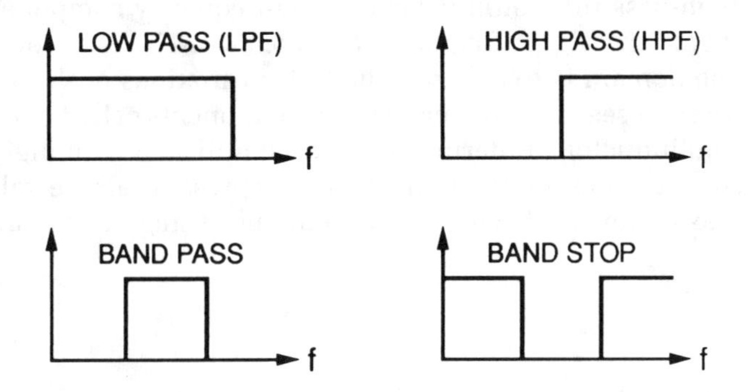

Cut-Off Frequency:

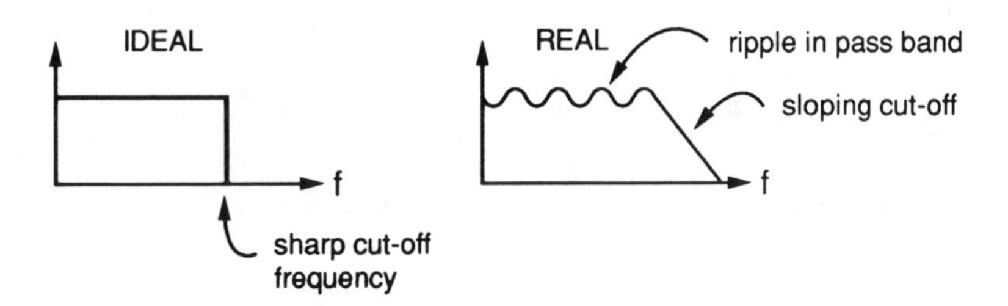

FILTERS (cont'd)

Filters are very useful in electronics when it is necessary to change the spectral shape of a signal. In hi-fi, filters are used in loudspeakers so that the speakers designed for low frequencies, called *woofers*, will receive only low frequencies at their input. The output from the hi-fi amplifier then passes through a low-pass filter before going to the woofer. Similarly, a high-pass filter is used to protect the speaker designed for high frequencies, called a *tweeter*, from the large energies of low frequency audio signals. Such filters used for loudspeakers are called crossover networks.

A filter affects not only the frequency characteristics of a signal, but also the actual shape of the waveform in the time domain. A low-pass filter passes the lower frequency components in the signal and filters out the higher frequency components. The effect of this on the time-domain waveform is to smooth out the faster variations in the waveform shape. A low-pass filter then is a waveform smoother. A high-pass filter eliminates the low-frequency components in the signal, leaving only the higher frequency components. The effect on the waveform in the time domain is to enhance the faster variations of the waveform.

Filters, we shall see later, are made from components called reactors, such as capacitors and inductors. Filtering can be accomplished with digital signals by performing calculations on the numbers that represent sample values of the waveform. For example, the calculation of a moving average accomplishes low-pass filtering.

Filters (cont'd)

Loudspeaker Crossover Network:

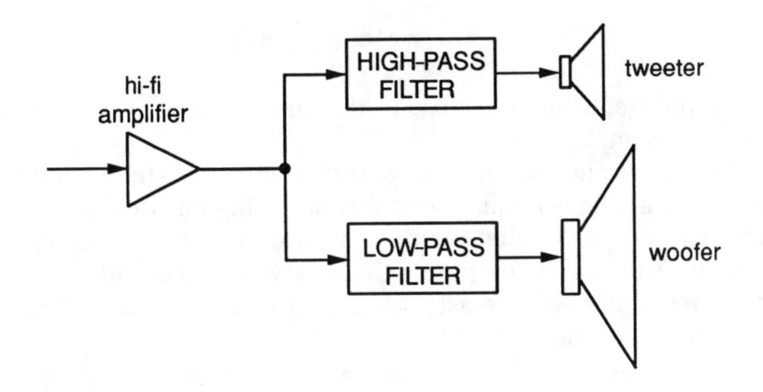

Smoothing:

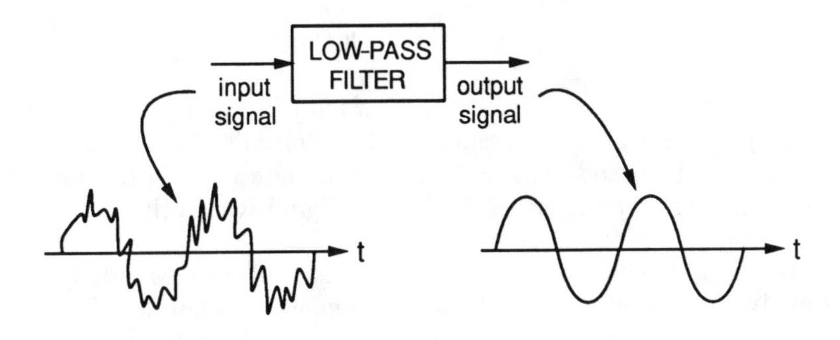

IMPULSE RESPONSE

The frequency response of a circuit $H(f)$ is important, since, if it is known, the spectrum of the output signal $O(f)$ can be obtained simply through multiplying the input spectrum $I(f)$ by the frequency response:

$$O(f) = H(f)\, I(f)$$

In effect, the frequency response of the circuit molds or shapes the spectrum of the input signal.

We saw earlier that one way to determine the frequency response of a circuit is by a single frequency at a time, using an oscillator as input to the circuit. The output of the circuit is measured and plotted as a function of frequency as the input frequency is slowly changed and swept through the range of frequencies of interest. The output of the oscillator ideally is constant as a function of frequency.

The preceding method is tedious and time consuming. Another method to determine the frequency response of a circuit is based on signal processing techniques. The fequency response of a circuit equals the output spectrum divided by the input spectrum, or

$$H(f) = \frac{O(f)}{I(f)}$$

If the input spectrum were flat or constant with frequency, the output spectrum would equal the frequency response of the circuit. The signal with the flat spectrum could be used as input to the circuit, and we would obtain the output signal. A spectrum analysis of the output signal would then produce the frequency response of the circuit.

The signal that has a flat spectrum is the *impulse*. An impulse is an extremely sharp spike. Its spectrum contains equal amounts of all frequencies within its bandwidth. The output of a circuit with an impulse as its input is called the *impulse response* of the circuit. A spectrum analysis of the impulse response then gives the frequency response of the circuit.

A sharp hand clap is a close approximation of an impulse. So, acousticians enjoy wandering around concert halls clapping their hands to obtain a rough idea of the acoustic characteristics of the hall.

Impulse Response

Frequency Response:

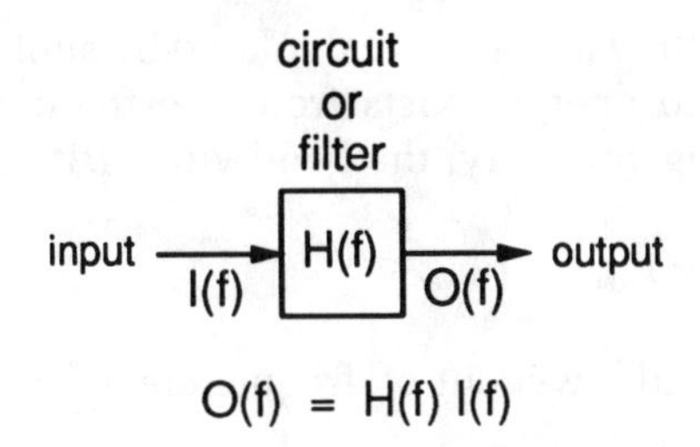

Impulse Response:

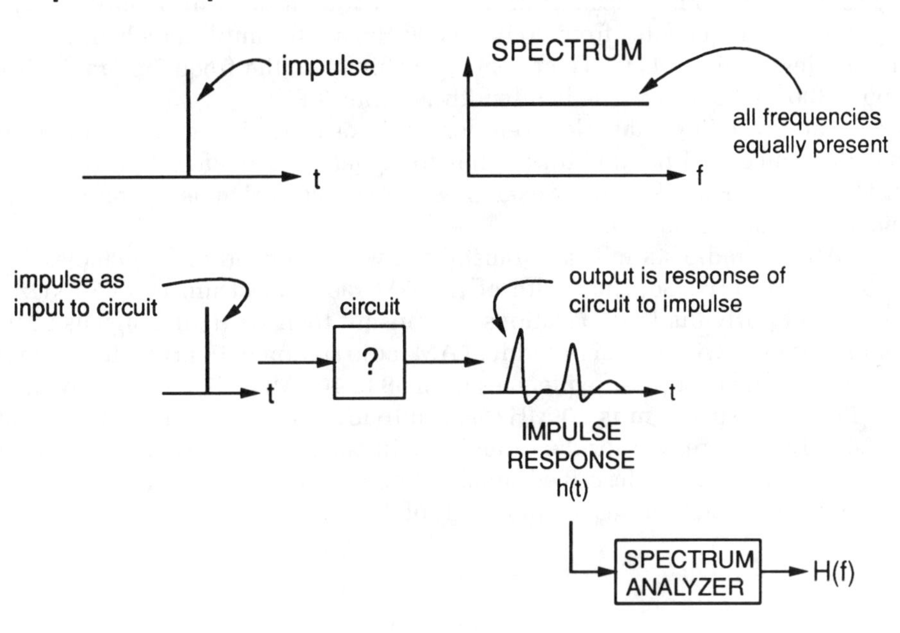

BANDWIDTH

Nearly all real signals have spectra that cover a finite range or band of frequencies. The width, in frequency, of this band of frequencies is called the *bandwidth* of the signal.

Mathematically, the bandwidth of a signal is calculated by subtracting the frequency below which little or no energy exists from the frequency above which little or no energy exists. Algebraically, the bandwidth *BW* is

$$BW = F_{\text{upper}} - F_{\text{lower}}$$

where F_{upper} and F_{lower} are the upper and lower cut-off frequencies of the spectrum of the signal.

A communication channel or circuit has a frequency response, and so it has a bandwidth associated with that response. In this case, the bandwidth represents the width of the band of frequencies passed by the channel or circuit. Similarly, any communication device has a bandwidth. A hi-fi amplifier typically passes frequencies from 20 to 20,000 Hz, so the amplifier's bandwidth is a trifle less than 20 kHz. Telephone speech covers the frequency range from about 300 to 3,300 Hz. The bandwidth is about 3 kHz.

The preceding examples were for audio or acoustic signals and devices, but the concept of bandwidth applies to signals and devices that operate in any medium. The next set of examples are for radio signals operating in the electromagnetic medium.

All AM radio stations are transmitted within a band of frequencies from 550 to 1,600 kHz. So, the width of the AM radio spectrum is 1,050 kHz. A number of individual radio stations all transmit their particular signals in this band. The bandwidth of an individual AM radio station is 10 kHz. The FM radio spectrum band occupies frequencies from 88 to 108 MHz. So, the bandwidth of the FM radio spectrum is 20 MHz. An individual FM radio has a bandwidth of 200 kHz. The reason for the differences in bandwidth between AM and FM radio stations will be described much later. The signal broadcast over the air by a television station has a bandwidth of 6 MHz.

Bandwidth

Definition:

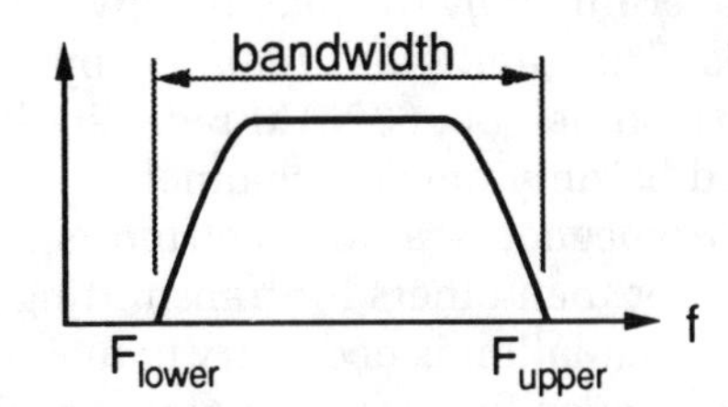

$$\text{bandwidth} = F_{upper} - F_{lower}$$

Examples:

Signal	Bandwidth
telephone speech	3,000 Hz
hi-fi stereo	20,000 Hz
AM radio station	10 kHz
FM radio station	200 kHz
TV station	6 MHz

BANDWIDTH (cont'd)

Nearly all communication media are limited in the bandwidth that they can pass, and so the number of communications channels that can be carried by the medium is determined by the total bandwidth and the bandwidth of the individual channels. The bandwidth required by a TV channel is 600 times that of an AM radio channel. So, 600 AM radio stations could be placed in the spectrum space used by only one TV channel.

Since the electromagnetic radio spectrum is not limitless and some frequency bands are better than others for transmitting commercial radio and TV signals, some form of regulation is necessary to maintain order in the allocation of radio frequencies for specific uses. Economic considerations also arise when deciding which types of signals will be transmitted or carried over a specific width of spectrum. For example, one TV channel might occupy a bandwidth that otherwise could be used to carry 1,500 telephone circuits.

The upper and lower frequencies that determine the bandwidth do not occur abruptly, and there will usually be some energy below and above the band of frequencies that determine the bandwidth. However, the energy in this *out-of-band* range of frequencies will usually be insignificant. The upper and lower cut-off frequencies are frequently defined as the frequency at which the amplitude spectrum has decreased by a factor of two.

Bandwidth (cont'd)

Channels:

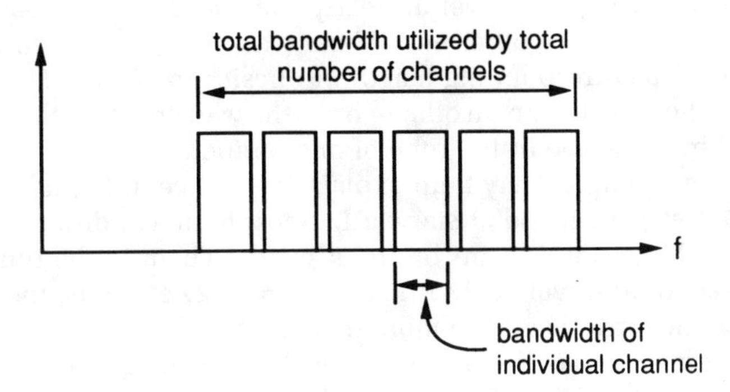

Bandwidth Determination:

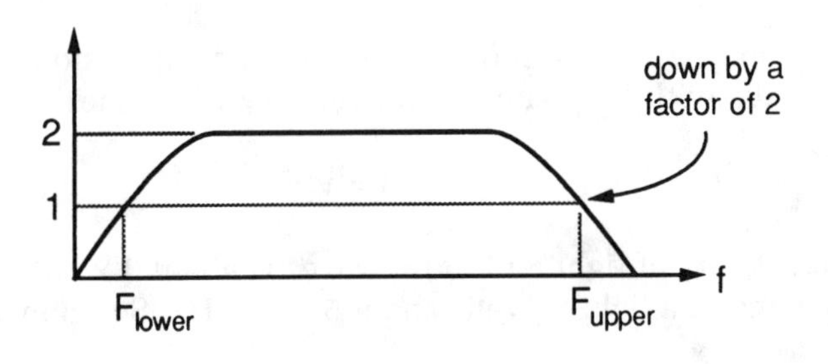

WAVELENGTH

Waves usually travel at some *velocity* over physical distance. For example, sound waves travel at the speed of sound from one end of a room to the other end. Electrical signals travel at nearly the speed of light from one end of an electrical wire to the other end. If the wave were frozen in time, the period would have a corresponding length in physical distance. This distance of one period is the wavelength. In other words, the wavelength is the physical distance traveled by the wave in the time of one period.

Some examples may help explain this concept. Sound travels at a speed of 1,127 feet per second at standard atmospheric conditions. For a sine wave of 1 kHz, the period is 1 ms or 10^{-3} seconds. Then, in the time of one period, the wave would travel 1,127 ft/s $\times$ 10^{-3} s = 1.127 ft. Thus, the wavelength of a 1 kHz sound wave is a little more than 1 foot.

The mathematical expression for the wavelength λ (Greek lambda) of a wave with a velocity v and period T is

$$\lambda = vT$$

Since frequency is the reciprocal of period, the expression for the wavelength of a sine wave with frequency F and velocity v becomes

$$\lambda = v/F$$

The velocity of light and radio waves is about 3×10^8 meters per second. The frequency of a light wave is about 5×10^{14} Hz. So, the wavelength of light is

$$\lambda_{light} = (3 \times 10^8)/(5 \times 10^{14}) = 0.6 \times 10^{-6} \text{ meters}$$

The distance of 10^{-6} meters is called a *micron*, abbreviated μm. Light is not a single frequency, but covers a range, or *spectrum*, of frequencies all with corresponding wavelengths. The *visible light spectrum* ranges from 0.4 to 0.7 μm for colors ranging from ultraviolet to infrared.

The term *angstrom* (abbreviated Å) is used for the extremely small wavelengths, characteristic of x-rays and other very high frequency waves. One angstrom equals 10^{-10} meters.

Antennas used in radio usually have lengths proportional to the wavelength of the radio wave. Hence, radio engineers usually express frequencies as their equivalent wavelengths. The lengths of organ pipes are proportional to the wavelengths of their resonant frequencies.

Wavelength

Definition:

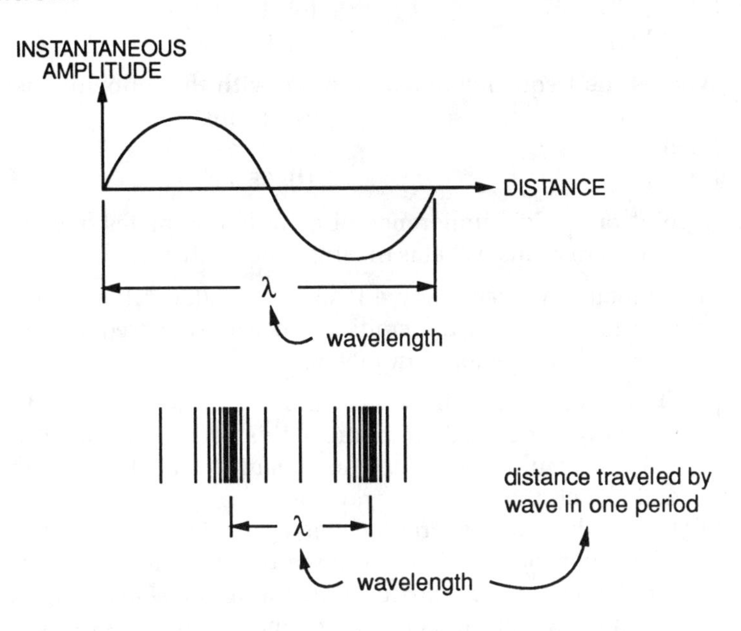

Equations:

$$\lambda = vT$$

$$\lambda = \frac{v}{F}$$

where λ = wavelength, v = velocity, T = period, F = frequency

$v_{sound} \approx 1{,}100$ ft/sec

$\lambda_{1\,kHz} \approx 1$ ft

$v_{light} \approx 3 \times 10^8$ meter/sec
or 186,000 miles/sec

$\lambda_{light} \approx 0.6\ \mu m$

PROBLEMS

1.1. What is the period of a sine wave with the following frequencies?
(a) 1 kHz
(b) 2 MHz
(c) 5 kHz
(d) 5 GHz
(e) 100 Hz
(f) 0.5 Hz

1.2. What is the frequency of a sine wave with the following periods?
(a) 0.5 s
(b) 5 s
(c) 2 μs
(d) 10 ms
(e) 2 sec
(f) 25 μs

1.3. A point on the circumference of a circle completes one-quarter of a full revolution in 25 ms. What is the frequency with which the circle is turning?

1.4. The velocity of a radio wave is about 300,000 kilometers per second. A cellular radio system transmits radio waves at frequencies near 1 GHz. What is their wavelength in meters?

1.5. (a) On a piece of graph paper, carefully draw one period of sine wave. Then draw the third harmonic of this sine wave but with an amplitude 1/3 the amplitude of the fundamental. Add together the two waves and sketch the resultant waveform.
(b) On another piece of graph paper, again sketch this resultant waveform. Then plot the fifth harmonic with an amplitude which is 1/5 that of the fundamental. Add together these two waveforms. Note that the resultant waveform is becoming more and more like a square wave.

Chapter 2: Electricity

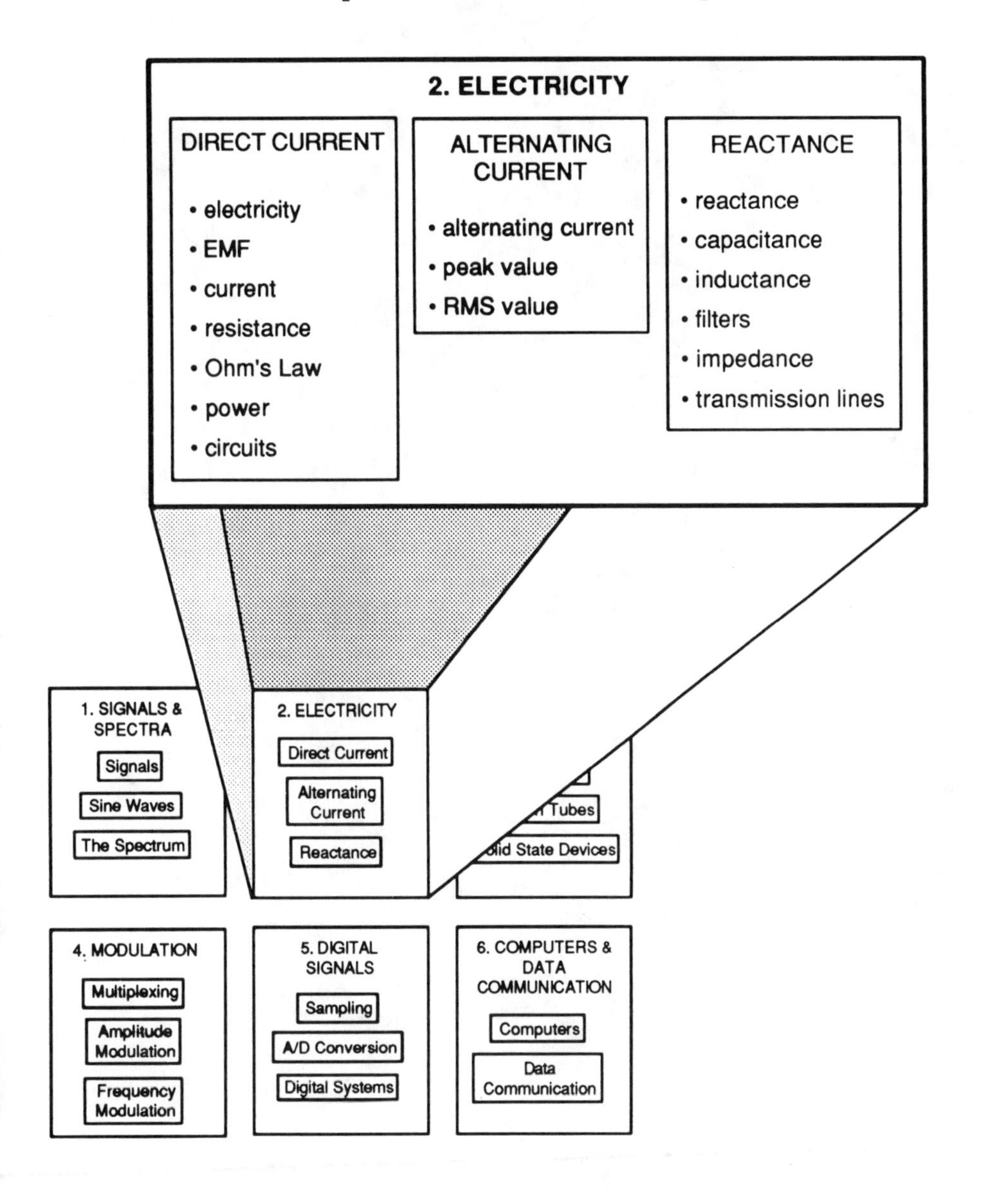

2

Introduction to Electricity

Nearly all signals used in modern communication systems are electrical at one time or another. So, it is important to understand electricity in general and more specifically the various effects encountered in electrical circuits with time-varying signals.

Two effects, *capacitance* and *inductance*, occur in electrical circuits with time-varying signals. These two effects can occur on a stray basis and will limit the bandwidth of electrical transmission circuits. These effects can also be deliberately introduced into electrical circuits to shape and control the spectrum of signals, thereby creating filters of various kinds.

To understand the effects of capacitance and inductance, we need to know about electrical circuits and how they are created from various circuit elements connected in *series* and in *parallel*. *Electricity* is the flow of electrons in an electric circuit. The *electrons* are caused to flow by an *electromotive force* that is measured in *volts*. The number of electrons flowing per second is the *current*, which is measured in *amperes*. The flow of electrical current is opposed by *resistance* and the effects of capacitance and inductance.

This module begins with an explanation of electricity in terms of an analogy to the flow of marbles closely packed together in a pipe and caused to flow in response to a constant force. A constant electromotive force in an electrical circuit causes a constant current that is called *direct current*, or dc. A time-varying force causes an *alternating current*, or ac. The electromotive force, the current, and the opposition to the flow of current are related by *Ohm's law*.

Series and parallel circuits are described. The effects of capacitance and inductance are explained in terms of their action on time waveforms and also their effects on the different frequency components present in a signal. The module ends with the practical application of capacitance and inductance to the understanding of the bandwidth limitations of transmission lines and the use of these two effects to create filters.

Direct Current

ELECTRICITY

Electricity refers to the flow of charged particles through some medium. If the medium supporting the flow has an abundance or surplus of charged particles, it is called a *conductor*. If there is an absence of surplus particles, the medium cannot support any flow and is called an *insulator*. Some media are better conductors than others so that a range exists from the best conductors to the best insulators. Media with qualities that lie in between are called *semiconductors*. During the early days of electricity, these charged particles were called electrons, from which we have the term "electricity," meaning the flow of electrons.

The flow of charged particles does not occur by itself. An electron-moving force is needed to cause the electrons to flow, or to move. This force is called an electromotive force, or EMF for short. The EMF is measured in volts. An EMF causes the surplus charged particles in a conductor to push upon each other, thereby causing a flow of charged particles. The number of charged particles flowing in a unit amount of time is called the current. The current is measured in amperes.

Some EMFs are constant with respect to time and cause the flow of a constant current. Such a constant current is called direct current (dc). Other EMFs vary with respect to time, alternating the direction of their force and also the direction of the current. Such a current is called an alternating current (ac). One example of an alternating current is a current that varies with the shape of a sine wave, or varies sinusoidally.

Certain effects encountered in electricity depend on whether the current is dc or ac. The ac-dependent effects vary with frequency, thereby becoming a means to create filters of various kinds and characteristics.

Electricity

Electricity:

The flow of electric charge

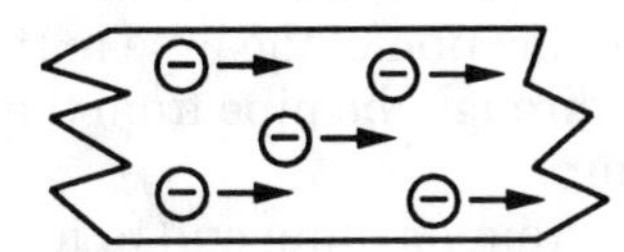

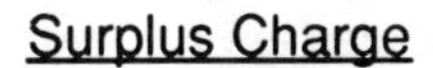

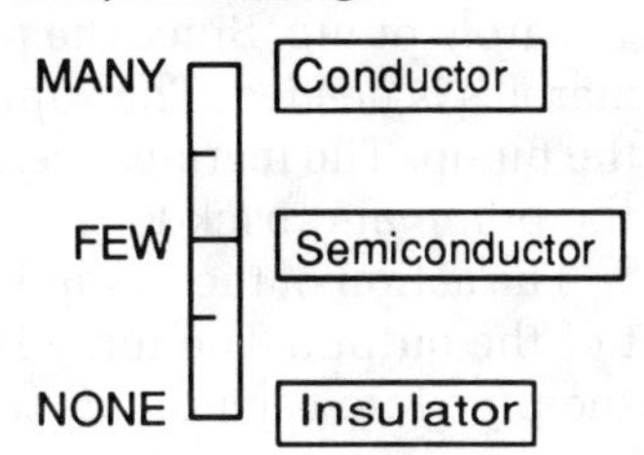

Electromotive Force (EMF):

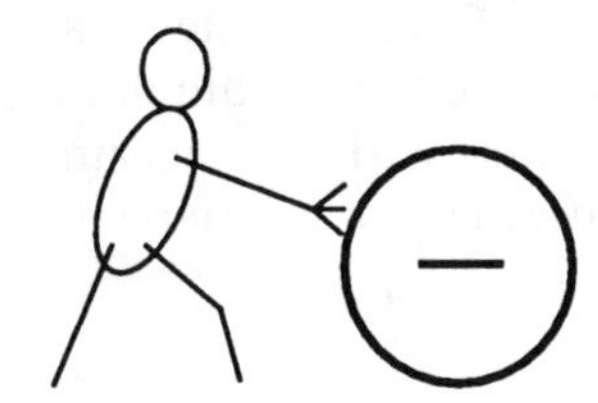

Current:

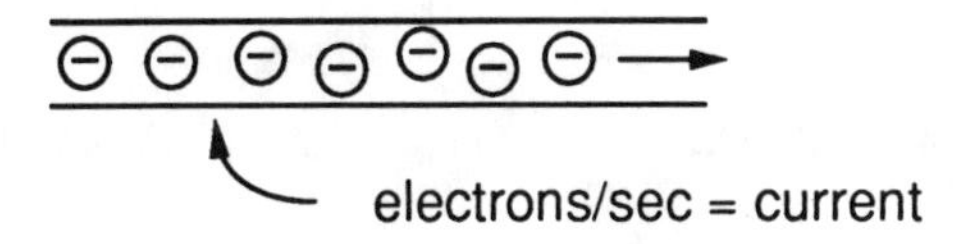

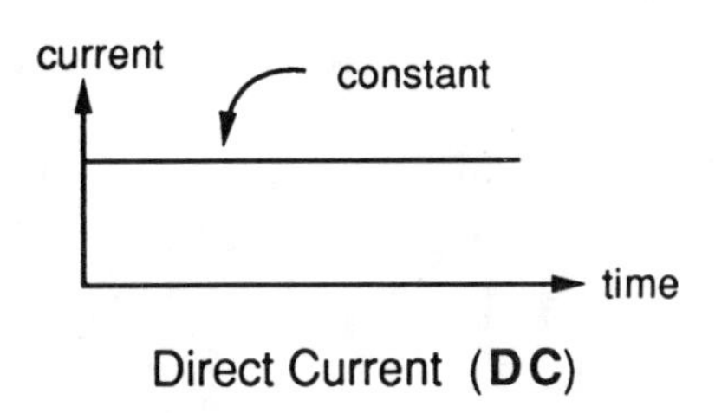

Direct Current (**DC**)

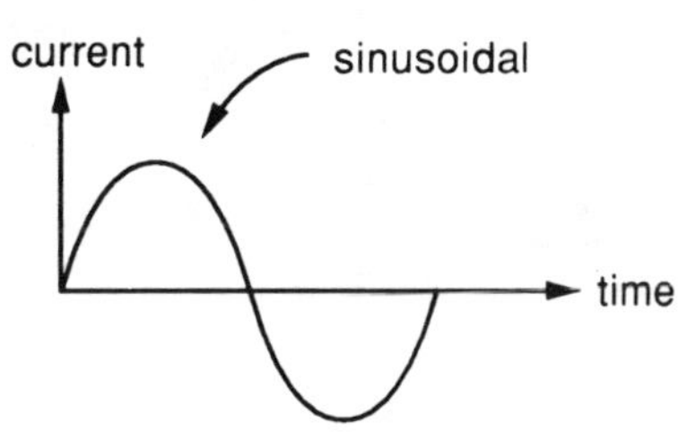

Alternating Current (**AC**)

ELECTRICITY ANALOGY

Electricity is frequently explained through the use of analogies. The analogy that we will use here is the flow of marbles in a frictionless pipe. The marbles are densely packed together in the pipe so that the marbles all touch each other.

The marbles are caused to move through the pipe by the force generated by a marble pump. Since the pump cannot create marbles, a continuous supply of marbles is needed. The supply is a return connection of the pipe to the input to the pump. The marbles then make a circuit through the pipe from the output of the pump and back to the input to the pump.

The action of the pump is to take marbles from its input and to force them out of the output. The force is sufficient to push the densely packed marbles in the pipe into each other, causing a continuous flow of marbles through the pipe.

A paddle wheel can be placed in the pipe in such a way that the marbles passing through the pipe cause the wheel to turn. The more friction that the wheel offers to the flow of marbles, the fewer marbles that flow in the circuit per unit of time, assuming that the force of the marble pump remains constant. If the pump force increases, more marbles will flow per unit of time, assuming that the friction or resistance to the flow of marbles offered by the wheel remains constant.

Electricity Analogy

Marbles:

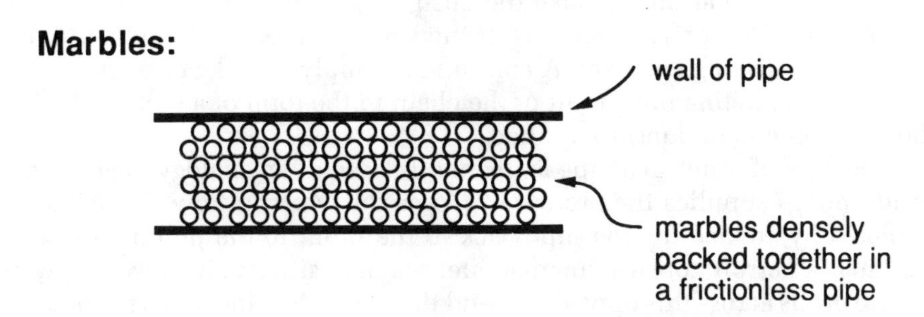

Marble Circuit:

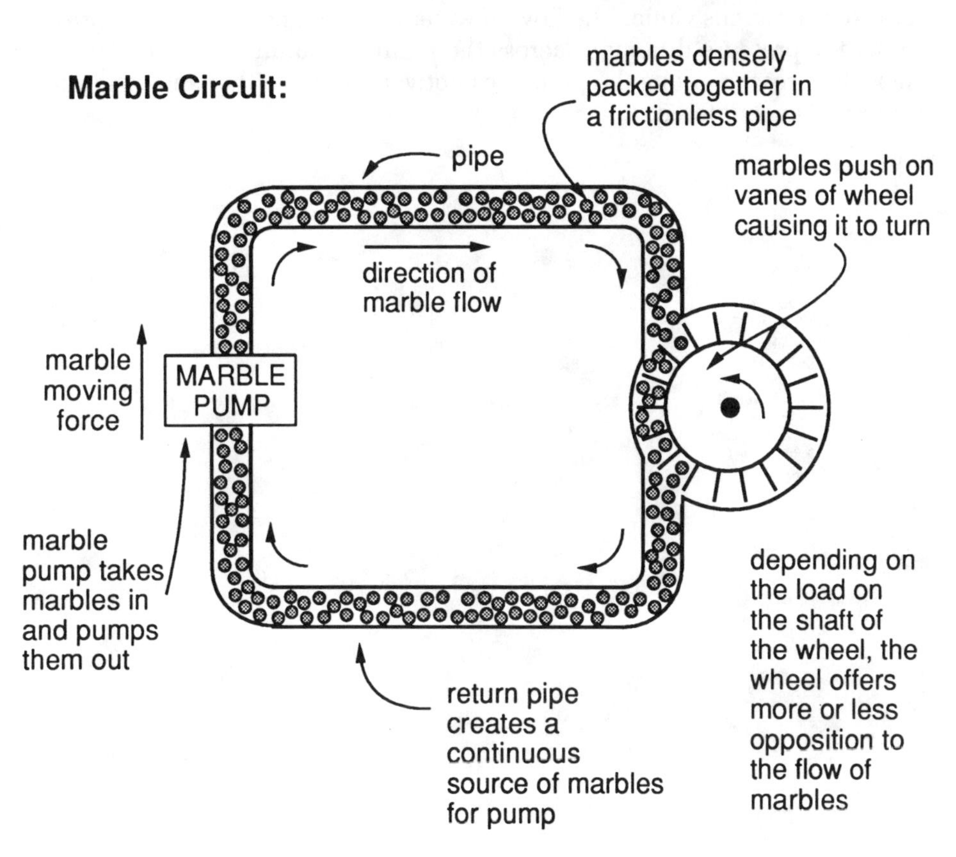

ELECTRICITY ANALOGY (cont'd)

An analogy that might make the need for a complete circuit clearer is a bicycle chain. The driving force is carried by the links in the chain and is created at the driving sprocket. A continuous supply of links is needed and is created by completing the circuit of the chain in the form of a loop. If the loop is broken, force is no longer supplied.

The flow of water in a pipe is frequently used as an analogy to electricity. A water pump supplies the pressure to force the water to flow. The circuit is completed by connecting the pipe back to the input to the pump. The water encounters a narrow constriction that offers an opposition to its flow. The water pressure drops across this opposition, and the pump then increases the pressure back to its previous value. The flow of water is caused not by absolute pressure but by the pressure difference across the pump. Assuming a frictionless pipe, the water pressure along the pipe does not vary unless either an opposition or a pump is encountered.

Analogy (cont'd)

Bicycle Chain:

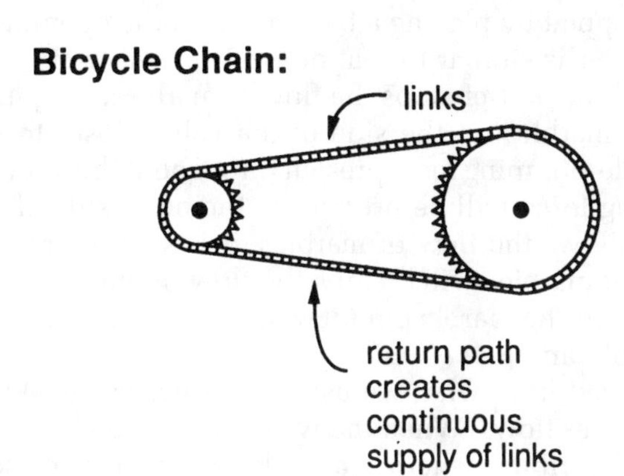

Water Flow:

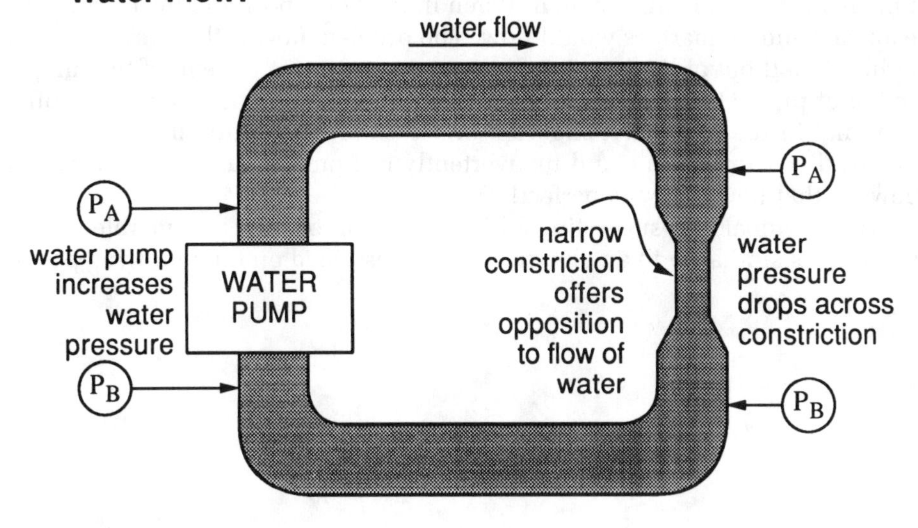

SWITCH ANALOGY

The flow of marbles can be stopped by placing a barrier across the opening in the pipe. The effect of the barrier is similar to that of a valve used to stop the flow of water through a pipe. If the barrier stops the flow of marbles, the pump will still attempt to push the marbles on the side of the valve closer to the pump, and there will be a marble-pumping force present throughout this portion of the pipe. No marble-pushing force will be present on the other side of the barrier. If the marble switch allows the flow of marbles, it is said to create a *closed circuit* of marbles. If the marble switch stops the flow of marbles, it is said to create, in effect, a break in the marble circuit. Such a break is called an opening in the circuit, or simply an *open circuit*.

This terminology is reversed from what we use for a water valve. When the water valve is closed, no water flows. When the water valve is open, water flows. All this is similar to describing whether a bridge is open or closed, namely, open to car traffic but closed to boat traffic.

The confusion in terminology can be eliminated by designing a section of hinged pipe as a marble switch. When in the "off" position, the circuit will be opened and no marbles would flow. The problem now is that we expect the marbles to fall out of the opening in the open portion at the end of the hinged section of pipe. One way to prevent this from happening is to make the pipe "special," in the sense that should a break occur it would seal itself at both ends so that no marbles could inadvertently roll out. What all this confusion shows is that no analogy is perfect!

The rationale for suggesting a hinged section of pipe as a marble-circuit switch is its similarity to an electric switch, described on the following pages.

Switch Analogy

Marble-Circuit Switch:

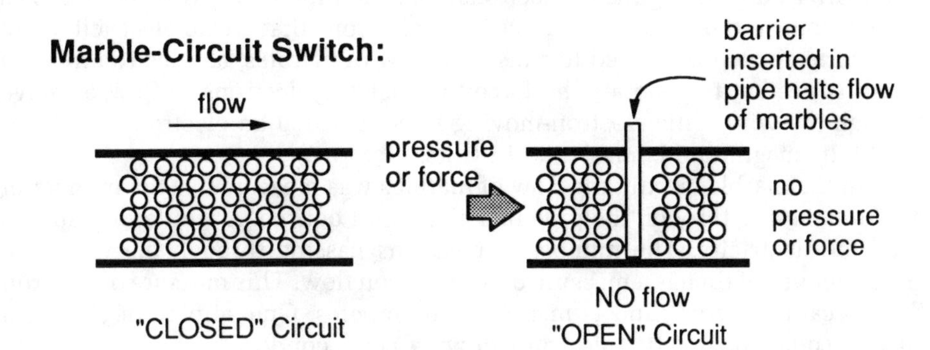

Terminology Confusion:

OPEN Circuit = no flow of marbles (or electric current)

CLOSED Circuit = flow of marbles (or electric current)

Hinged-Pipe Marble Switch:

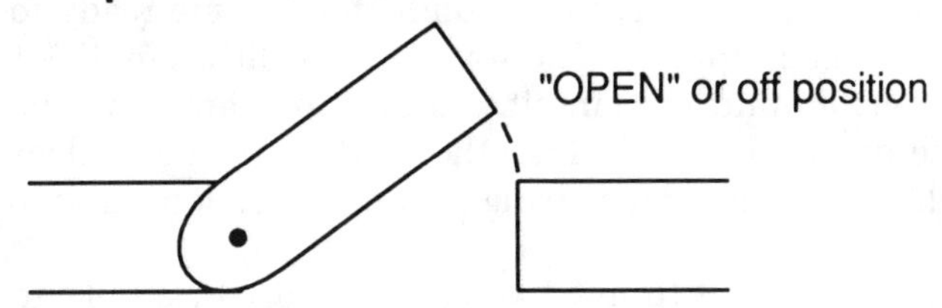

Electric Switch:

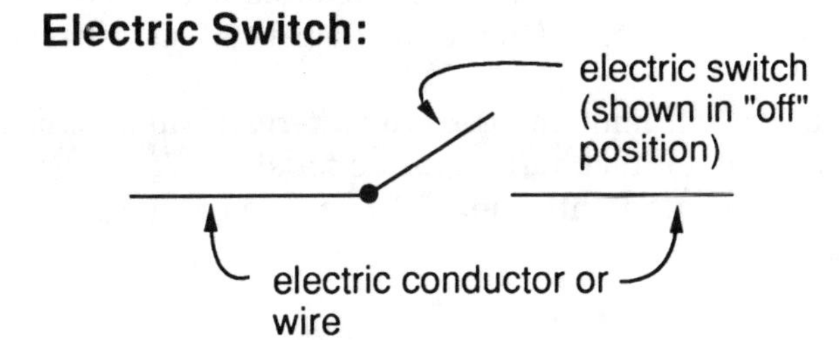

ELECTRIC CIRCUIT

The marbles are analogous to electrons, and the pipe that carries the marbles is analogous to the wire that conducts the electrons that create electricity. Just as a marble pump is needed to cause the flow of marbles, an electron-moving pump is needed to generate the force to cause the electrons to flow or move through the wire. This electron-moving force is called an electromotive force (EMF). Its magnitude is measured in volts.

In the marble circuit, the flow of marbles was measured as the number of marbles passing through the pipe in a given unit of time. A similar measure is used for electricity. The number of electrons passing through the wire in a given amount of time is a measure of the electron flow. This measure of electron flow is called current and is measured in amperes. One ampere of electrical current equals 6.28×10^{18} electrons flowing per second.

The flow of electricity can be stopped by breaking open the wire through the use of a *switch*. An electric switch opens or closes the circuit, according to whether the switch is in the "off" or the "on" position. With a marble-circuit switch, a barrier that does not conduct marbles is used to interrupt the flow of marbles in the marble circuit. An electric switch interrupts the flow of electrons in the electric circuit by placing a barrier in the form of an insulator in the circuit. The insulating barrier is the "cutting" of the wire by the action of the electric switch.

Electrons are always present in a conductor and are ready to flow when an electromotive force is applied. The source of the electromotive force simply "pumps" electrons around the circuit. No electrons are generated or lost by the EMF source or in the circuit. Like the marbles in the analogy above, the electrons are the means for transferring a force, or energy, from one place to another.

Electricity does useful things like causing heat in an electric heater or creating light in an electric light bulb. Such devices offer opposition or resistance to the flow of electrons when placed in an electric circuit. These devices are shown symbolically as a zig-zag line, and they are called electrical resistance, or simply a *resistor*.

One source of a constant electromotive force is a *battery*. A battery creates a constant current, what we call a direct current, or dc for short. A dc current does not change its value or direction with time. It is nontime varying.

Electric Circuit

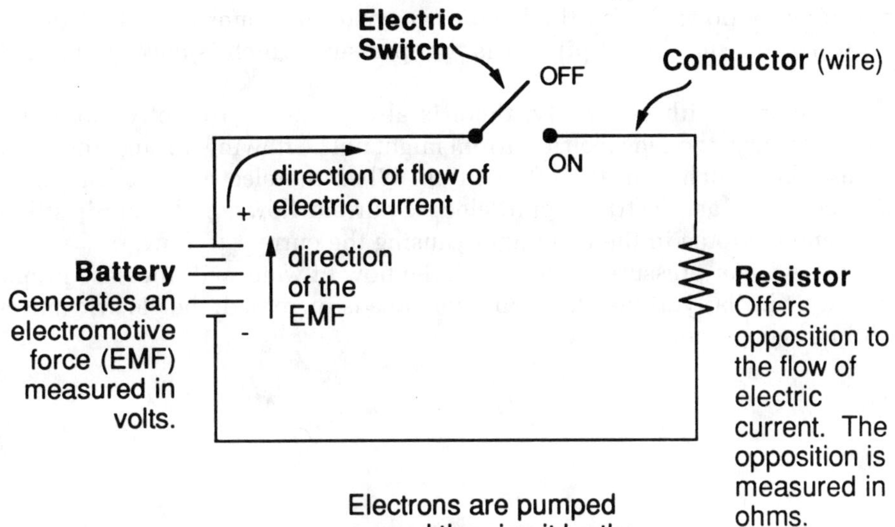

Electrons are pumped around the circuit by the EMF generated by the battery.

The EMF source does not generate electrons and hence needs a continuous supply to pump. This supply comes from the completed return circuit.

CURRENT FLOW

There is always water pressure in a pipe even though the valve is turned off and no water flows. When the valve is opened, the water flows and there is still water pressure in the pipe. It is that pressure which is making the water flow.

Similarly, with electricity, there is always an electromotive force in a conductor, even though electric current might not be flowing through the circuit because the switch is in the off position. When the electric switch is turned on to complete an electric circuit, electric current flows, and there is still an electromotive force in the conductor causing the current to flow.

If the water pressure is increased, the flow of water will increase proportionately. Similarly, if the electromotive force is increased, the electric current will increase proportionately.

Current Flow

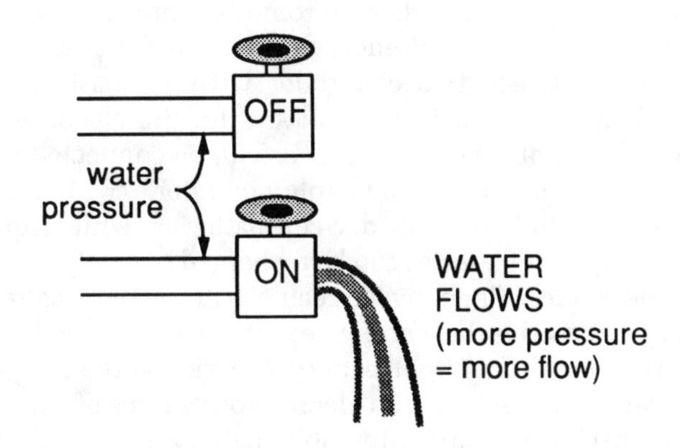

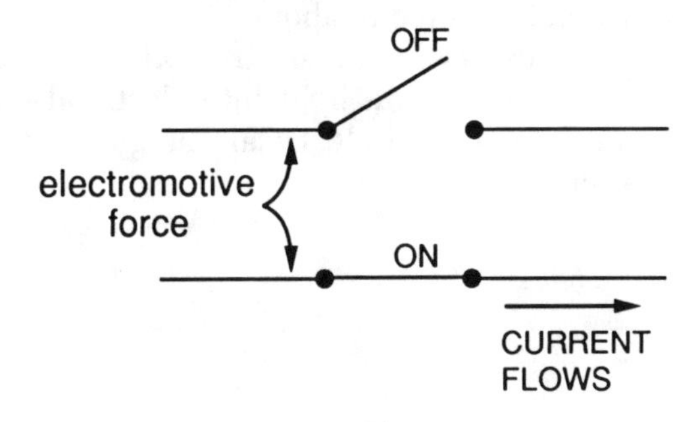

EMF SOURCES

The most standard source of a dc electromotive force is a battery. A battery is a device that converts chemical energy into electrical energy. The conversion of energy occurs between two electrodes within a basic unit called a *cell.* The specific chemicals used in the cell and for the electrodes determine the electromotive force that is generated. Cells can be connected together in series to obtain larger voltages, as will be explained later.

A chemical paste is used in dry-cell batteries, while liquids are used in wet-cell batteries. Some batteries can be recharged by applying an electromotive force at the electrodes. Such rechargeable batteries are called *secondary* or *storage batteries.* Batteries that cannot be recharged are called *primary batteries.*

The conventional flashlight battery is a zinc-carbon dry-cell battery. Its electrochemical action generates an electromotive force of 1.5 volts at the terminals of each cell. The standard automobile battery is a lead-cell storage battery consisting of six cells connected together internally in series. Each individual cell generates an electromotive force of about 2 volts, and the storage battery itself generates a total electromotive force of about 12 volts.

A number of other devices can create dc electromotive forces. A solar or photovoltaic cell converts the energy from light into electrical energy. A dc generator converts mechanical force into electrical energy. A thermocouple converts heat into electrical energy.

EMF Sources

Battery:

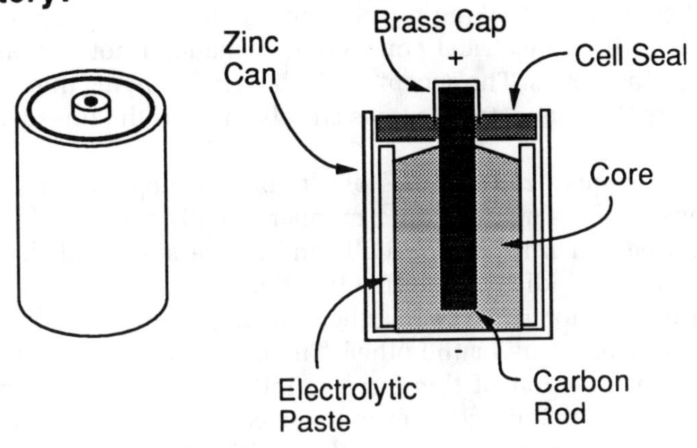

Zinc-Carbon Dry Cell

Other Sources:

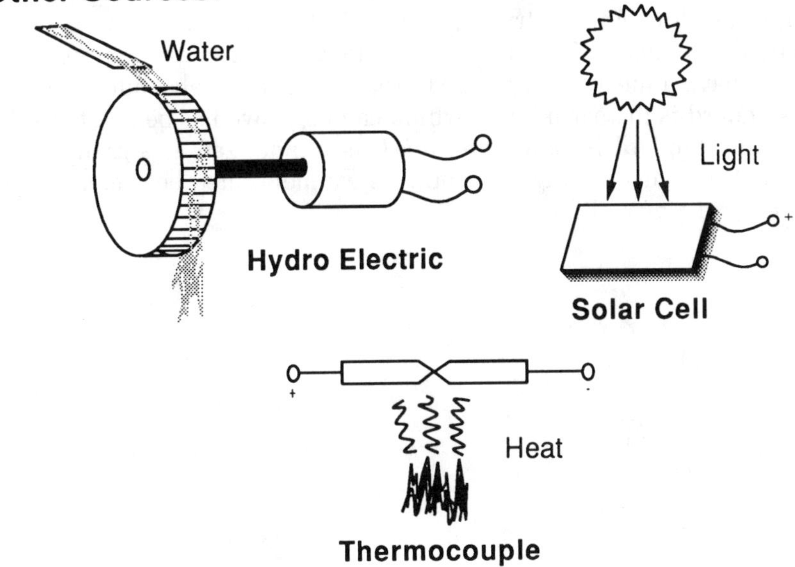

Hydro Electric

Solar Cell

Thermocouple

SYMBOLS

Symbols are used in electric circuit diagrams to depict various devices and circuit elements. Straight lines depict electrical conductors. If two conductors cross each other without making an electrical connection, a little bump is sometimes drawn. If an electrical connection is made, a dot is drawn at the point where the lines cross. The accepted symbols for showing that no connection occurs where the connectors cross is simply to cross the lines without dot or bump.

A battery is symbolized as pairs of alternating long and short lines. A battery can consist of a single cell or a number of cells connected together in series. A single pair of lines will usually indicate a single-cell battery, and multiple pairs of lines indicate a multicell battery.

An electromotive force can be in either one direction or the other, causing the current to flow one way or the other. Thus, an electromotive force has a direction, and the direction of this force is indicated by the *polarity* of the battery. The force is in the direction from the negative to the positive terminal. By convention, the longer line represents the positive terminal, and the shorter line represents the negative terminal.

An EMF is somewhat like water pressure in the sense that the pressure difference between two points, rather than absolute pressure, determines the rate of water flow. For electricity, it is the EMF difference between two points in a circuit that determines the current. However, in some circumstances, it is useful to have a measure of the EMF relative to some absolute standard. One such standard is the earth itself, which can be viewed as being at absolutely 0 volts. This standard is called *ground.* It is symbolized as a group of parallel horizontal lines decreasing in length. As we mentioned before, a resistance is depicted as a zig-zag line.

Symbols

Conductors:

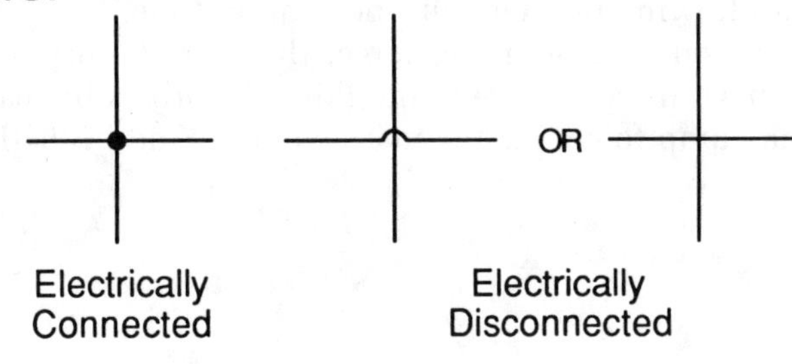

Battery:

+
E Multicell
-

+
Single Cell
-

Resistance:

Ground:

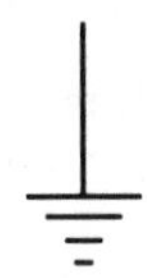

OHM'S LAW

In the marble analogy, let E represent the force generated by the pump, R represent the friction offered by the paddle wheel, and I represent the number of marbles flowing in the circuit per unit of time.

For a constant pumping force, the marble current decreases in direct proportion as the resistance to the flow of marbles increases. Mathematically, if E is the pump force, I is the marble current, and R is the resistance, then,

$$I = E/R$$

or, alternatively,

$$E = IR$$

A similar relationship exists in an electrical circuit between the electromotive force E in volts, the current I in amperes, and the resistance R in ohms. This relationship is called Ohm's law, and is

$$E = IR$$

If the electromotive force E, commonly called the voltage, and resistance R are known, then the current I can be calculated from

$$I = E/R$$

Similarly, if the voltage and current are known, then the resistance can be calculated from

$$R = E/I$$

The accepted abbreviation for ampere is A, volt is V, and the Greek uppercase omega Ω is the abbreviation for ohm. The term "amps" also is frequently used as an abbreviation for amperes.

The various prefixes for powers of ten apply to voltage, current, and resistance. So, a voltage of 10 mV is the same as 10^{-2} volts, a current of 1 mA is the same as 10^{-3} amperes, and a resistance of 1 MΩ is the same as 10^{6} ohms.

Ohm's Law

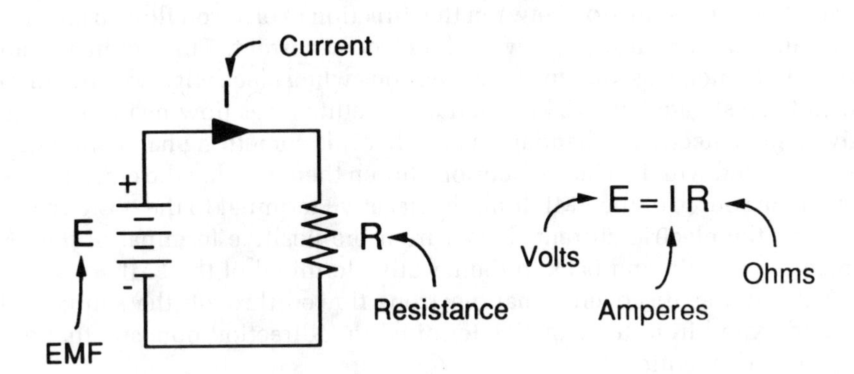

Other Forms:

Given the EMF and resistance, the current is:

$$I = E/R$$

Given the EMF and current, the resistance is:

$$R = E/I$$

DIRECTION OF CURRENT FLOW

There is frequent confusion between the direction of electron flow in an electric circuit and the direction of flow of the electric current. This confusion arose because not much was known about electrons when electricity was first discovered and investigated. It was known that something was flowing in a conductor to give rise to electrical phenomena and that this something had a direction of flow associated with it. The convention chosen then was that the current flowed through the source of the EMF from the negative terminal to the positive terminal. Thus, the electric current flowed from the positive terminal of the EMF, through the circuit, and back to the negative terminal of the EMF.

Later it was discovered that electrons flowed through the source of the EMF and exited from the negative terminal in a direction opposite that of the established convention.

Electrical engineers have become accustomed to the old convention and actually find it useful in analyzing electric circuits. We will follow the old convention here: electric current in a circuit flows in a direction through the EMF source from the negative to the positive terminal. One way of remembering this is to visualize electricity as the flow of positively charged particles, and these positively charged particles flow from the positive terminal of the EMF source.

When an electric current flows through an opposition or resistance, the electromotive force or voltage decreases. Another way of stating this is that an electric current flowing through a resistance creates a voltage across the resistance in such a direction as to oppose the flow of current.

The direction or polarity of a voltage is indicated by the use of plus (+) and minus (–) as labels on the terminals of the EMF, or by an arrow in the direction from negative (–) to positive (+).

Current Flow

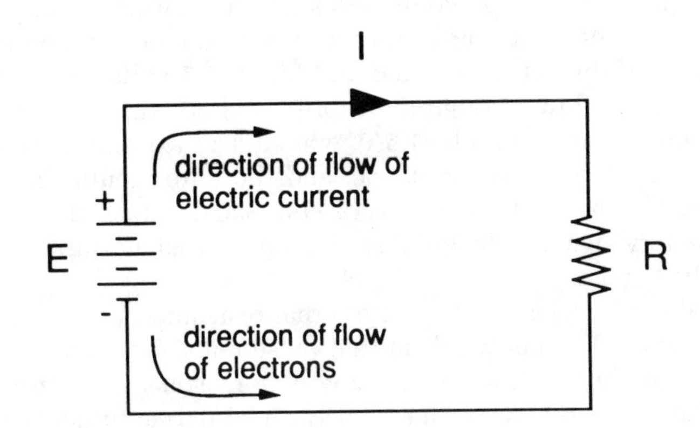

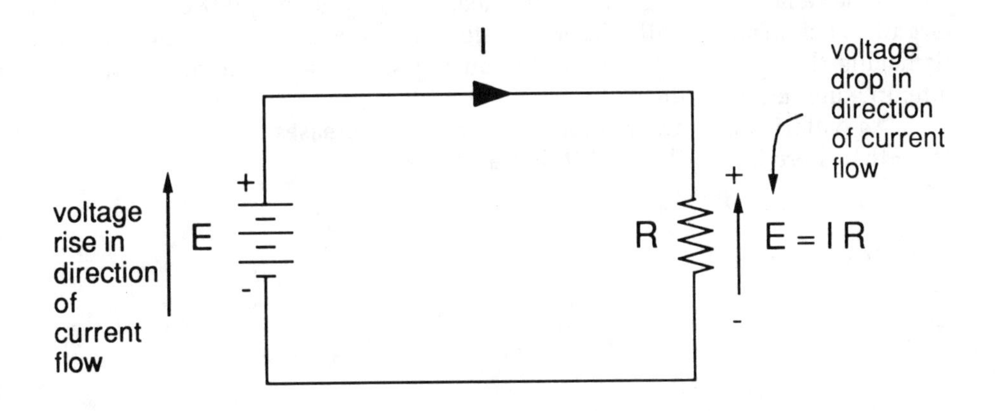

BATTERIES

An interesting question is, what would happen if a wire were directly connected across the terminals of a battery? In theory, a wire is a perfect conductor with no resistance. So, if the battery created an EMF of 1.5 volts, we would expect an infinite current to flow through the short-circuited wire, since the current, according to Ohm's law, would be 1.5/0, which is an extremely large number. Such a large current could cause havoc, and the wire would disappear in a puff of smoke. However, if a wire were connected across the terminals of a flashlight battery, we would quickly determine that nothing catastrophic happened! Why?

In the real world, a battery has an internal resistance, as do wires, though usually quite small. The net result is that these resistances combine to limit the current flowing in the short-circuited wire to reasonable amounts, particularly for a flashlight battery, which has a fairly high internal resistance, typically about 0.3 ohms. Incidentally, an automobile battery is quite a different matter, since it has a fairly low internal resistance, typically about 0.007 ohms. A short circuit across an automobile battery would cause large currents to flow, possibly damaging the battery, and certainly causing sparks and much heat. Do not attempt such an experiment!

As batteries age, their internal resistance increases until it becomes so large that there is very little EMF left at the terminals.

Batteries

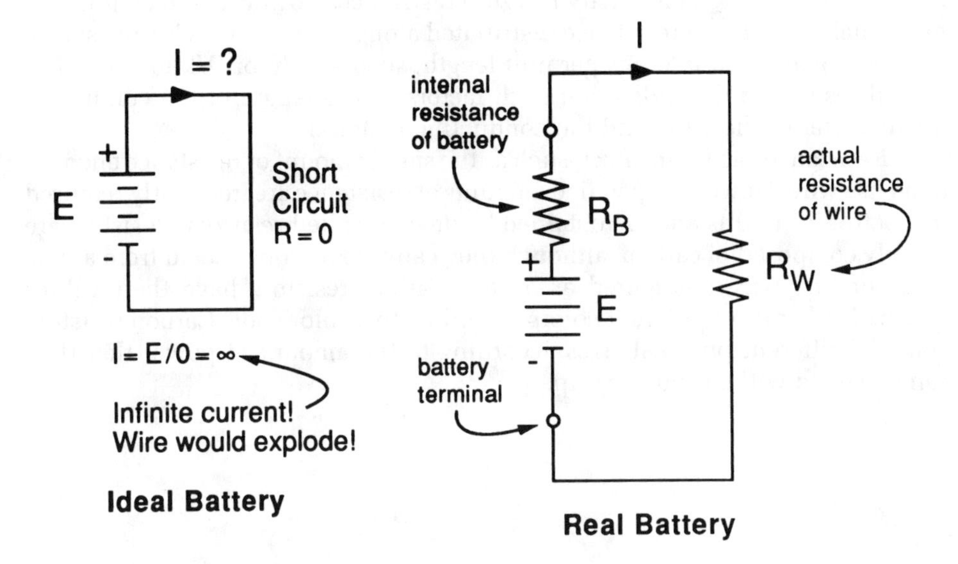

A real battery has internal resistance that limits the flow of current.

RESISTANCE

An ideal conductor of electricity has zero resistance. However, real conductors have small amounts of resistance distributed along their length. This resistance is usually expressed in ohms per unit length, such as Ω/ft or Ω/kilometer. The overall resistance depends upon such factors as the length of the conductor, the diameter of the wire, and the conducting material.

Resistance can be an effect, such as the small amount of resistance encountered in a wire. However, specified amounts of resistance are frequently required in electronic circuits and are obtained by devices called *resistors*. Resistors are usually composed of carbon, although they can also be constructed from a long piece of fine wire configured as a coil. Carbon resistors have their values encoded as bands of different colors according to a color code. Carbon resistors come in different physical sizes according to the amount of power that they can dissipate without burning up.

Resistance

Resistance Effect:

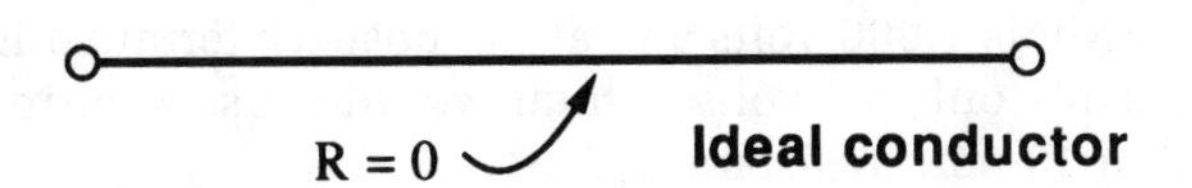

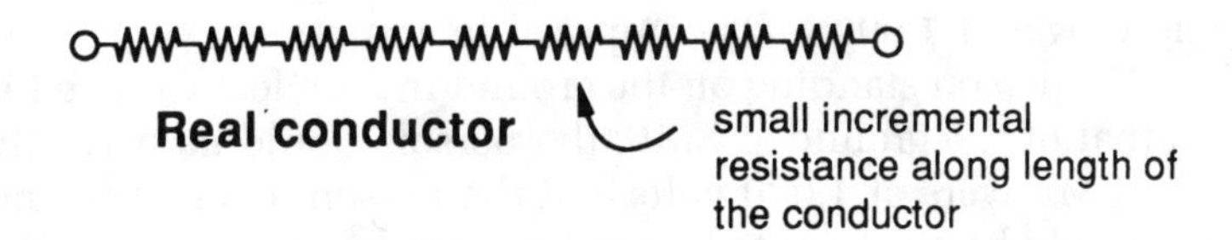

Resistor:

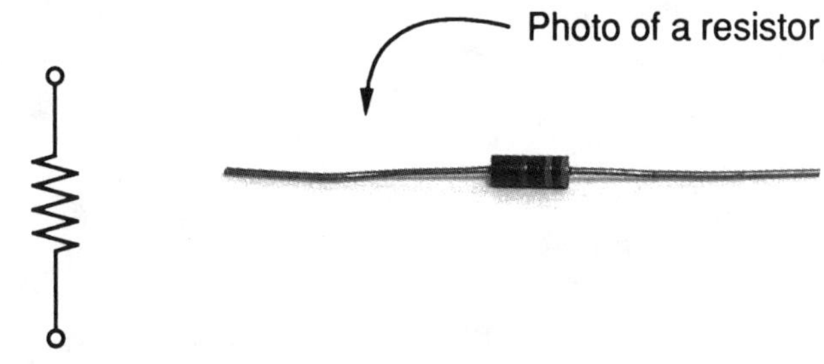

GROUND

It is the EMF across an opposition that causes an electric current to flow. The voltage at some point in an electric circuit can have a value compared to an absolute reference of 0 volts. Suppose the voltage at the negative terminal of a source of an EMF is 1,000 volts and at the positive terminal is 1,010 volts. The EMF difference is only 10 volts, which would cause a current of 1 ampere to flow through a 10-ohm resistance.

The absolute reference of 0 volts is called ground and is symbolized by a series of parallel lines decreasing in length. If in our example we connect the negative terminal of the EMF to ground, the same 10 volts is still across the 10-ohm resistance, causing a current of 1 ampere to flow. However, the EMF is no longer at 1,000 volts compared to ground.

If a person standing on the ground in bare feet were to touch the negative terminal of the grounded EMF, the person would be perfectly safe. However, if the EMF were at 1,000 volts and the person touched the negative terminal, they would be electrocuted since they would have completed a circuit through their body to ground. Grounding is frequently done for reasons of electric safety, which will be explained more thoroughly later in this chapter.

Ground

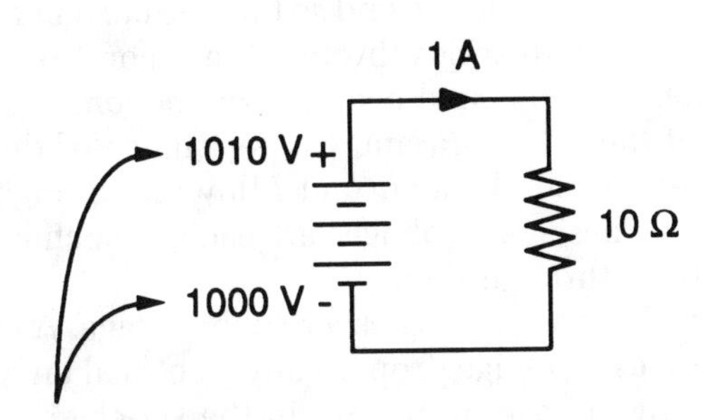

The EMF difference across the source is (1010 V - 1000 V) = 10 V. This 10 V causes a current of 1 A to flow through the resistance of 10 Ω.

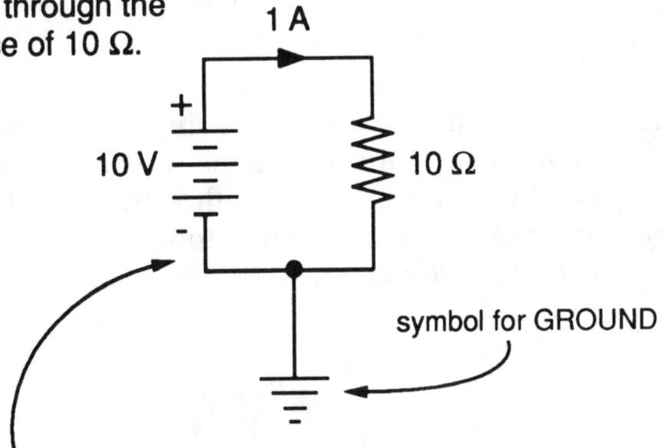

GROUND respresents an absolute EMF of 0 V. The effect in this example is to anchor the EMF at the negative terminal of the source to 0 V. There is still a 10 V EMF across the 10 Ω resistance causing a current of 1 A to flow.

SERIES CIRCUITS

Two resistances can be connected end to end so that the current flowing through the first resistance then flows in series through the second resistance. Such a connection of two resistances is called a *series connection.*

If the resistance of the first is R_1 and the resistance of the second is R_2, then the voltages created in each by a current I flowing through each of them are $E_1 = IR_1$ and $E_2 = IR_2$. These two voltages are each in such a direction as to oppose the flow of current through them.

A battery or other EMF source creates a rise in voltage. A current flowing through a resistance creates a voltage drop. In any electrical circuit, the voltage rises must be balanced by voltage drops. So, in the series circuit, the sum of the two voltage drops across the two resistors must equal the EMF, E, generated by the source. Here, we have

$$\begin{aligned} E &= E_1 + E_2 \\ &= IR_1 + IR_2 \\ &= I(R_1 + R_2) \end{aligned}$$

Since $E = IR$, $R_1 + R_2$ can be viewed as a single resistance. So, resistances in series add to give a series equivalent single resistance. In other words, R_1 and R_2 can be replaced by a single resistor with value $R_1 + R_2$.

The total voltage E across the two resistors in series divides proportionately across each resistor. Stated mathematically, the voltage across R_1 is

$$E_1 = \frac{R_1}{R_1 + R_2}E$$

This can be proved, since the voltage E_1 across R_1 is simply $E = IR_1$ and the current I is the voltage E divided by the total resistance in the circuit, namely, $R_1 + R_2$, or $I = E/(R_1 + R_2)$. Simple substitution gives the desired result. Similarly, it can be shown that the voltage across R_2 is

$$E_2 = \frac{R_2}{R_1 + R_2}E$$

A series connection of two resistors is called a *voltage divider.*

Series Circuits

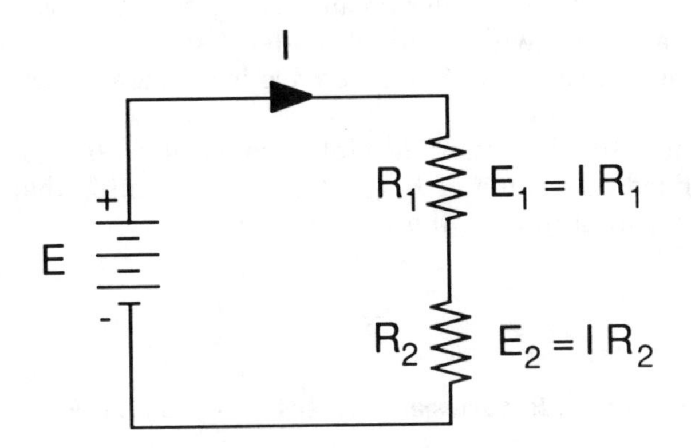

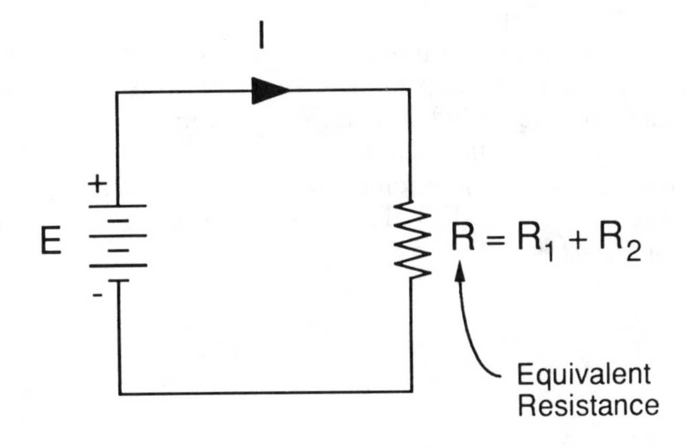

POTENTIOMETER

The principle of the voltage divider creates a way to vary a voltage. This is accomplished by a resistor with a variable slider that can choose or "tap" any intermediate value of resistance. Such a device is called a *potentiometer* or a *rheostat.*

The tap divides the total resistance into two portions, R_1 and R_2. The sum of R_1 and R_2 is fixed at a value of $R_{TOT} = R_1 + R_2$. Assuming that no, or very little, current flows from the tap, the voltage at the tap E_2 is

$$E_2 = \frac{R_2}{R_{TOT}} E$$

So, as the slider is moved, R_2 varies and causes E_2 to vary. A potentiometer has three terminals.

The resistance could be carbon material or the inherent resistance of a long coil of thin wire. A potentiometer made from a coil of wire is called a *wire-wound potentiometer.*

The tap can be connected to one of the other terminals to short out that portion of the resistance. This produces a variable resistance across the two terminals as the slider is varied. A *variable resistance* is shown symbolically as a resistor with an arrow through it.

Variable resistors and potentiometers are used in electric circuits to control or vary the intensity of a light bulb, the volume of a hi-fi amplifier, or the brightness of a TV screen.

Potentiometer

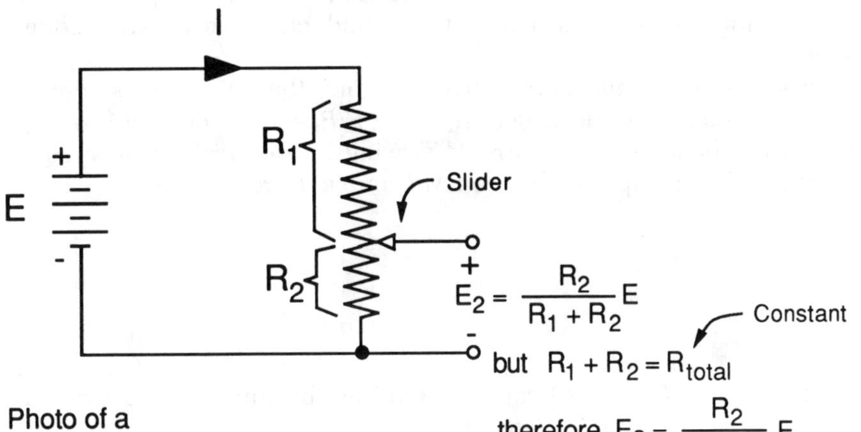

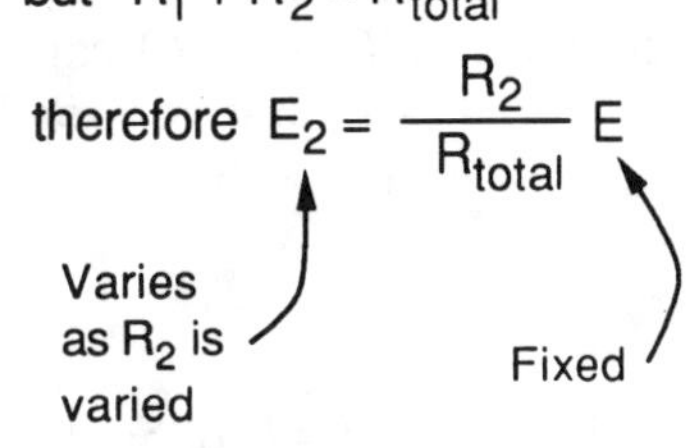

Photo of a potentiometer

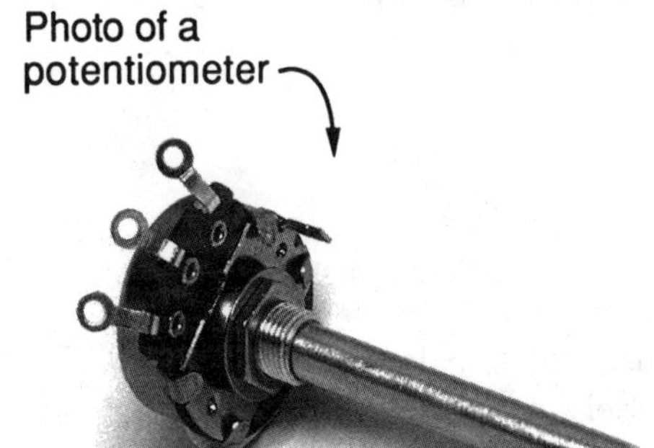

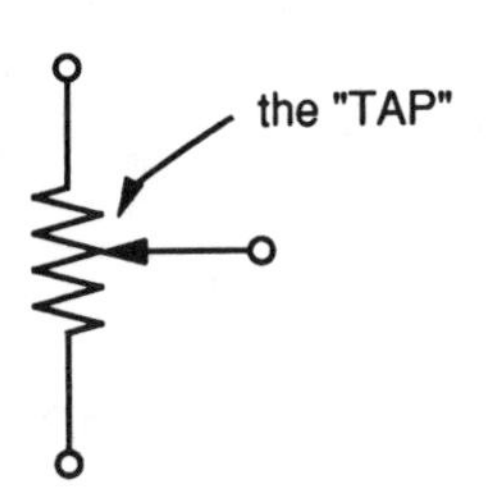

Variable Resistance

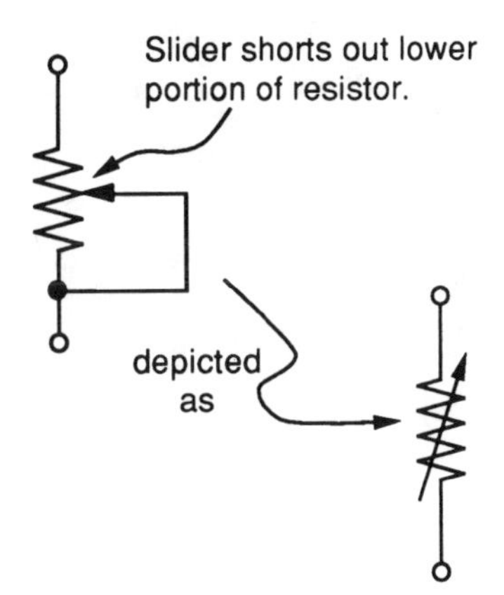

PARALLEL CIRCUITS

Two resistors can be connected in parallel so that the two top conductors are connected together and the two bottom conductors are likewise connected together.

In a *parallel connection* of two resistors, the voltage E is across each resistor. So, the current flowing in R_1 is $I_1 = E/R_1$ and the current flowing in R_2 is $I_2 = E/R_2$. The sum of the currents flowing in each resistor must equal the total current being supplied by the EMF source. Here, we have

$$I = I_1 + I_2$$
$$= E/R_1 + E/R_2$$
$$= E(1/R_1 + 1/R_2)$$

The term $(1/R_1 + 1/R_2)$ can be simplified by finding the least common denominator. The expression for I then becomes

$$I = E\frac{R_1 + R_2}{R_1R_2}$$

but, by Ohm's law, $I = E/R$. So, the *equivalent* single resistance of two resistors connected in parallel is $R_1R_2/(R_1 + R_2)$.

Connection of two resistors in parallel is called a *current divider.* The current through each resistor is inversely proportional to the resistance. In effect, the total current divides through each resistor, with most of the current flowing through the path of least resistance.

Parallel Circuits

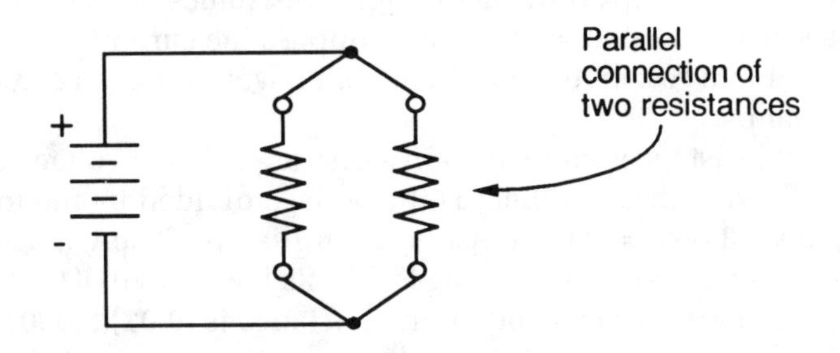

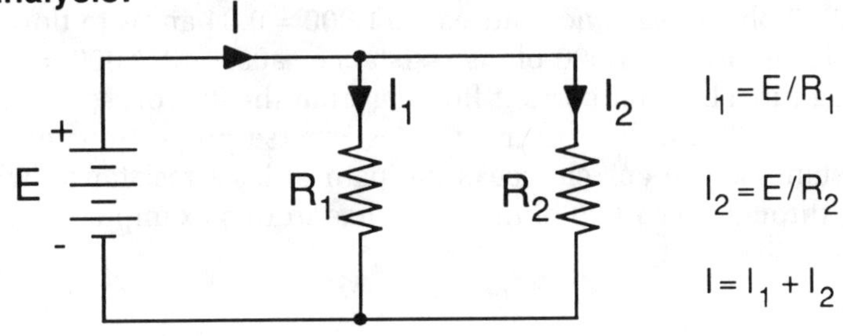

$$I_1 = E/R_1$$

$$I_2 = E/R_2$$

$$I = I_1 + I_2$$

$$R = \frac{R_1 R_2}{R_1 + R_2}$$

Equivalent resistance

EXAMPLES

A current of 0.1 amps flowing through a resistance of 1,000 ohms would create a voltage of $0.1 \times 1{,}000 = 100$ volts to oppose the current flow. A 12-volt battery connected to a resistive load of 200 ohms would draw $12/200 = 0.06$ amps or 60 milliamps.

Two resistances connected in series are called a voltage divider. The total current flowing through them is the voltage divided by the total resistance, or $40/(1{,}000 + 3{,}000) = 0.01$ amps or 10 milliamps. The voltage drop across the 1,000-ohms resistance, according to Ohm's law, is $(0.01)(1{,}000) = 10$ volts. The voltage drop across the 3,000-ohms resistance is $(0.01)(3{,}000) = 30$ volts. Thus, the 40 volts divides proportionately across the two resistances.

Two resistances connected in parallel are a current divider. The 40 volts across the 1,000-ohms resistance causes $40/1{,}000 = 0.04$ amps to flow through it. The 40 volts across the 4,000-ohms resistance causes $40/4{,}000 = 0.01$ amps to flow through it. The total current flowing from the 40-volt source is $0.04 + 0.01 = 0.05$ amps. Thus, the current divides inverse proportionately through the two resistances. The current seeks the path of least resistance, with most of it flowing through the 1,000-ohms resistance in this example.

Examples

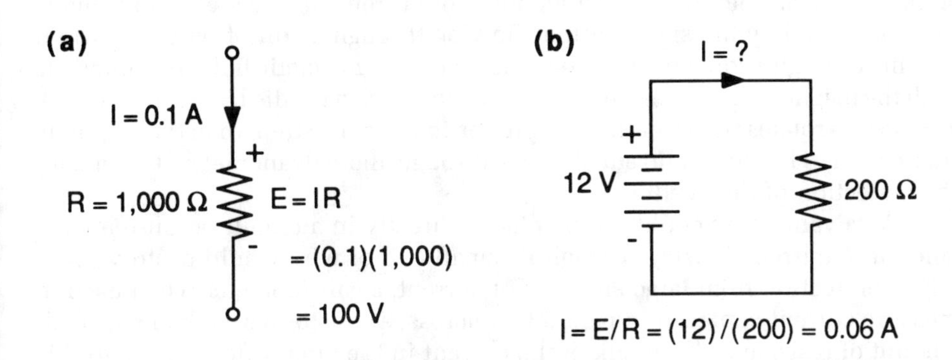
(a)
I = 0.1 A
R = 1,000 Ω
E = IR
= (0.1)(1,000)
= 100 V
(b)
I = ?
12 V
200 Ω
I = E/R = (12)/(200) = 0.06 A

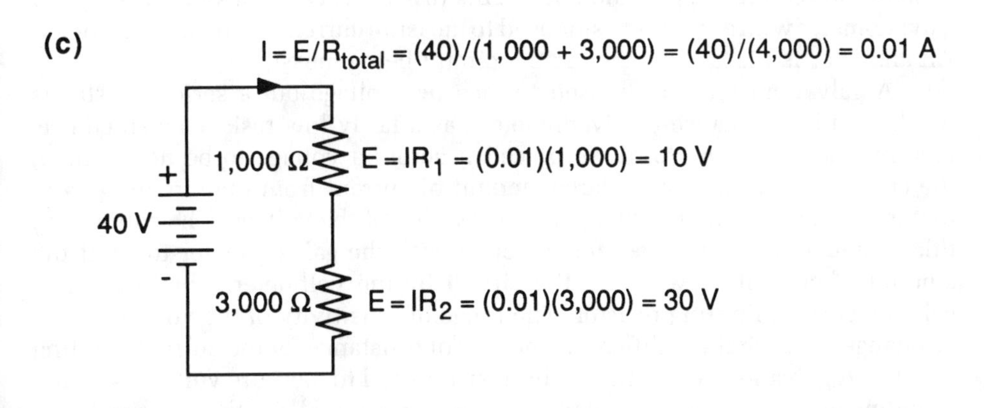
(c)
I = E/R_total = (40)/(1,000 + 3,000) = (40)/(4,000) = 0.01 A
1,000 Ω
E = IR_1 = (0.01)(1,000) = 10 V
40 V
3,000 Ω
E = IR_2 = (0.01)(3,000) = 30 V

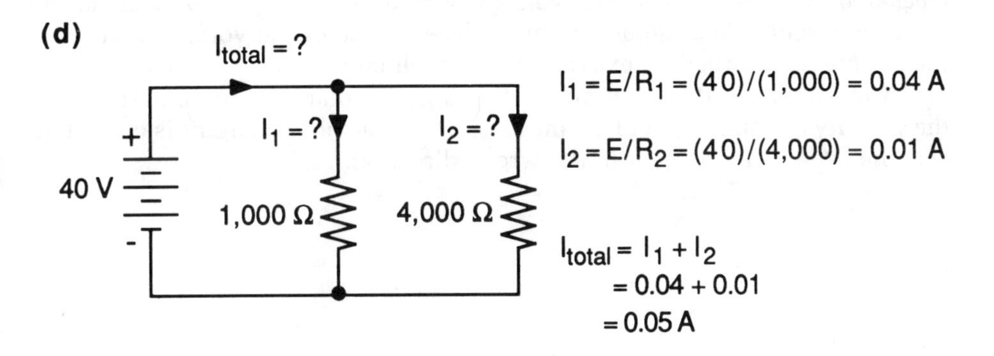
(d)
I_total = ?
I_1 = ?
I_2 = ?
40 V
1,000 Ω
4,000 Ω
I_1 = E/R_1 = (40)/(1,000) = 0.04 A
I_2 = E/R_2 = (40)/(4,000) = 0.01 A
I_total = I_1 + I_2
= 0.04 + 0.01
= 0.05 A

MEASUREMENT

A *galvanometer* measures small amounts of current in terms of the magnetic field produced by the small current flowing through a coil of wire. A current flowing through a coil creates a *magnetic field*. This magnetic field then interacts with the magnetic field created by a magnet to which a needle has been attached. The needle rotates with a deflection proportional to the strength of the magnetic interaction. The more current flowing through the galvanometer, the greater the deflection of the needle.

A galvanometer could not be placed directly in a circuit because a large amount of current flowing through its small coil of wire would destroy it. To afford protection from large amounts of current, a galvanometer when used to measure current is placed in parallel, or across, a resistor with a known, small amount of resistance. The bulk of the current in the circuit flows through this resistor, protecting the galvanometer. This resistor is called a *shunt resistor*. A galvanometer with a shunt resistor used to measure current is called an ammeter. An ammeter is placed in series in the circuit being measured.

A galvanometer can be used to measure voltage but a series resistor is needed. This is because a galvanometer has a fairly low resistance since it is merely a coil of wire. So, if we placed it across a voltage to be determined, the coil would draw a fairly large amount of current from the circuit, greatly disturbing the circuit and giving a false reading of the voltage. The solution to this problem is to put a resistor in series with the galvanometer to limit the amount of current drawn from the circuit by the voltmeter. This resistor is called a current-limiting resistor. The range of sensitivity of the voltmeter can be changed by switching different amounts of resistance for the current-limiting resistor. A galvanometer with a series resistor used to measure voltage is called a voltmeter. A voltmeter is placed across, or parallel to, the voltage being measured. A voltmeter measures voltage by sampling a small amount of current from the circuit. The higher the internal resistance of the voltmeter, the less the circuit is disturbed by extracting too much current.

Currents have direction and voltages have polarity. It is important that the polarity of connection of an ammeter or voltmeter in a circuit is correct, or else the needle can be bent in the wrong direction.

Measurement

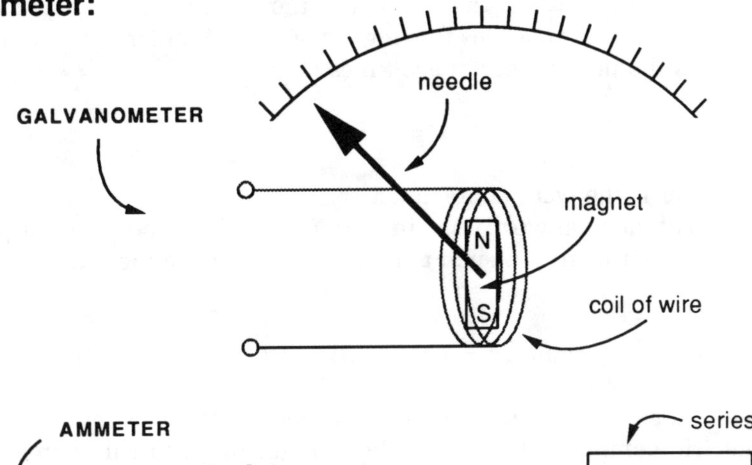

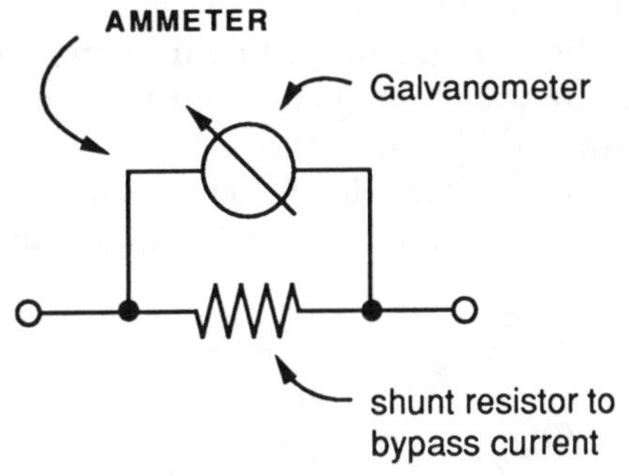

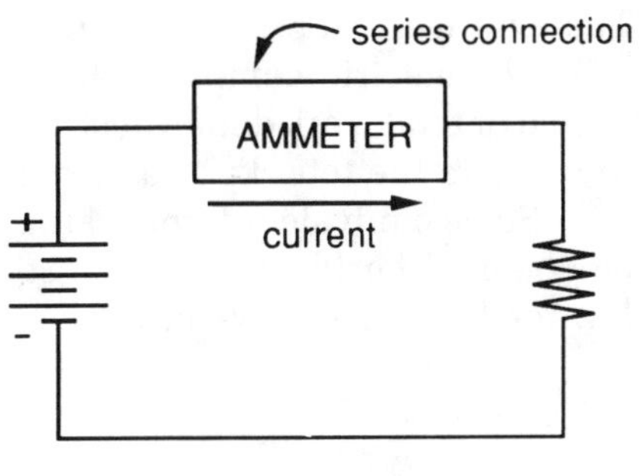

Voltmeter:

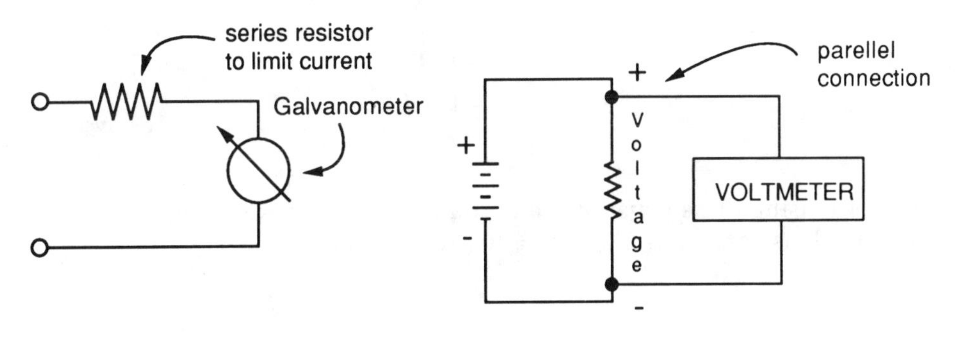

POWER

Power is a measure of the rate at which work is being done. In the marble analogy, a measure of power is the product of the force being exerted by the pump times the number of marbles flowing per second. Similarly, power in an electric circuit is the product of voltage times current,

$$P = EI$$

The unit of power is the watt, abbreviated W.

A measure of the total work done in time T hours is the power multiplied by time. For an electric circuit, such a measure would have the units of *watt-hours*

$$\text{Energy} = \text{Total work} = PT$$

The total work done in a specified amount of time is called *energy*.

The electric company charges for the amount of electrical energy consumed or total work done. Electric meters are watt-hour meters and are designed to measure the total kilowatt-hours consumed by the user.

Some equivalent forms of the equation for power in an electric circuit can be obtained fairly easily by substituting in the equation $P = EI$, using Ohm's law for E or I. Here, we have

$$\begin{aligned} P &= EI \\ &= E\,(E/R) \\ &= E^2/R \end{aligned}$$

Also,

$$\begin{aligned} P &= EI \\ &= (IR)I \\ &= I^2R \end{aligned}$$

By using these equations, a current of two amps flowing in a resistance of 100 ohms consumes $2^2 \times 100 = 400$ watts of power.

Power

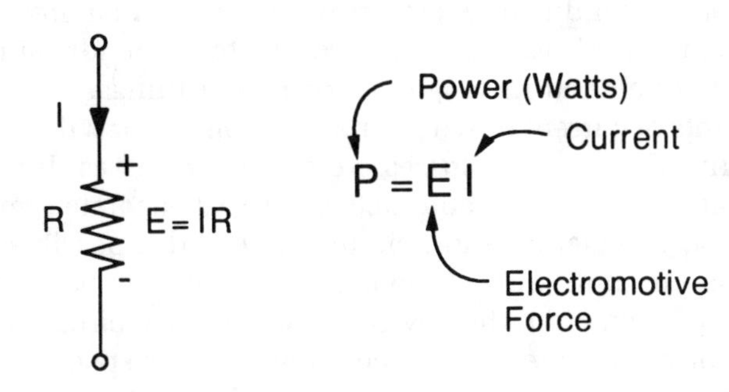

Other Forms:

If current and resistance are known:

$$P = EI = (IR)I = I^2R$$

$E = IR$

If voltage and resistance are known:

$$P = EI = E(E/R) = E^2/R$$

$I = E/R$

Energy:

Energy = Total work done in time T = P x T

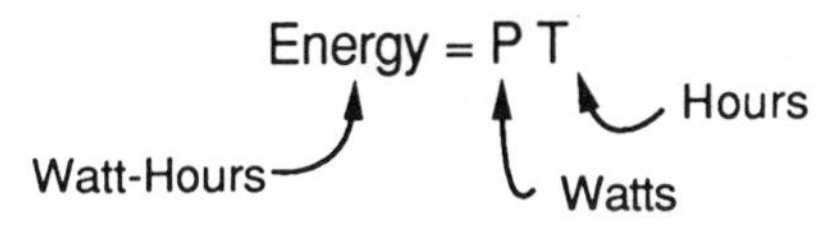

PROTECTION

As we saw earlier, no conductor is perfect in the sense of having absolutely no resistance. The amount of resistance of a conductor depends on its cross-sectional area, its material, and its length, among other things.

Large amounts of current flowing through a conductor will generate power losses in the form of heat. Such losses are called I^2R losses. If there were an excessive amount of current in a conductor, the heat generated might melt the insulation to cause a dangerous short circuit along with a possible fire. For this reason, devices are used to prevent too much current from being drawn. If the current exceeds some amount, the device causes an open circuit. Such devices are called *fuses* or *circuit breakers*, depending on their specific principles of operation, and are connected in series with the circuits to be protected.

A fuse consists of a short piece of metal through which the current flows. The metal is narrowed a precise amount, so that for a specified amount of current, enough heat is generated in the narrow path to cause it to vaporize, opening the circuit and giving protection. Once this happens, the fuse must be replaced with a new one.

A circuit breaker can be reset if it senses excessive current and trips. Slow protection is given by a strip of two different metals bonded together so that as the combination heats up it bends. When the bimetallic strip bends a specified amount in response to excessive current, it trips the breaker and the circuit is opened. Fast protection is given by an electromagnetic mechanism.

Problems of excessive current and high losses in conductors are important considerations in electric power distribution systems, but are rarely encountered in communication system applications. This is because the amount of current is typically quite small in communication circuits.

Protection

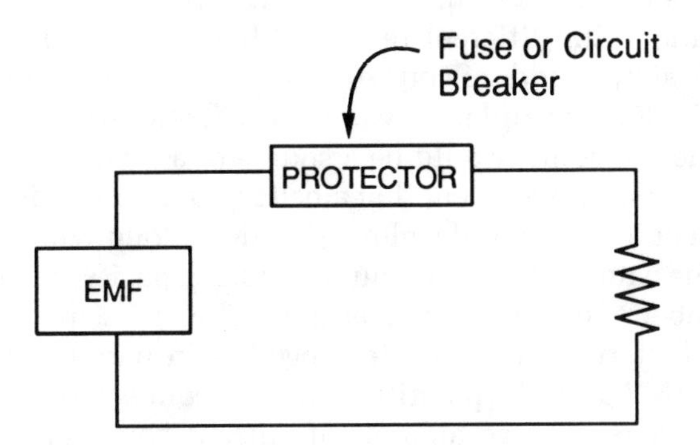

Fuse:

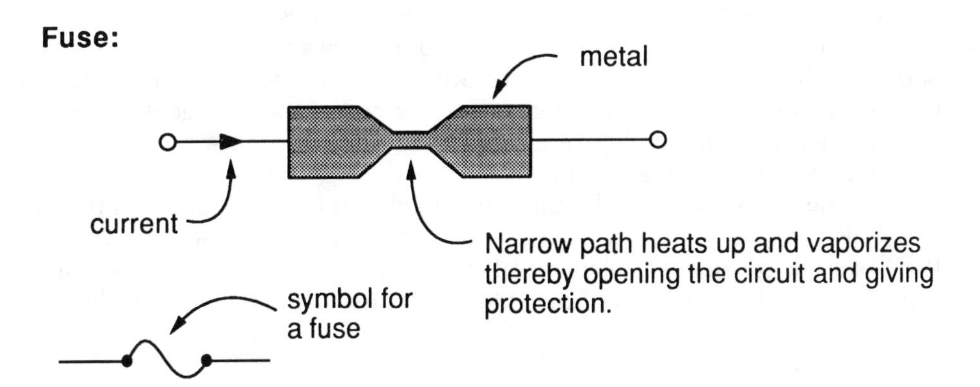

Circuit Breaker:

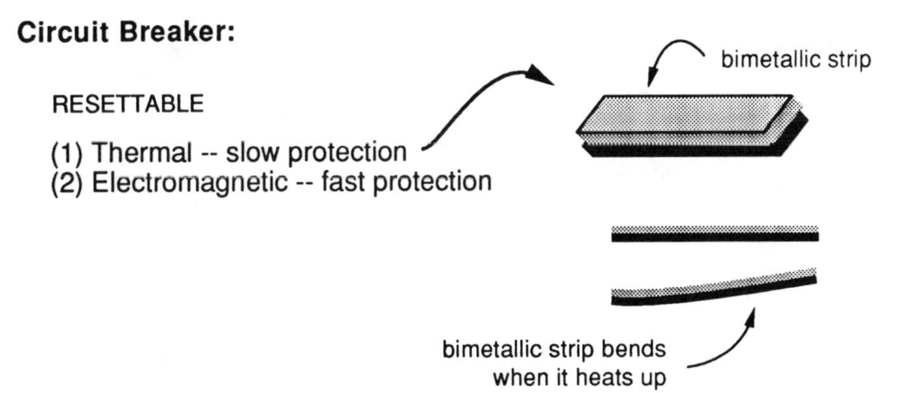

SERIES AND PARALLEL EMFs

More than one source of an electromotive force can be connected together in either series or parallel with different results. When connected in series, the voltages of the EMFs simply add, of course, taking into account the polarity of the individual EMFs. For example, if two 1.5-volt batteries were connected together in series, the net result would be a source of a 3-volt EMF.

The maximum amount of current that a battery can deliver is determined by the electrochemical action that develops the EMF along with the size and the construction of the battery. This maximum current capacity can be increased by connecting a number of batteries together in parallel, since parallel currents are additive. When batteries are connected together in parallel, each battery must have the same EMF and like polarities must be connected together. If the EMFs were not the same, a large amount of current could circulate in the parallel circuit created between the batteries. The net EMF of identical batteries connected in parallel is the same as one of the batteries alone.

The internal resistances of a number of batteries connected in series are additive. This can create a fairly large drop in the voltage at the terminals of the series connection of the batteries if the current flow is large. Connecting batteries together in parallel reduces the internal resistance of the parallel combination, and so reduces this effect.

As mentioned previously, the symbol used so far for a battery is that for one consisting of a number of cells connected together in series. A cell is the basic electrochemical element of a battery. A single-cell battery can be symbolized as a single pair of lines. Multiple pairs will indicate multiple cells.

Series and Parallel EMFs

Series:

E_1

E_2

$E = E_1 + E_2$

Series connected EMFs add algebraically.

Parallel:

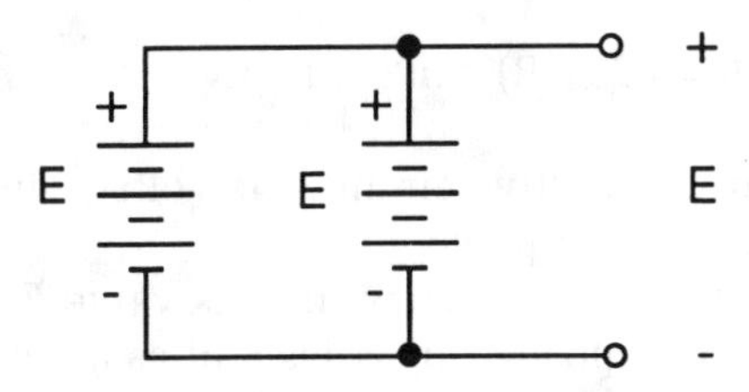

Parallel connected EMFs must be identical in voltage.

Cells:

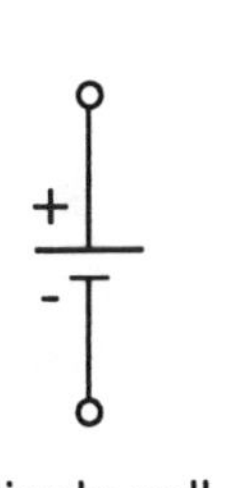

Single cell battery

Multi cell battery

Alternating Current

WAVEFORMS

The electric current produced by a battery is constant or nonvarying in both direction and magnitude. Electric current flows directly, in a single direction, and is called *direct current*, or simply dc.

A different type of current, which repeatedly changes its direction and magnitude, also exists. Such a current is called an *alternating current*, or ac.

An alternating current is produced by an alternating electromotive force. If such an alternating EMF were plotted as a function of time, it would be changing in amplitude and polarity peak. One type of alternating EMF is a sine wave $e(t)$ with a maximum amplitude E. An alternating EMF is depicted symbolically as two overlapping circles, or as one period of a sine wave enclosed within a circle.

A sinusoidal alternating EMF produces a sinusoidal alternating current flowing through a resistive circuit. If the value of the resistance is R, by Ohm's law, the alternating current $i(t)$ would then be

$$i(t) = e(t)/R = (E_{\text{peak}}/\text{R}) \sin(2\pi Ft)$$

where F is the frequency. In a purely resistive circuit, the voltage and current have the same phase and are said to be in phase.

The average value of a sinusoidal ac voltage with peak value E_{peak} is zero, since two alternate half cycles have opposite polarities and as a result cancel each other. The peak-to-peak value of the ac voltage is $2\ E_{\text{peak}}$.

The frequency of electric power in North America is 60 Hz. European power has a frequency of 50 Hz.

Alternating Current

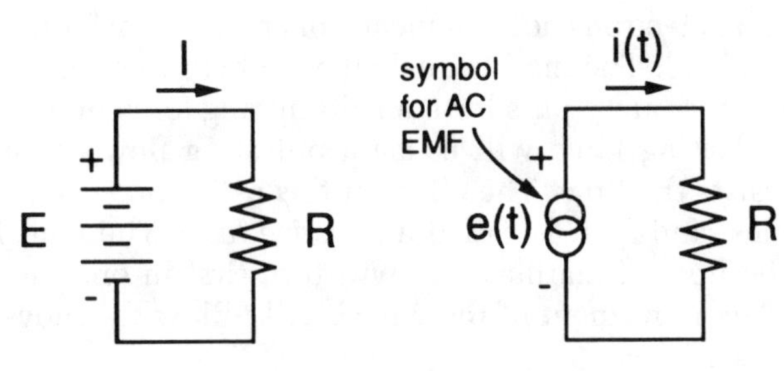

Direct Current
E & I are constant
with respect to time

Alternating Current
e(t) & i(t) vary sinusoidally
with respect to time

DC Waveforms:

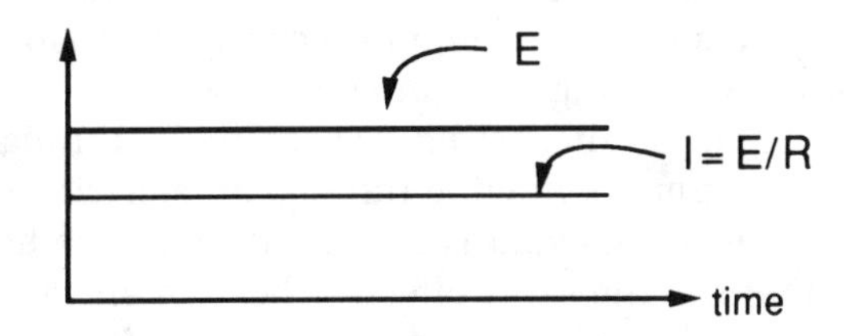

AC Waveforms:

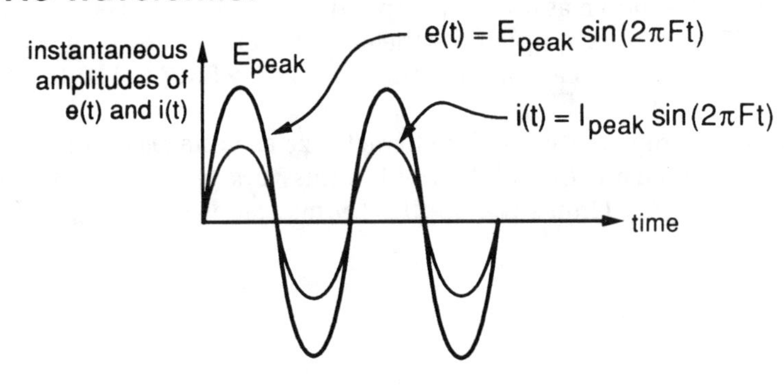

EFFECTIVE VALUE

A sinusoidally varying current pulsating to and fro through a resistive load performs useful work. The electrons act as a means of energy transfer from the source of the EMF to the load. The marble analogy may help to clarify this.

Assume that the marble pump puts forth an alternating force in the shape of a sine wave. This pulsating force will cause a pulsating flow of marbles through a wheel that resists the flow. The effect of this resistance is a load on the movement of marbles, and so it is called a *resistive load.* The wheel will be caused to turn by the flow of marbles, but will turn first in one direction and then in the other. The movement of the wheel will follow the movement of the marbles.

The net rotation of the wheel is zero because the movement in one direction cancels the movement in the other. However, a fan attached to the wheel would swing to and fro, performing useful work. So, work does occur with an alternating current.

This should not be a surprising revelation. Taking one step forward and then one step backward is nearly as tiring as walking straight ahead, even though no net distance is walked in the first case. A useful measure then would be the effective value of the energy expended by stepping backward and forward in terms of the energy expended by walking straight ahead.

Useful work also occurs when an alternating current flows through an electric circuit. If the alternating current is flowing through an electric heater, it will get hot just like it does if a direct current is flowing through the heater. There is then a direct current that has the same effective heating effect as the alternating current. This effective value of an alternating current is called its *root mean square*, or rms, value.

If a sine wave has a peak or maximum amplitude of E_{peak}, the rms value of the sine wave can be shown as 0.707 multiplied by E_{peak}. In the case of electric power, the familiar 110 volts supplied by the electric company at an electrical outlet is the rms value. The peak value of this alternating EMF is 110/0.707 or 155 volts.

The rms values are usually used to characterize ac voltages and currents. The power of an ac device is the product of the rms voltage multiplied by the rms current, or $P = E_{rms} I_{rms}$. Ohm's law applies to rms values so that $E_{rms} = I_{rms} R$.

Effective Value

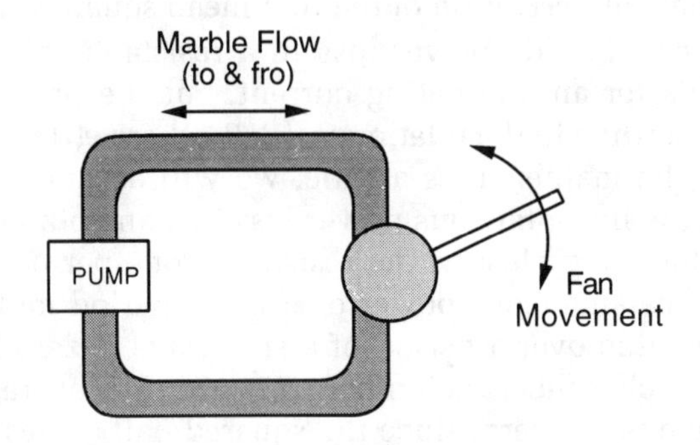

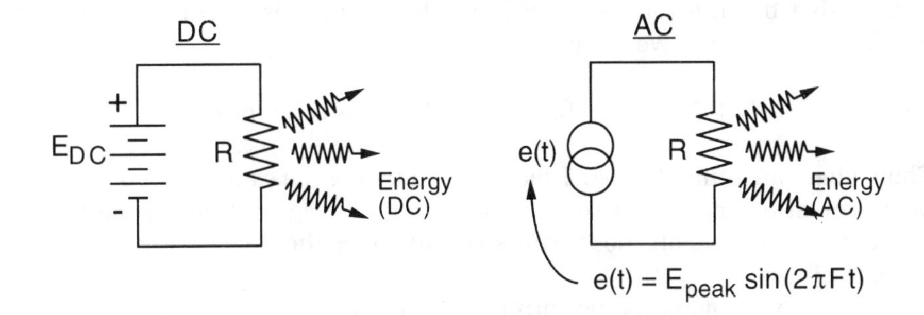

For the same effective energy, Energy DC = Energy AC.

$$E_{DC} = \frac{E_{peak}}{\sqrt{2}} = 0.707\, E_{peak}$$

DC voltage with same effective energy of sinusoidal AC voltage with peak amplitude E_{peak}

Called: Root Mean Square (RMS)

$$RMS = \frac{Peak}{\sqrt{2}} \qquad \text{(for sine wave only)}$$

ROOT MEAN SQUARE

The mathematical derivation of the root mean square value of a sine wave will be treated here. The dc power loss in a resistance of R ohms is E_{eff}^2/R. This relation holds for an alternating current, but the power varies continuously with time according to the relations $e^2(t)/R$, where $e(t)$ is the sinusoidal voltage. To simplify the mathematics a little, we will assume a resistive load of one ohm so that the instantaneous power loss in a one-ohm resistor is simply $e^2(t)$.

The total energy lost in the resistor in one period would be the average value of the instantaneous power over a full period multiplied by the period. The average value over a period of a sine wave is zero because the polarities of each half cycle cancel each other. However, the average value of the square of a sine wave is not zero, since the squared half cycles are both positive.

The effective value of the alternating current will be the value of a direct current that has the same energy loss. The energy loss of this direct current would be $(E_{eff})^2 T$. So, we have

$$(E_{eff})^2 T = \text{ac Energy loss} = [\text{Average value of } e^2(t)]\, T$$

Then, after canceling the two T factors, and taking the square root of both sides of the equation, E_{eff} is the square root of the average (or mean) value of $e(t)$ squared. Shortening all these words simply gives the root mean square, or E_{eff} becomes E_{rms}.

The key to actually performing these calculations for a sine wave is determining the average value of the square of a sine wave with amplitude E_{peak}. This calculation involves determining the area under the curve of $E_{peak}^2 \sin^2(t)$ for one period. We can show that the area under the squared sine wave is $E^2 T/2$. So, the average value over T is $E_{peak}^2/2$, and the root mean square is $E_{peak}/\sqrt{2}$, or $0.707E_{peak}$.

The concept of root mean square value can be extended to any waveshape. For example, a square wave has a rms value that is simply the maximum amplitude of the square wave.

RMS of Sine Wave

DC

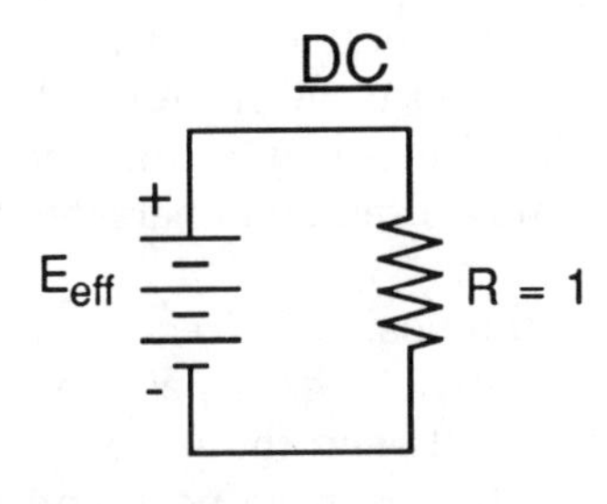

Power = $(E_{eff})^2$

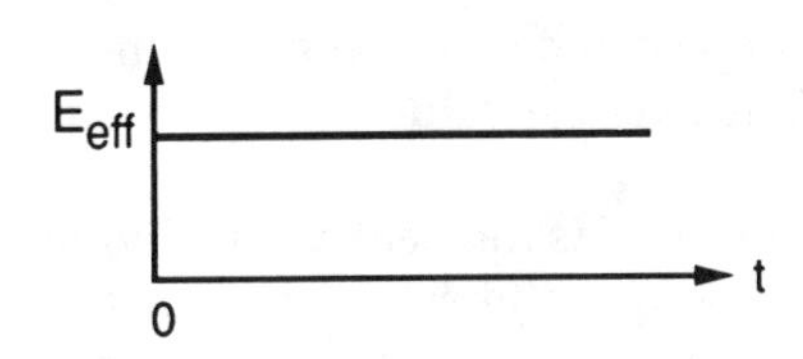

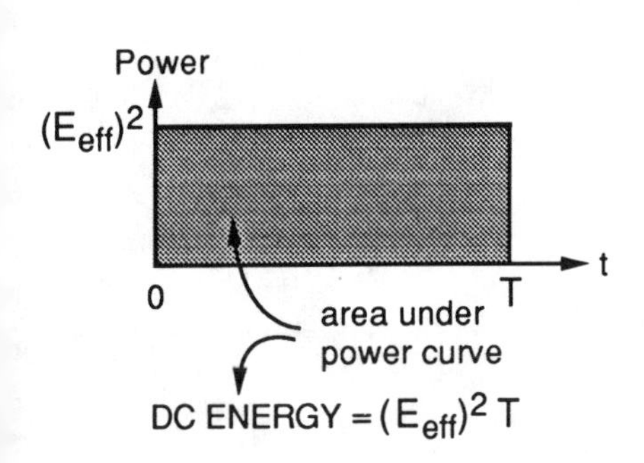

AC

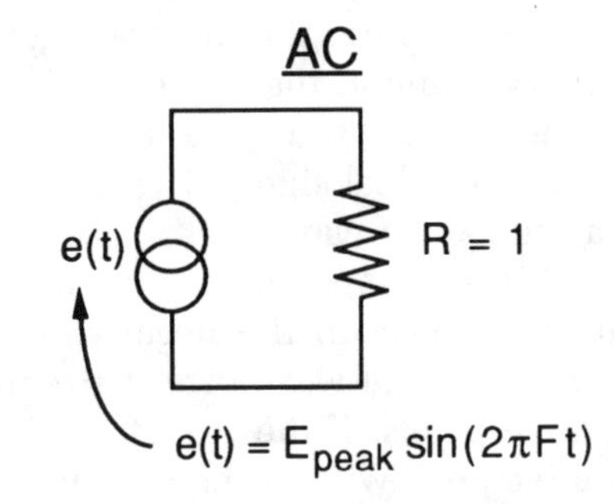

Instantaneous Power = $e^2(t)$

$= (E_{peak})^2 \sin^2(2\pi Ft)$

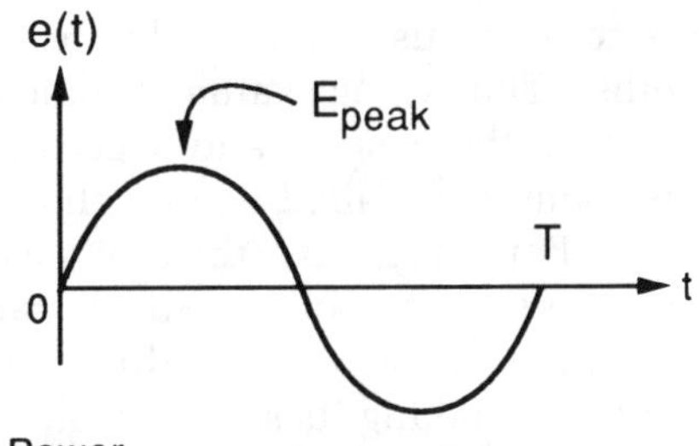

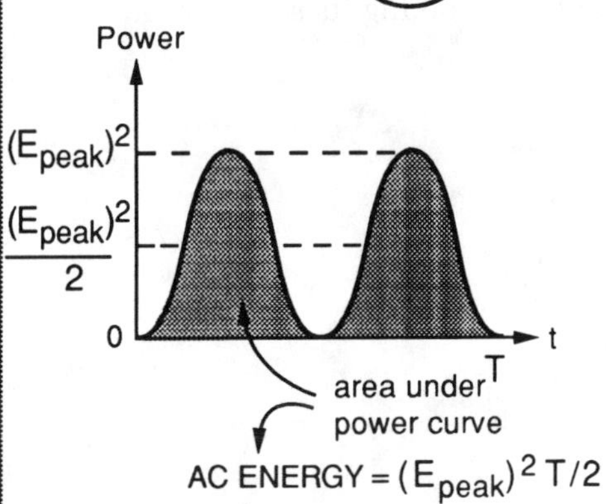

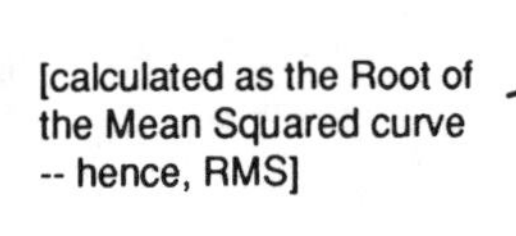

[calculated as the Root of the Mean Squared curve -- hence, RMS]

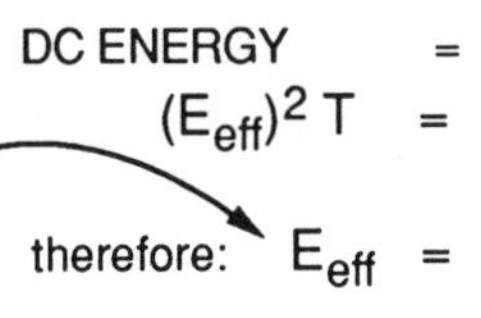

$$\text{DC ENERGY} = \text{AC ENERGY}$$

$$(E_{eff})^2\, T = (E_{peak})^2\, T/2$$

$$\text{therefore: } E_{eff} = \frac{E_{peak}}{\sqrt{2}}$$

WAVEFORM MEASUREMENTS

An electromotive force or current can have any waveshape. The waveshape can be as simple as the sine wave generated by the power company and delivered to every home at the power socket in the wall. Or, the waveshape can be as complex as the sound of a symphony orchestra converted by a microphone into an electrical signal. Every waveshape can be measured and characterized in a number of ways.

A waveshape will have a maximum peak in the positive direction and a minimum peak in the negative direction. There also will be a peak-to-peak measure for the waveform that equals the distance between the positive and negative peaks. If the waveform is not symmetric, the positive peak and the negative peak will not be the same. For a sine wave, the positive and negative peaks are indeed the same, but a sharp pulse will have no negative peak at all.

A waveshape will have an average value measured over some specified interval of time. The average value of a sine wave measured over a full period is zero because alternate half cycles have opposite polarities that cancel each other. The average value of some other waveshape can be anything, depending on how the positive and negative areas add. The average value of a waveshape is sometimes called its dc value.

For a sine wave, the root mean square (rms) value is the peak value divided by the square root of two. For any other waveshape, the rms value must be calculated by actually finding the average value of the square of the waveshape and then taking its square root.

Waveform Measurements

Peak Value:

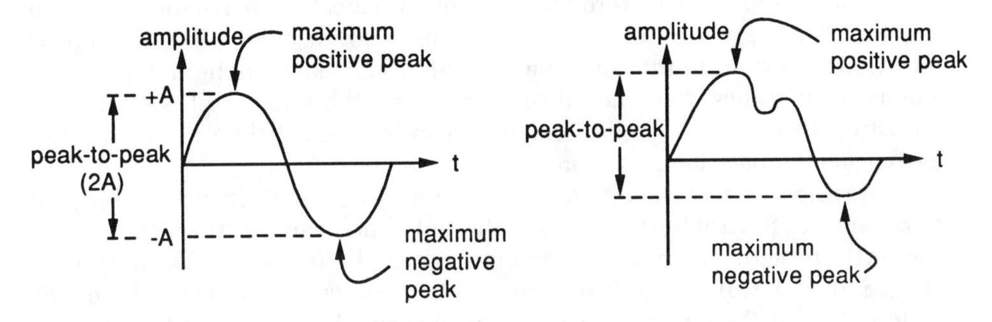

Average Value:

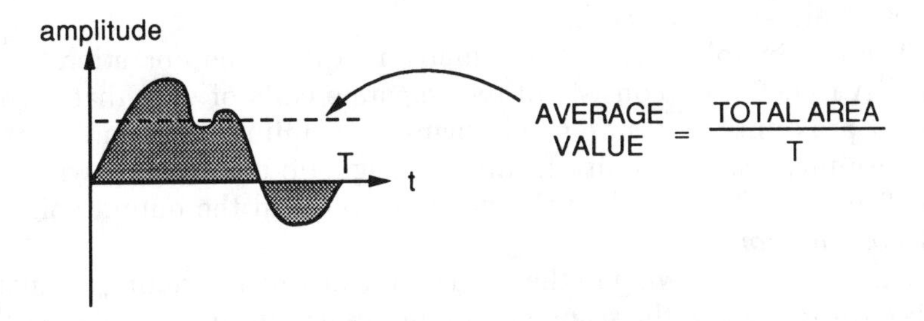

RMS Value:

For sine wave:

$$E_{RMS} = \frac{E_{peak}}{\sqrt{2}}$$

For any waveform:

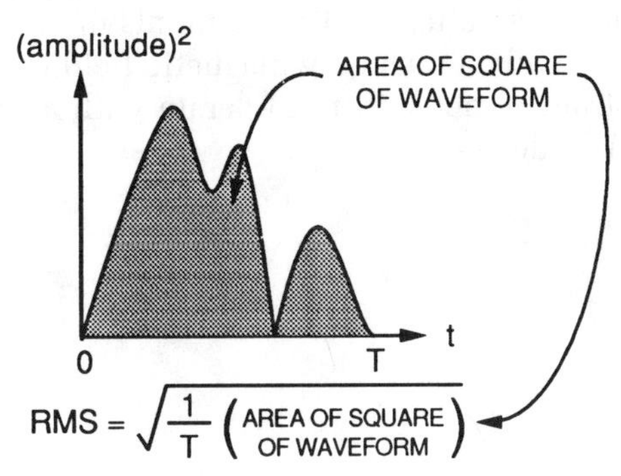

$$RMS = \sqrt{\frac{1}{T}\left(\text{AREA OF SQUARE OF WAVEFORM}\right)}$$

TRANSFORMERS

There are some basic principles of electromagnetism that are important to understanding some electrical devices and phenomena.

A dc current flowing through a coil of wire creates a magnetic field about the coil. The magnetic field has a north and a south pole, depending on the direction of current in the coil. The strength of the magnetic field depends on, among other things, the amount of current and the number of coils of wire. The strength of the magnetic field will also increase if the coil surrounds a ferromagnetic material as its core.

Moving a magnet along, or near, the axis of a coil of wire induces an electromotive force at the terminals of the coil. The polarity of this EMF changes sign as the direction of movement of the magnetic field changes. The magnitude of the induced EMF depends on, among other things, the strength of the magnetic field and the speed with which it changes. It is essential to remember that the magnetic field must be changing to induce a voltage in the coil. As might be expected, an ac current flowing in a coil will create an alternating magnetic field about the coil.

These principles of electromagnetism explain the operation of a *transformer*. A transformer consists of two separate coils of wire that are in close physical proximity to assure good magnetic coupling between the two coils. The close physical proximity facilitates magnetic coupling between the two coils. The input coil is called the *primary coil*, and the output coil is called the *secondary coil*.

An ac current flowing in the primary creates an alternating magnetic field that is close enough to the secondary to induce a voltage across the secondary's terminals. The alternating magnetic field created by the primary appears to be the same in effect as if a magnet were being moved through the secondary to induce a voltage at the terminals of the secondary.

Only a changing magnetic field can induce an EMF in a coil of wire. So, a transformer will not operate with a dc signal. A transformer is exclusively an ac device.

Transformers

Electromagnetic Principles:

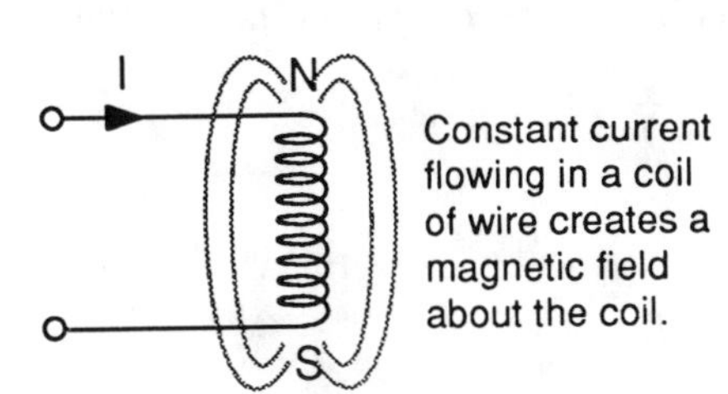

Constant current flowing in a coil of wire creates a magnetic field about the coil.

Moving a magnet (or magnetic field) through or across a coil of wire creates an EMF at the terminals of the coil.

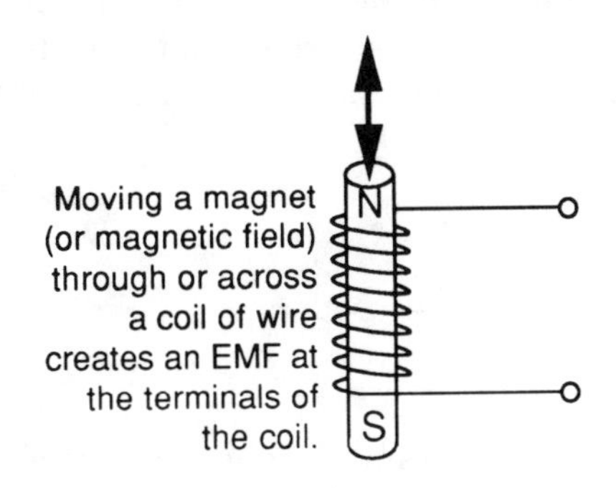

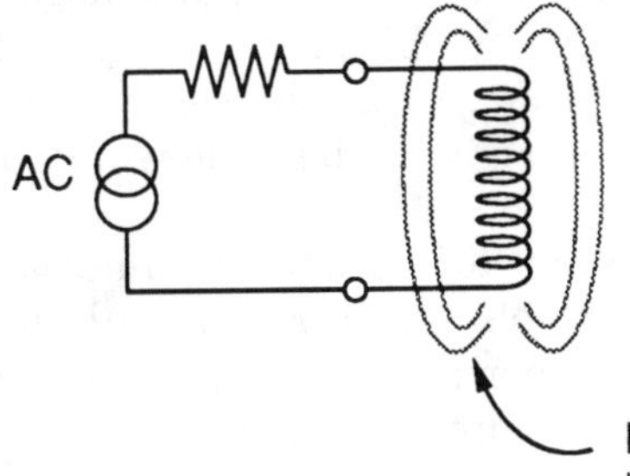

An alternating current flowing in a coil of wire creates an alternating magnetic field about the coil.

Magnetic field changing in both polarity and magnitude.

The Transformer:

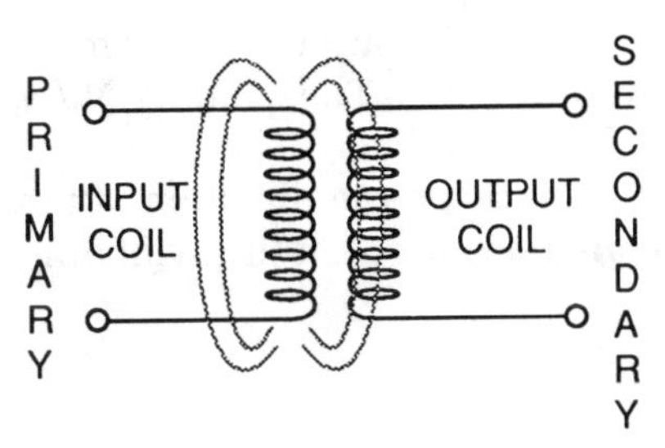

Alternating current at input coil creates an alternating magnetic field that affects the output coil thereby creating an alternating EMF at the terminals of the output coil.

Note: A changing magnetic field is required to create an EMF at the secondary. Hence, tranformers will not work with unchanging DC.

TRANSFORMERS (cont'd)

The strength of the magnetic field and the corresponding induced voltage depends on the number of coils of wire in the primary and in the secondary. If N_p and N_s are the number of turns in the primary and in the secondary, respectively, and if E_{in} and E_{out} are the rms input and output voltages, then,

$$E_{out}/E_{in} = N_s/N_p$$

or

$$E_{out} = (N_s/N_p)E_{in}$$

As a result, by choosing the appropriate number of turns it is possible to change ac voltages with a transformer. The quantity of N_s/N_p is called the transformation ratio, n. If n is greater than one, the output voltage will be greater than the input voltage. If n is less than one, the output voltage will be less than the input voltage. In this way, ac voltages can be stepped up or down. A transformer with a transformation ratio greater than one would be accordingly called a step-up transformer.

The transformer is called a passive device since it cannot generate power. Its input power $E_{in}I_{in}$ must equal its ouput power $E_{out}I_{out}$, with the exception of internal power losses in the transformer itself. This equality can be used to derive a relation for the current transformation properties of a transformer, namely,

$$I_{out} = (N_p/N_s)I_{in}$$

In other words, if the output voltage increases, the output current must decrease proportionately to keep the output power constant. The quantity N_p/N_s is called the turns ratio.

The polarity of the output EMF depends on the direction of the winding of the coils of wire. Small dots are sometimes used on the symbol for a transformer to indicate the polarity.

Transformers (cont'd)

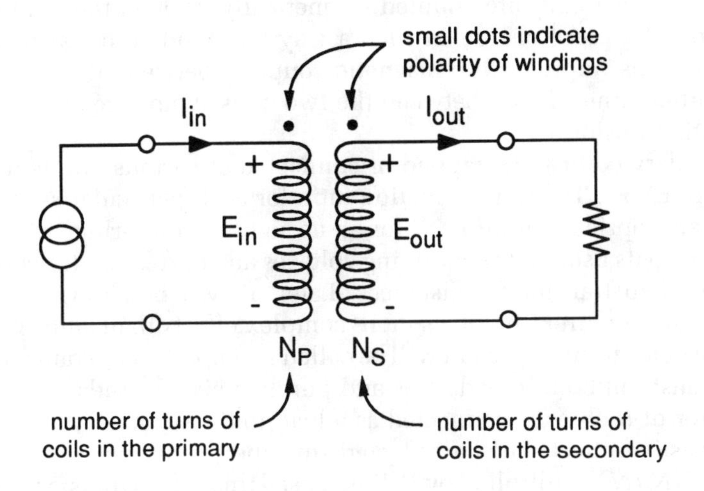

Transformation Ratio

$$\frac{E_{out}}{E_{in}} = \frac{N_S}{N_P} \quad \text{OR} \quad E_{out} = \frac{N_S}{N_P} E_{in}$$

$$I_{out} = \frac{N_P}{N_S} I_{in}$$

Turns Ratio

$$\boxed{P_{in} = P_{out}}$$

Passive
Device

TRANSFORMERS (cont'd)

Some transformer coils are coupled magnetically through the air. Others are wound around a piece of iron in such a way that the two coils share the same iron core. This improves the magnetic coupling between the two coils. One or two vertical lines drawn between the two coils symbolize an iron-core or ferromagnetic transformer.

The secondary coil can be tapped or connected at various points to give multiple output EMFs. The transformation ratio for each particular secondary coil gives the appropriate output EMF for an ac signal at the primary.

In addition to its uses for transforming voltages and currents, a transformer can also be used to transform resistances. Later, it will be shown that the "resistance" to an ac current involves more complex effects than simply resistance. The resistance to an ac current will be called an *impedance*. Transformers can scale or transform both impedances and purely resistive loads.

If a resistor of R ohms is connected as a load to the secondary of a transformer, the resistive effect will be reflected into the primary as if it were a resistive load of $(N_p/N_s)^2$ multiplied by R. When used this way, a transformer can transform a resistor or an impedance from one value to another. Transformers are used this way in vacuum-tube amplifiers to match the impedance required by the output power tubes to the impedance of the loudspeaker.

The electromagnetic concepts that are basic to the operation of the transformer were discovered in the 1830s by the American Joseph Henry and the Briton Michael Faraday. In honor of their contributions to electricity, two basic units are named after them: the henry and the farad. The American William Stanley applied transformers with ac to electric power distribution.

Transformers (cont'd)

Core Types:

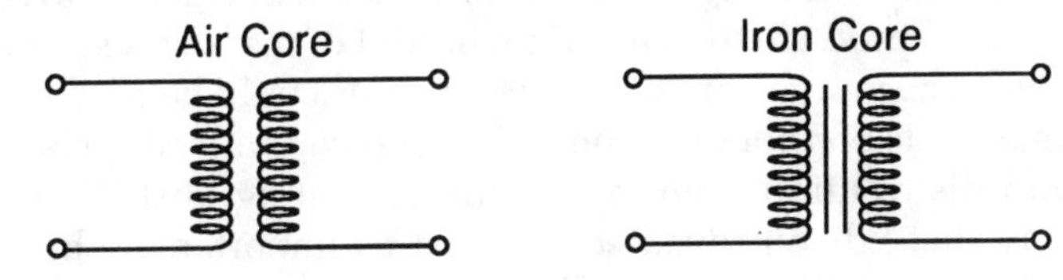

Multiple Taps:

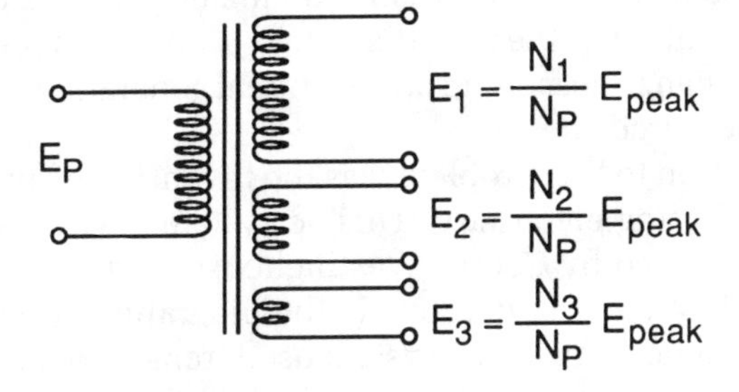

Matching Transformer:

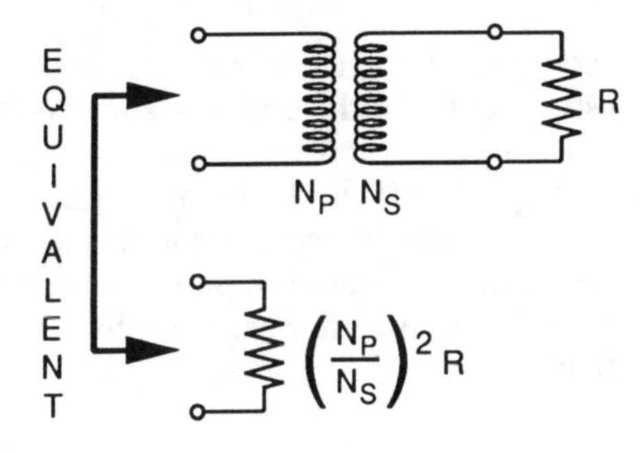

ELECTRIC POWER

On October 21, 1879, Thomas Alva Edison invented an electric light bulb that was different from any other previous source of electric light. Edison's invention had a filament with a high resistance. This meant that relatively little current was needed to produce the same light output of the low-resistance light sources. This smaller current meant that electric wires could be much smaller in diameter, which made possible an economical and centralized system of power generation and distribution. The system for generating, distributing, and metering electricity that Edison invented was just as important as his invention of the high-resistance electric light bulb.

Light bulbs were connected in parallel across a dc source of 110 volts in Edison's system. The individual currents drawn by each bulb added and so created the need for ever-increasing diameters of wire to keep the I^2R losses reasonable. Eventually, the wire diameter became impractically large, the distribution system had to cease, and an additional generating station and distribution system was required.

The solution to this problem was the invention of an ac system for generating and distributing electricity. This ac system was invented by Nikola Tesla and commercialized by George Westinghouse. A very good account of the life of Tesla and his many inventions is the biography, *Tesla: Man Out Of Time,* by Margaret Cheney. Tesla's ac system used transformers so that higher voltages could be used in the distribution system and then stepped down to a 110-volt rms value at the customer's premises. The use of the higher distribution voltages meant that the currents could be lower, leading to much lower I^2R losses and more practical wire diameters. Tesla invented the ac motor that made the system useful practically.

A roaring battle occurred between the dc system of Edison and the ac system of Tesla and Westinghouse. In the end, the superiority of the ac system defeated the dc system, and all electric power is now ac. Electric power in the United States has a frequency of 60 Hz and a nominal rms voltage of 110 volts. In actuality, the nominal 110 volts is usually about 120 volts. Most electric appliances are rated in terms of the amount of power in watts that they require. The current can then be easily calculated. For example, a 1,200-watt hair dryer requires 1,200/120 = 10 amperes.

Electric Power

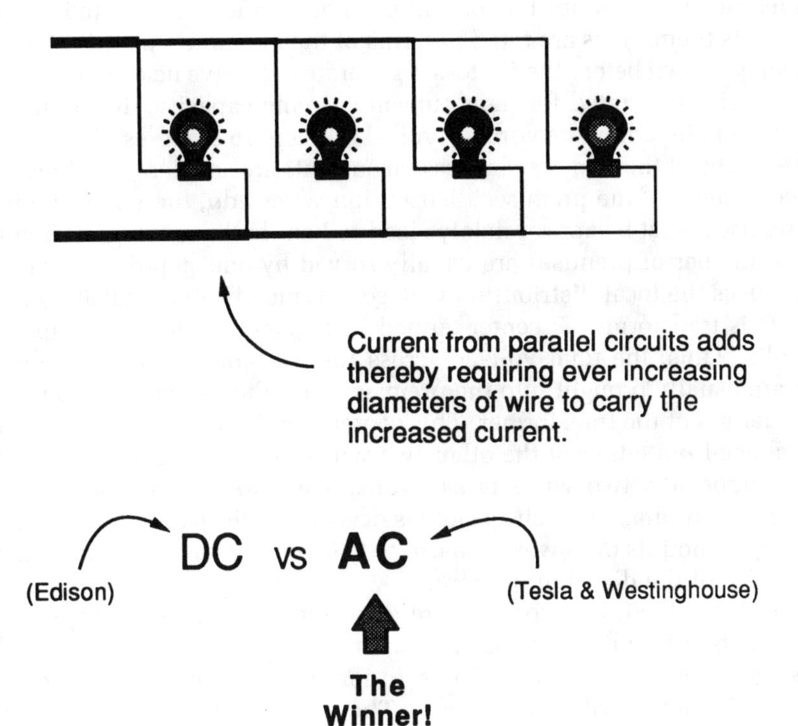

ELECTRIC POWER DISTRIBUTION

The electric outlets in our homes and businesses are all connected in parallel. The outlets themselves are rated in terms of the maximum permissible current that can be drawn before the I^2R losses generate excessive heat. The distribution wires are likewise rated for their current-carrying capacity, depending on the diameter and length of the wire. A typical rating is 15 amperes. Fuses or circuit breakers protect the circuits if excessive currents are accidentally drawn. Since all the currents of the premises distribution wires add, the wires coming into the premises must be appropriately sized to handle the much greater currents.

A number of premises are usually served by one step-down transformer that reduces the local distribution voltage of typically about 2,000 volts to 120 volts. This transformer is center-tapped so that each side of the tap delivers 120 volts. Thus, the total voltage across the transformer is 240 volts. Three wires are usually brought into the premises with the center wire connected to the center tap of the transformer. This center wire is grounded at the premises. The voltage from either of the other two wires to ground is 120 volts, and the voltage across the two wires is 240 volts. The 240 volts is used within the premises for appliances such as clothes dryers and electric cooking ranges that draw large amounts of power. The use of 240 volts reduces the current, and so too the size of the distribution wire.

The step-down transformers are connected in parallel with each other. The currents drawn by each transformer are additive, and after a while the I^2R losses again become excessive. The solution is the use of further transformers with even higher distribution voltages. The nationwide grid that interconnects utility companies and enables them to share power loads operates at 750,000 volts.

Color-coded wires are used within the premises to ensure safety. For a 120-volt circuit, the black wire is the "hot" 120-volt circuit (black for death), and the white wire is the ground (white for life). Although the wiring of 120-volt circuits is conceptually simple, the risks to life and property are quite high, and the installation of such circuits should be performed only by licensed electricians.

Power Distribution

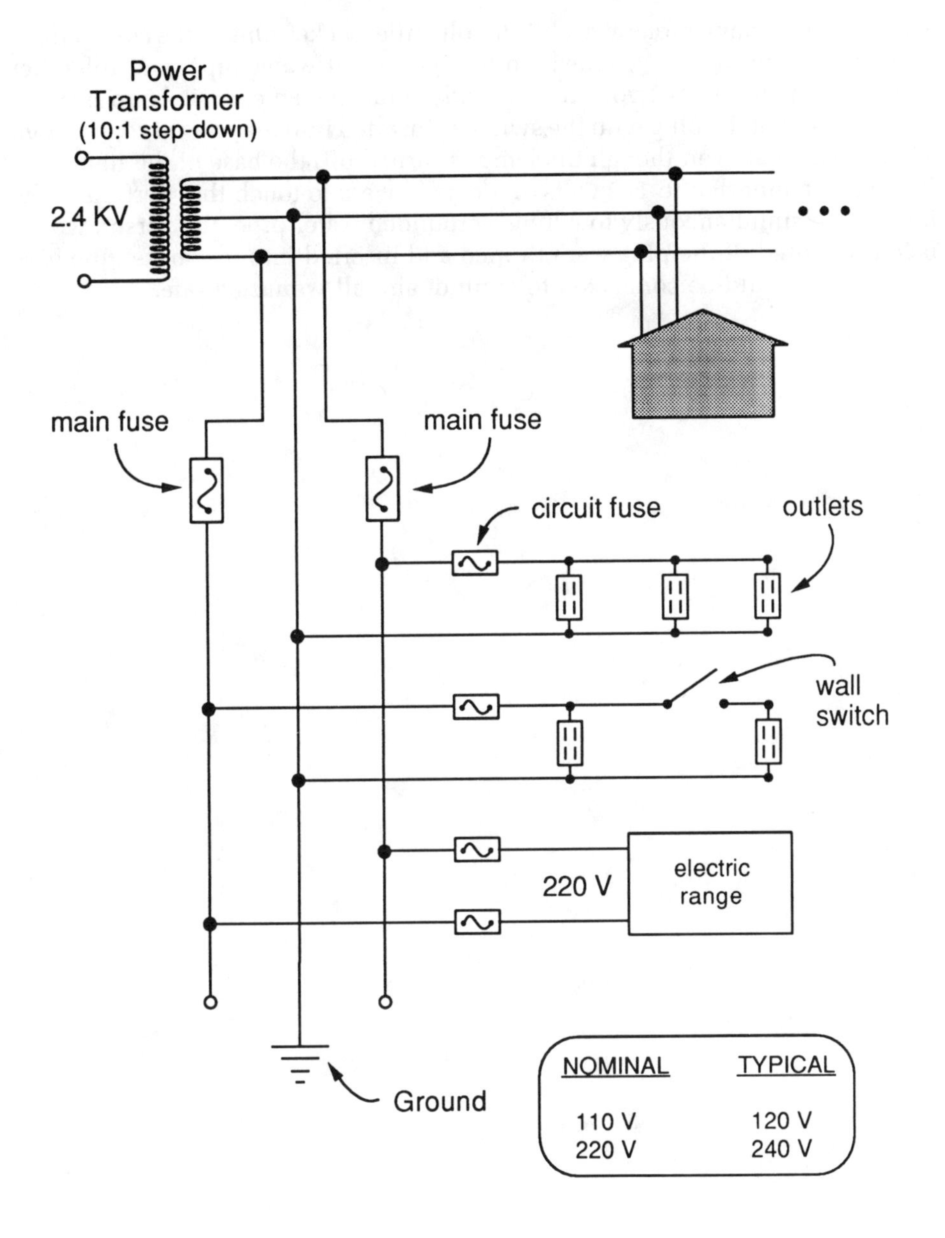

ELECTRIC POWER SAFETY

There are two connections at each 120-volt outlet socket. One of the connections is to ground, usually by connection to a grounded water pipe, and the other connection is to the 120-volt line. Consider plugging an electric lamp into the socket such that the plug with the switch is inserted into the ground connection. This means that even though the lamp is turned off, the base of the light bulb is directly connected to 120 volts. If a person were to touch that portion of the bulb while simultaneously touching a grounded water pipe, the person would be electrocuted! If the plug were flipped and inserted into the socket, the base of the bulb would be connected to ground, and all would be safe.

Electric Safety

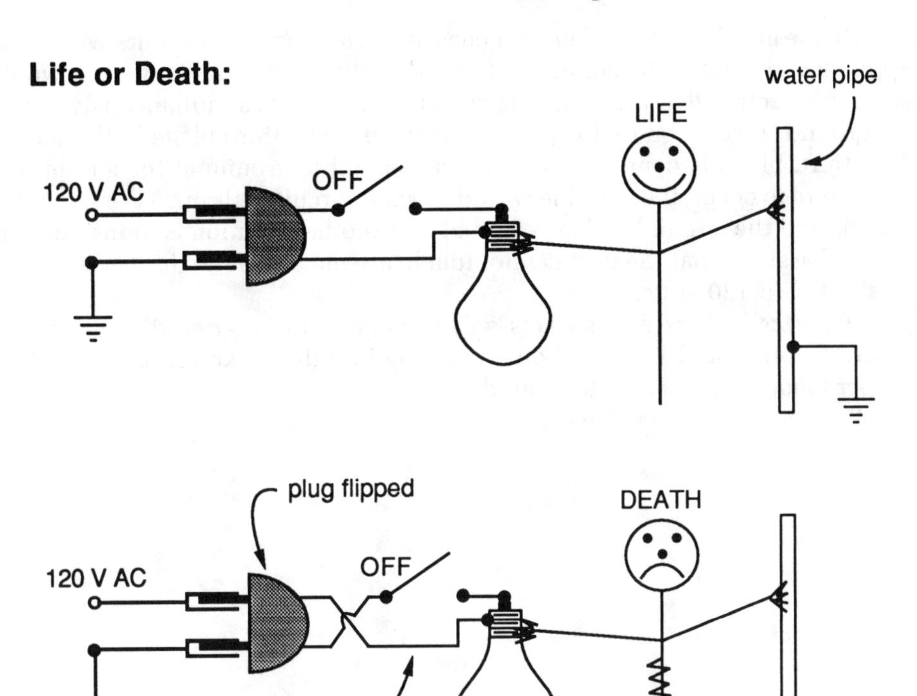

ELECTRIC SAFETY (cont'd)

Some appliances have a metal chassis or case. If a short occurs within the appliance, the 120 volts could be attached to the chassis, and a person would easily be electrocuted if the chassis were touched while simultaneously touching ground. A solution to this problem is the use of a third blade in the socket. This third blade is connected to the chassis and to ground at the socket. If a short were to occur, the 120-line would be short circuited through the grounded chassis and the circuit breaker would blow. Another solution is to insulate the metal chassis so that the user cannot touch any metal that might inadvertently be shorted to 120 volts.

Polarized plugs and sockets solve many of these ground problems. A polarized plug can be inserted only one way into the socket, assuring that the proper blade is connected to ground.

Electric Safety (cont'd)

Chasis Ground:

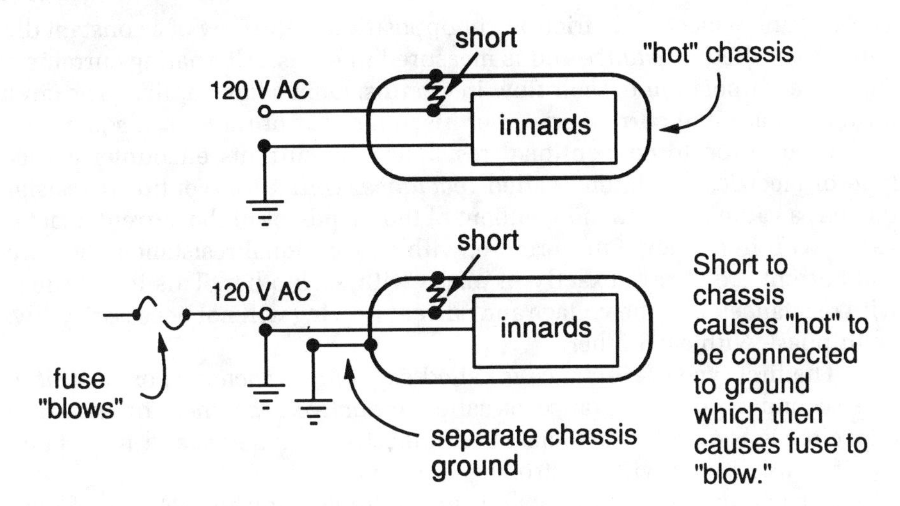

Polarized Sockets:

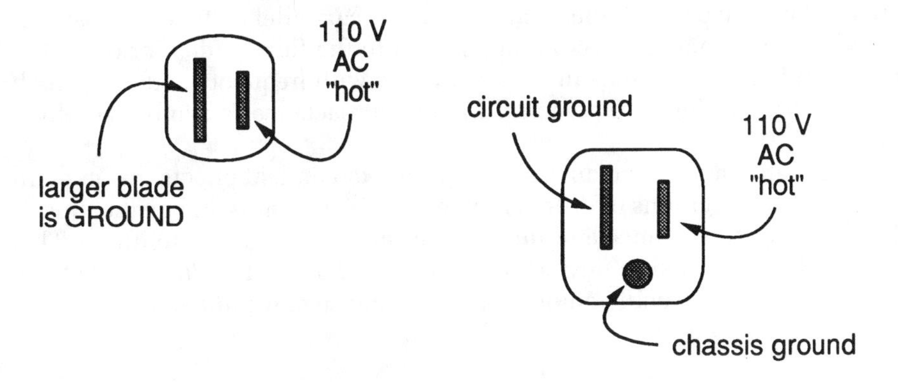

Reactance

SUMMARY

The amount of electrical "friction" or opposition to the flow of a constant direct current is called *resistance* and is measured in ohms. Alternating currents also encounter opposition to their flow in circuits. Ohm's law applies to ac circuits for the voltage and current on either instantaneous or root mean square bases.

In addition to conventional resistance, ac currents encounter a special type of electrical "friction" called *reactance*. Unlike conventional resistance that has a value which is independent of the frequency of the current, reactance varies with frequency. Furthermore, with conventional resistance, the voltage and current are always exactly in phase with each other. This is not the case with reactances, and the voltage and current flowing will not necessarily always be in phase with each other.

The first type of reactance is called *capacitive reactance*, or *capacitance*. The second type of reactance is called *inductive reactance*, or *inductance*. Capacitance is an effect associated with electrostatic fields, and inductance is an effect associated with electromagnetic fields.

Capacitive reactance offers an infinite opposition to the flow of dc. For other frequencies, capacitive reactance decreases inversely with frequency, offering very little opposition to very high frequencies. Capacitive reactance is symbolized by two parallel lines that represent the parallel plates of a capacitor.

Inductive reactance offers no opposition to the flow of dc. For other frequencies, inductive reactance increases linearly with frequency, offering much opposition to very high frequencies. Inductive reactance is symbolized by a coiled line.

Capacitance and inductance are frequency-dependent effects that explain the bandwidth limitations of different types of wire transmission media. Specified amounts of capacitance and inductance can be used to create filters. This is accomplished by using devices called *capacitors* and *inductors* that are constructed to have specific amounts of capacitance and inductance.

Reactance

Resistance:

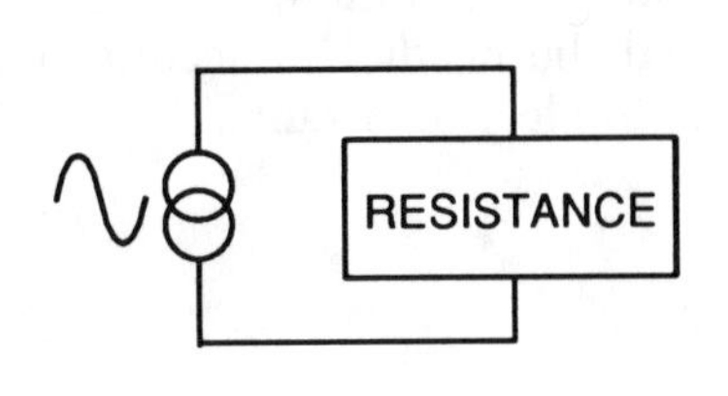

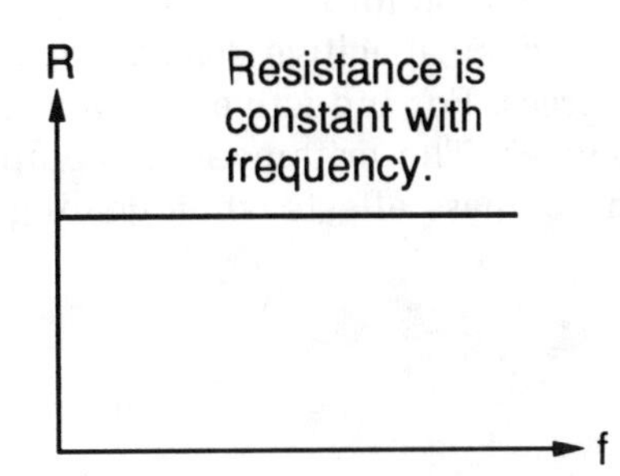

Capacitive Reactance:

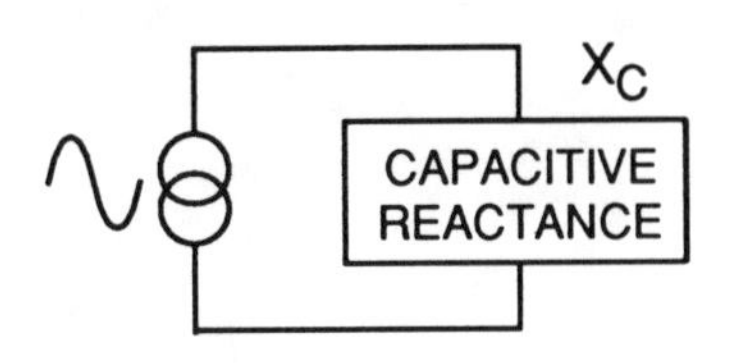

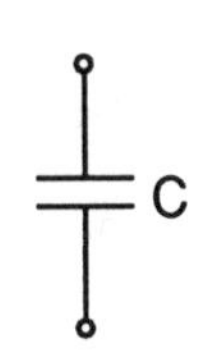

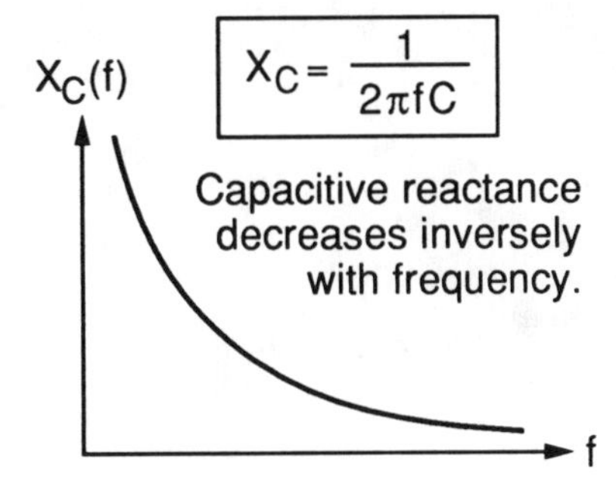

Inductive Reactance:

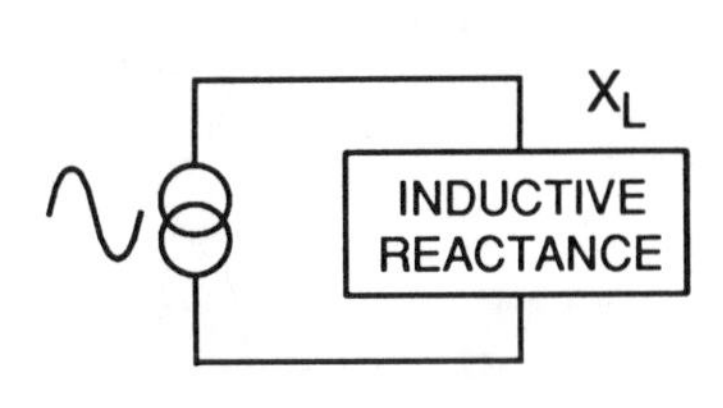

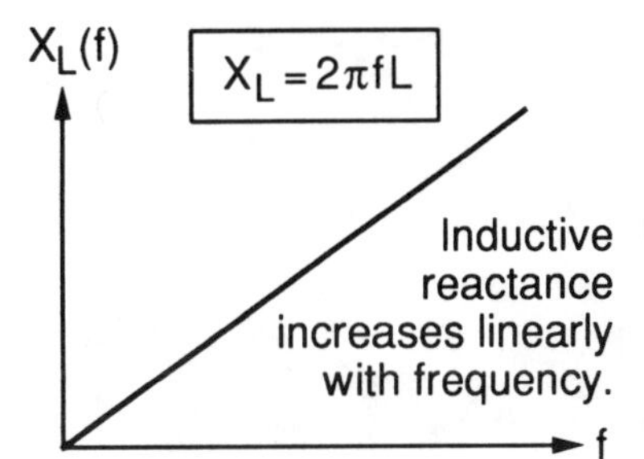

REACTANCE AND PHASE

The current through a resistance is in phase with the applied voltage. This is not the case for reactance.

For capacitive reactance, the current leads the applied voltage by 90 degrees. For inductive reactance, the applied voltage leads the current by 90 degrees. The arithmetic of complex numbers will be needed to account for these phase effects when dealing with reactance in electric circuits.

Reactance and Phase

Resistance:

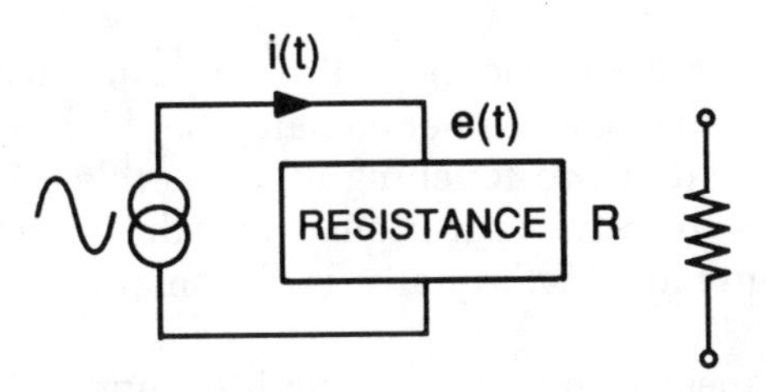

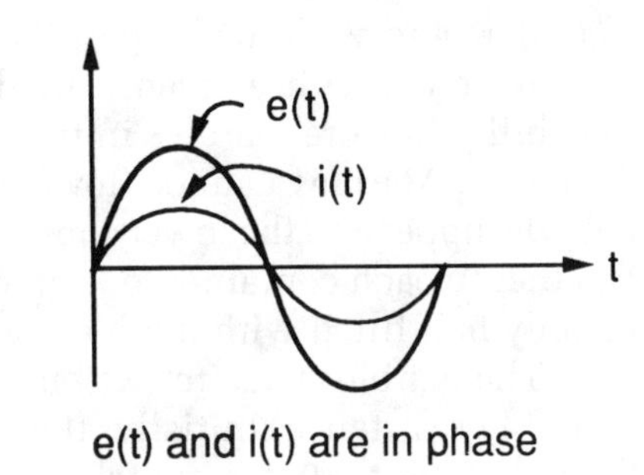

e(t) and i(t) are in phase

Capacitive Reactance:

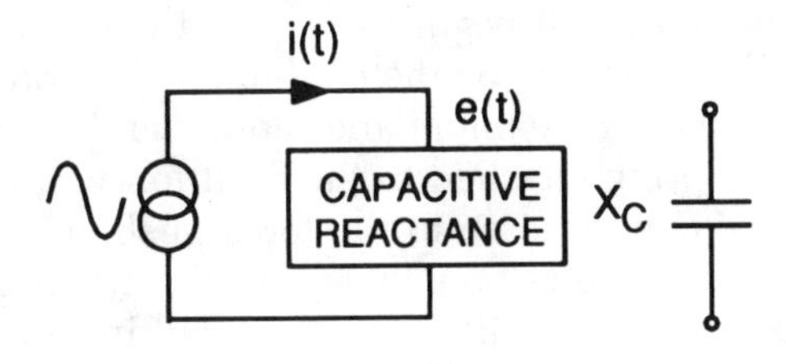

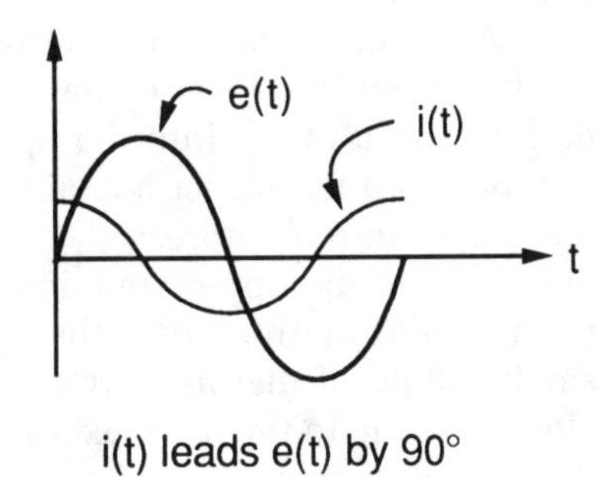

i(t) leads e(t) by 90°

Inductive Reactance:

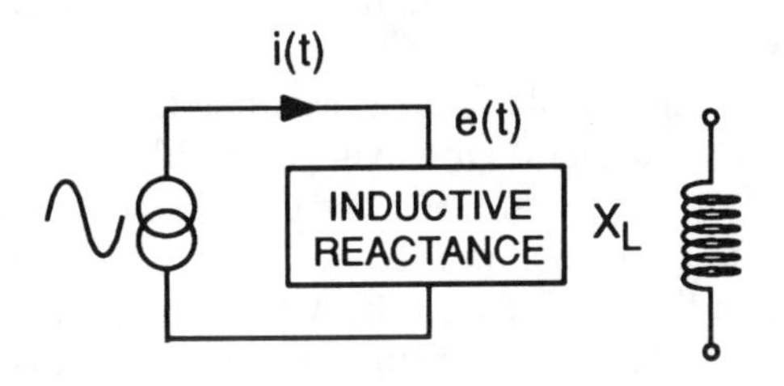

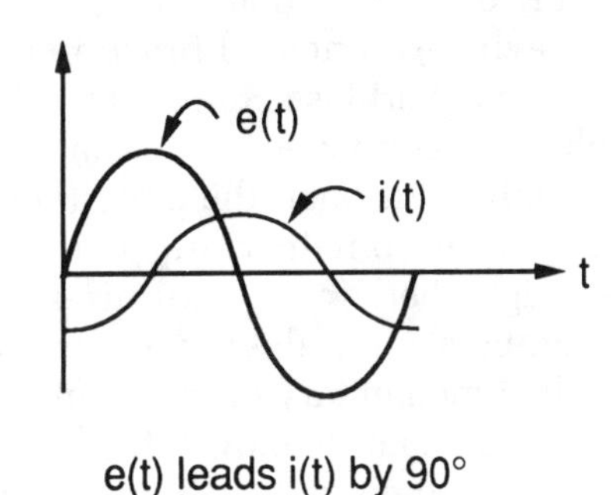

e(t) leads i(t) by 90°

CAPACITANCE ANALOGY

The marble circuit analogy will be used to help explain capacitive reactance. The analogy is not exactly precise, but capacitive reactance should be easier to understand with the help of the analogy.

A new circuit element called a marble capacitor will be invented with the ability to store marbles in two containers, an upper container and a lower container. Marbles cannot flow unless they are touching other marbles. So, both the upper and the lower containers will have a partition to push the marbles together in each container to keep them touching. Initially, each container will be only half filled with marbles.

The marble capacitor will be connected in a marble circuit in series with a marble resistance. Initially, the circuit will be open so that no marbles can flow. At time $t = 0$, the switch closes, and we can observe the behavior of the marble capacitor.

As soon as the switch closes, marbles will begin to flow in the circuit. Marbles from the lower container will be pumped by the marble-moving force of the marble pump into the upper container. As more and more marbles are pumped into the upper container, the spring attached to the partition will be more and more compressed, creating a counterforce to the force pushing the marbles into the upper container. As this counterforce increases, the net force pushing the marbles into the upper container will decrease, resulting in a smaller flow of marble current. Ultimately, the counterforce will become so close in value to the force exerted by the pump that the two forces will cancel, then no more marbles will flow, and the marble current ceases.

A plot of the marble current as a function of time will show an initial spurt of current that decays with time to zero. The shape of this curve is a decaying exponential function of time.

No marbles actually flow through the marble capacitor, since there is no physical connection between the two containers. However, a marble current does flow through the circuit, even if only momentarily, as the marbles from the lower container are pumped into the upper container. If we wait long enough, there will be no current flow in the marble circuit. A marble capacitor then blocks the flow of a direct current. The current flow that does occur is only a momentary or transient phenomenon.

The marble-moving counterforce across the marble capacitor is initially zero and then gradually builds up until it equals the marble-moving force of the pump. If the marble pump were connected directly to the marble capacitor, there initially would be no counterforce built up to oppose the marble pumping force of the pump and an infinite marble current would be generated. The insertion of a series resistance in the circuit prevents this infinite surge of current.

Capacitance Analogy

Marble Capacitor:

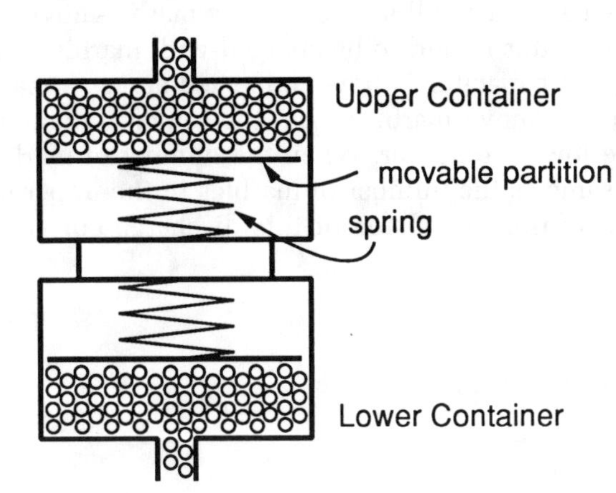

Note: Marbles can not flow until marble capacitor is connected to marble-conducting circuit and pump.

Charging:

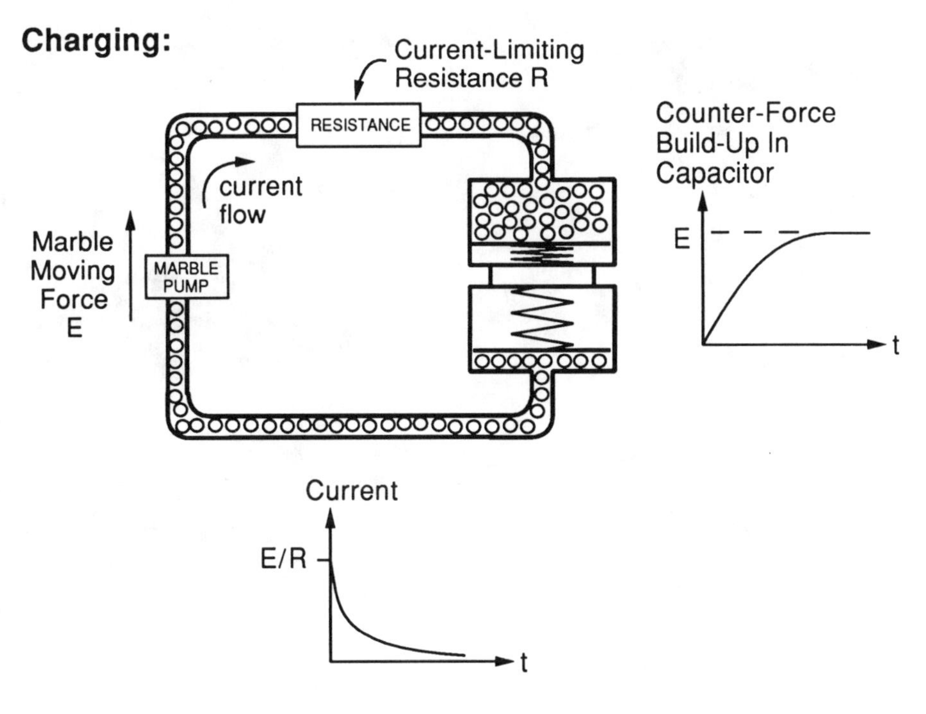

CAPACITANCE ANALOGY (cont'd)

After the flow of charging current has stopped, the marble capacitor can be removed from the circuit. Since there is no way for the marbles in the upper container to escape, the capacitor will save or store a marble-moving force. In this state, the marble capacitor is said to be charged with marbles.

If the charged marble capacitor is now connected across a marble resistance, the stored force will move marbles from the upper container through the circuit and into the lower container. When the number of marbles in the lower container is the same as the number of marbles in the upper container, the marble current will cease. The shape of this discharge current is also a decaying exponential.

Capacitance Analogy (cont'd)

Discharge:

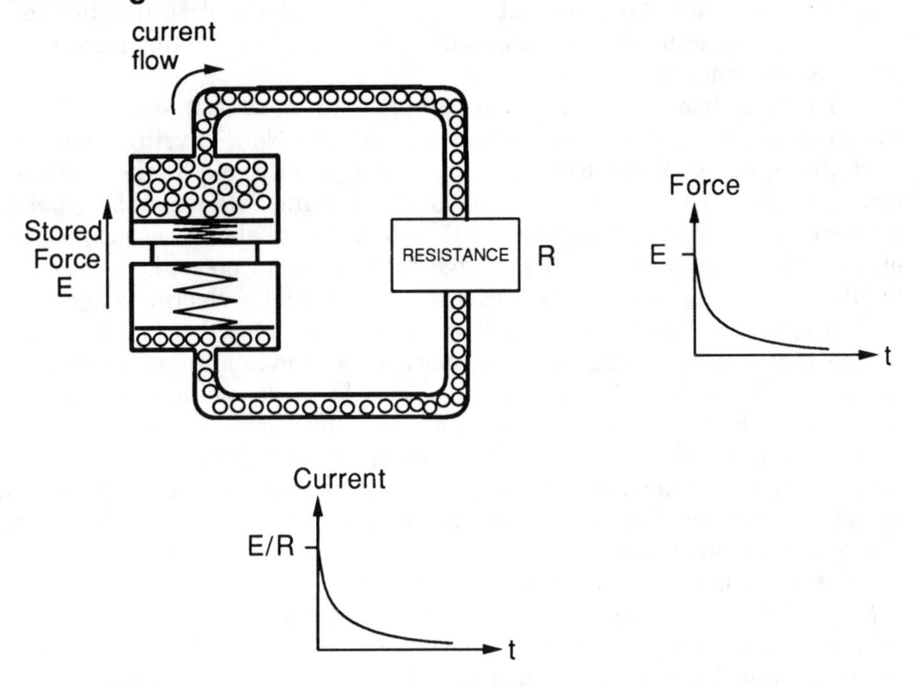

ELECTRICAL CAPACITANCE

Electrical capacitance is an effect that occurs whenever electric charge can accumulate and be stored between two conducting surfaces separated by an insulating material. An electrical capacitor is symbolized as two horizontal lines which represent the two sheets of nontouching conducting material that comprise a capacitor.

The two sheets, or plates, that comprise a capacitor are extremely large in surface area and so have a great excess of electrons. Normally, in an uncharged state, the number of electrons on each plate is the same. If the number of electrons were to become unbalanced, an electrostatic field would be generated between the two plates. This field would retain the imbalance because opposite electric charges attract. So, if a charge inequity were created, the capacitor would retain the inequity, acting as a storage device for electric charges.

A capacitor has an associated value of capacitance that is a measure of its capacity to store a charge. Capacitance is measured in *farads* and is formally defined as the ratio of the charge in *coulombs* to the voltage in volts. One coulomb of charge is equivalent to 6.28×10^{18} electrons.

An uncharged capacitor, with a capacitance of C farads, is connected in series with a resistance R and a source of an electromotive force E. A series switch prevents any flow of current until time $t = 0$ when the switch is closed. Since there is no charge stored on the capacitor to oppose any accumulation of electrons, the capacitor appears as if it were a short circuit, and a current E/R initially flows. However, as the current continues to flow, excess electrons from one plate are pumped onto the other plate, gradually building up an electric charge that opposes any further electron accumulation. When the charge is sufficient to create an EMF equal to the source E, current flow ceases.

A plot of the instantaneous current $i(t)$ as a function of time shows an initial current of E/R that gradually decreases toward zero at a decaying exponential rate given by

$$i(t) = \frac{E}{R} e^{-t/RC}$$

The e symbolizes the so-called *natural number*, which equals about 2.718. The current never really reaches zero, but very closely approaches zero. The current is said to approach zero asymptotically.

The voltage $v_c(t)$ across the capacitor is initially zero and then gradually increases, asymptotically approaching the value E. The equation describing the voltage rise is

$$v_c(t) = E(1 - \mathrm{e})^{-t/RC}$$

A series current-limiting resistance is needed in the circuit, since a capacitor looks like a short circuit to a sudden change in applied voltage and so initially draws an infinite amount of current.

Electrical Capacitance

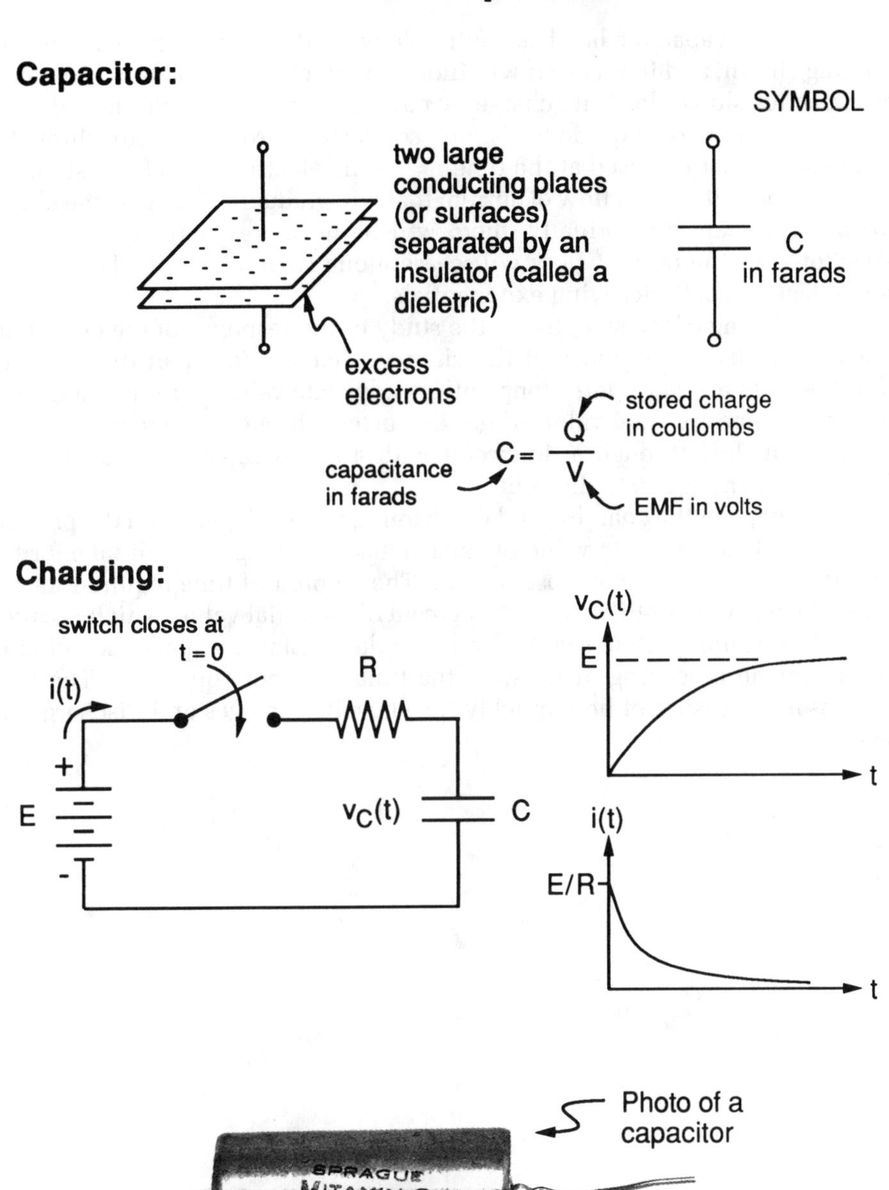

ELECTRICAL CAPACITANCE (cont'd)

Once the capacitor has been fully charged, it can be removed from the charging circuit, and in theory it will indefinitely retain its charge. In actuality, the capacitor slowly loses its charge by current leakage through the air.

The fully charged capacitor is next connected across a resistor through a series switch that is closed at the time $t = 0$. The electromotive force stored in the capacitor will cause a flow of current that will gradually dissipate the charge stored in the capacitor. Finally, there will be no more charge stored in the capacitor, and the current flow will cease along with any EMF. The current and voltage are both decaying exponentials.

The closing of the switch and the study of the response of the circuit are called the *transient response* of the circuit. After the transient dies out, the circuit reaches a *steady state* along with steady-state values for the voltage and current. The steady-state value of the dc current through a capacitor is zero. Any current flow through a dc circuit with a series capacitor is a transient phenomenon that quickly dies out to zero.

The shape of the charging and discharging curves depends on the product of R and C. The larger the value of capacitance, the longer it will take for the current to decay to a value close to zero. The amount of time required for the current to decay is 1/e-th or about 37 percent of its initial value, which is called the *time constant* of the circuit. For a simple resistance-capacitance circuit, such as in the preceding discussion, the time constant equals RC. The time constant is a measure of how quickly the capacitor charges and discharges in the circuit.

Capacitance (cont'd)

Discharge:

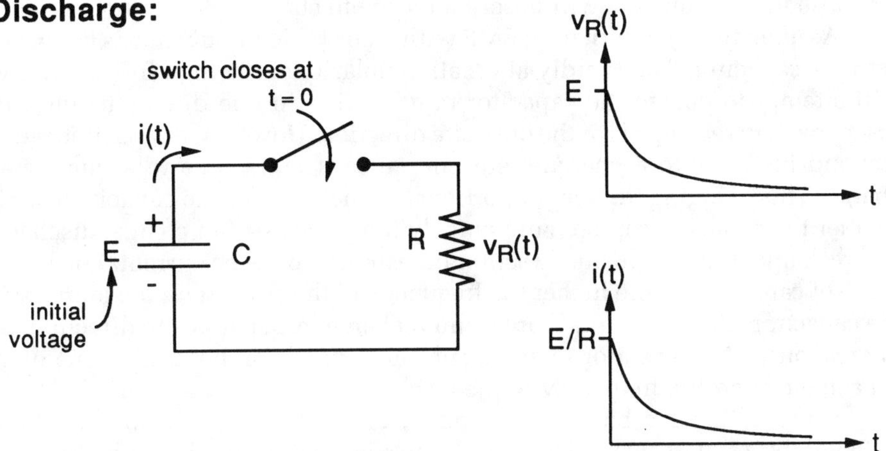

Time Constant:

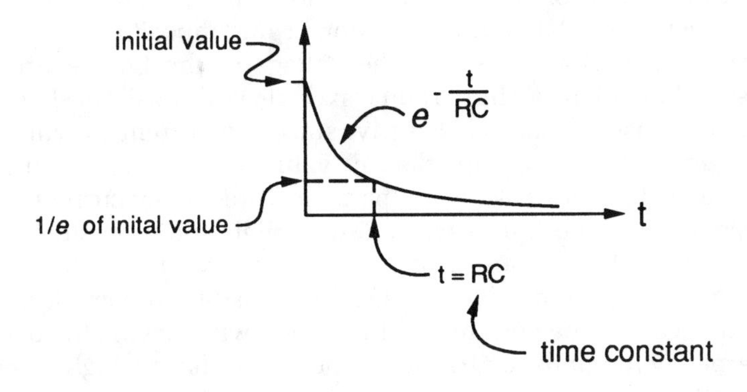

CAPACITANCE—AC EFFECTS

Our discussion of capacitance has so far concerned its transient effects in response to dc. Now, we will describe its ac effects.

Assume that an alternating EMF with a fairly high frequency is connected across a capacitor. The rapidly alternating polarity of the high frequency EMF will attempt to charge the capacitor rapidly, first in one direction, and then discharge it to recharge it in the opposite direction. However, a capacitor cannot respond instantly to a charging current, since it takes time to accumulate a charge. This charging time is proportional to the value of the capacitance. The smaller the capacitance, the more quickly the capacitor can charge, discharge, and recharge in the opposite direction in response to the ac current. For a given value of capacitance, the higher the frequency of the ac current, the more easily the capacitor can charge, discharge, and recharge in the opposite direction. So, the response of a capacitor to an ac current depends on both the value of the capacitance and the frequency of the ac current.

Perhaps this can be made clearer by an example. The source of the ac EMF will be a square wave. This avoids the complexity of a sinusoidally changing EMF. The square wave will cause switched dc charging effects as the capacitor is alternately charging in one direction, then suddenly charging in the opposite direction.

If the frequency of the square wave EMF is low, then the capacitor will have plenty of time to become nearly fully charged before the polarity of the applied EMF reverses to charge the capacitor in the opposite direction. The voltage waveform will rise quickly to the voltage of the EMF source as it periodically switches polarity. The current waveform will consist of short spurts of current with opposite polarities. Clearly, since the current flowing in the circuit is near zero most of the time, the rms value of the current will also be about zero. A capacitor does not easily "pass" low-frequency current.

If the frequency of the square wave EMF is high, the capacitor does not have time to become fully charged before the polarity of the applied EMF reverses. So, the voltage across the waveform is mostly quite small and will have a small rms value. The current would be high when trying to charge the capacitor alternately in opposite directions, and so will have a high rms value. A capacitor easily passes high-frequency current.

Capacitance—AC Effects

Low-Frequency Square Wave:

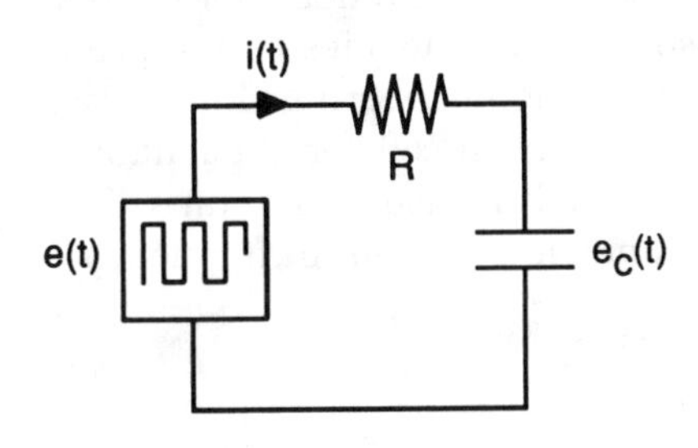

Capacitor charges quickly with respect to frequency of charging voltage, and current flowing in circuit is negligible.

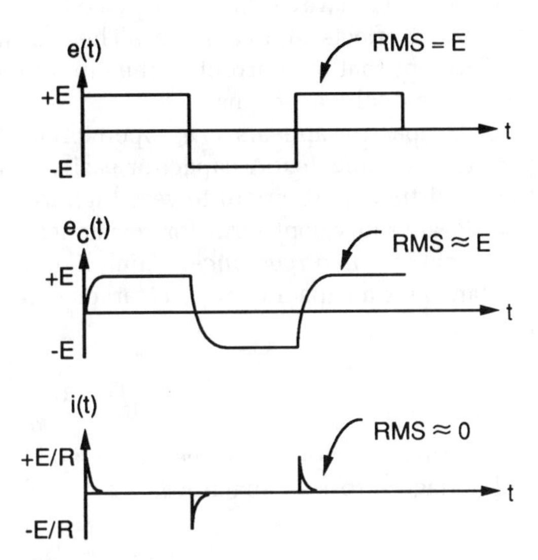

High-Frequency Square Wave:

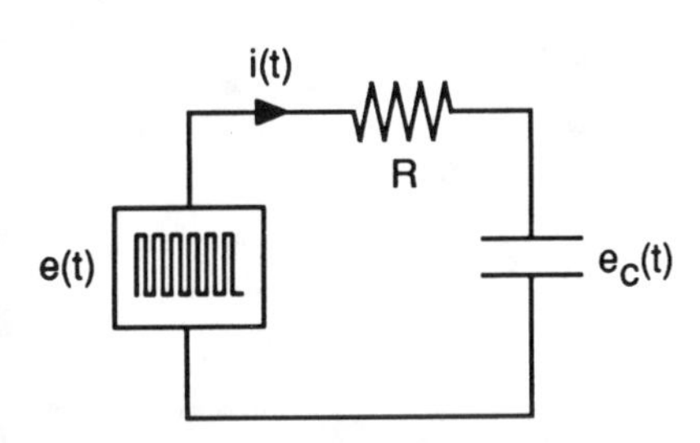

Capacitor never has time to charge fully, and hence current flowing in circuit is limited only by the resistance.

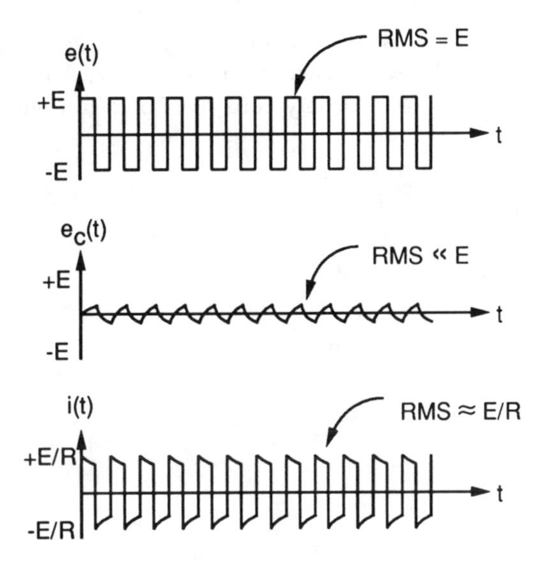

CAPACITANCE—FREQUENCY EFFECTS

We used a square wave EMF in the preceding discussion to simplify the frequency-dependent effects of a capacitor. These same effects occur with a sinusoidal EMF except that the current in the circuit and voltage across the capacitor are also sinusoidal waveforms.

A capacitor appears as an open circuit to dc current and does not readily pass low frequencies. A capacitor easily passes high frequencies and appears to be a virtual short circuit to very high frequencies. Remembering Ohm's law, $E = IR$, an equivalent term for resistance can be introduced for capacitance. This term is called *reactance*, which like resistance, is measured in ohms. The reactance for a capacitor with C farads capacitance to a current of frequency f hertz is

$$X_c(f) = 1/(2\pi fC)$$

Ohm's law applies to reactance, and the relationship for the rms current and voltage across a capacitor is

$$E_{\text{rms}} = I_{\text{rms}} X_C$$

Because the reactance depends on frequency, both the voltage and current would likewise vary with frequency. A capacitor offers infinite reactance to a dc signal and nearly zero reactance to a very high frequency signal.

Frequency Effects

Frequency Behavior:

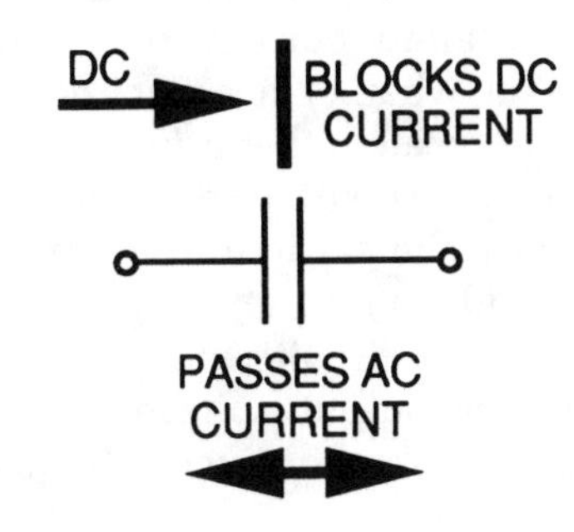

Reactance:

reactance is a function of frequency

reactance in ohms

$$X_C(f) = \frac{1}{2\pi f C}$$

frequency in hertz

capacitance in farads

$X_C(f = 0) = \infty$

behaves like an open circuit at low frequencies

$X_C(f)$

f

behaves like a short circuit at high frequencies

$X_C(f = \infty) = 0$

CAPACITANCE—FILTER EFFECTS

A capacitor can be used in an electrical circuit to block lower frequencies and to pass higher frequencies. Remember that no current actually flows through the capacitor itself, but, depending on the frequency of the current, the capacitor either allows or does not allow current to flow in the circuit in which it is connected. Depending on whether a capacitor is connected in series or in parallel, it can be used to create either a *high-pass filter* or a *low-pass filter.*

When connected in series with a resistance R, a capacitor creates a high-pass filter. This is because the capacitor blocks low frequency currents from reaching the resistance and causing a voltage to appear across the resistance. If the input voltage to such a circuit were held constant in amplitude as the frequency were varied, the output voltage taken across the resistor would increase as the frequency increased. Such a frequency response is associated with a high-pass filter.

When connected in parallel with a resistance, a capacitor creates a low-pass filter. This is because the capacitor shorts out the high-frequency currents to prevent them from ever reaching the resistor and creating a voltage across it. If the input voltage to such a circuit, including a current-limiting series resistance, is held constant in amplitude as the frequency is varied, the output voltage across the parallel capacitor-resistor combination decreases as the frequency increases. Such a frequency response is associated with a low-pass filter.

Capacitors can be used in hi-fi loudspeakers to protect the low frequency loudspeaker (called the woofer) from high frequencies and the high frequency loudspeaker (called the tweeter) from low frequencies. This type of filter is called a crossover network, since the signal crosses over from the woofer to the tweeter as the frequency increases.

Filter Effects

High-Pass Filter:

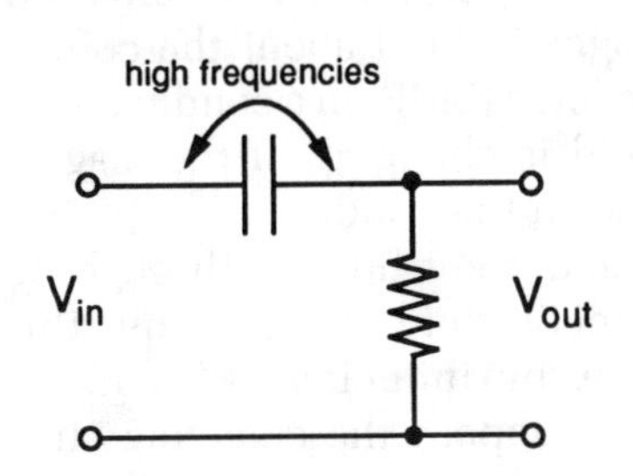

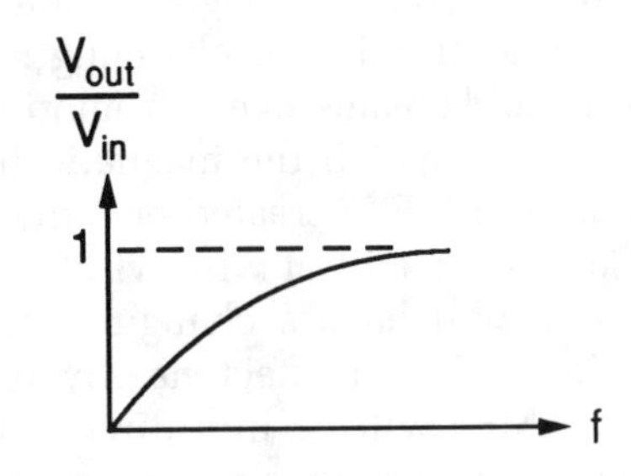

Low-Pass Filter:

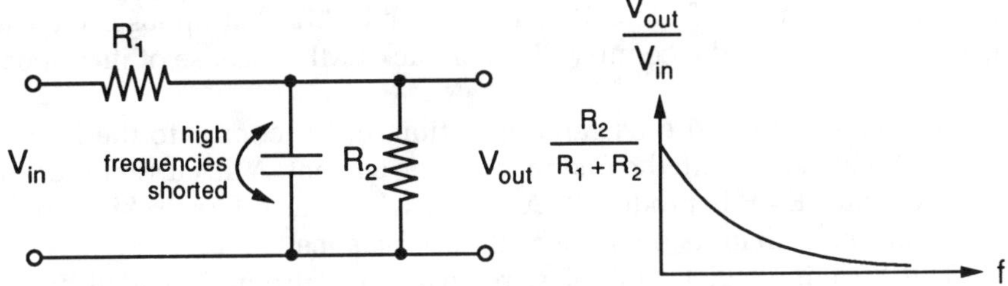

Loudspeaker Crossover Filter:

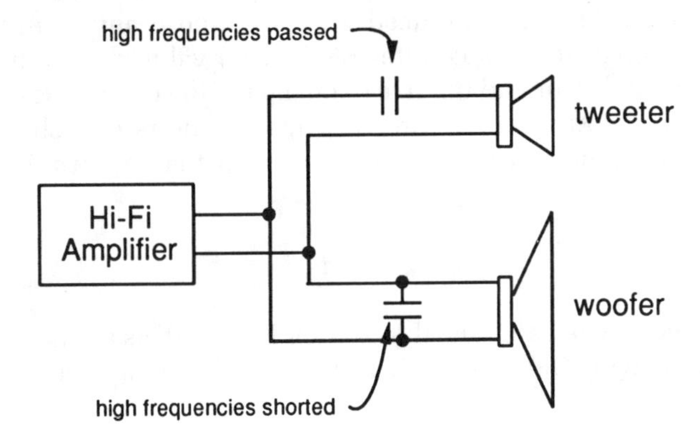

INDUCTANCE

There are two electromagnetic effects associated with electricity that are essential to understanding inductance. A constant dc current flowing in a coil of wire creates a constant magnetic field about the coil, and an alternating current creates an alternating or changing magnetic field about the coil. A changing magnetic field creates an electromotive force (EMF) in a conductor that depends on the rate at which the magnetic field is changing. If the magnetic field is constant, no EMF is created or induced in the conductor.

Consider a coil of wire with an ac current flowing through it. Each turn of the coil will have a changing magnetic field about it, and this changing magnetic field will affect nearby turns by inducing EMFs in them. These induced EMFs will be in a direction to oppose the changing current. In the same way that moving a magnet more quickly through a coil of wire creates a stronger EMF, the faster the ac current alternates, the faster will be the rate of change of the magnetic field and the stronger will be the value of the EMF induced in each turn of the coil. This induced EMF that opposes changes in the ac current is called a counter EMF, or back EMF, because of its opposition nature.

Hence, a coil of wire has an opposition, or "friction," to the flow of an ac current that varies with the frequency of the current. A dc current is constant, and no counter EMF is produced. A coil of wire thus offers no "friction" to a dc current, other than its own internal dc resistance.

The "friction" of a coil of wire to an ac current is called *inductive reactance*. The property associated with a coil that gives a measure of the strength of the counter EMF is called the *inductance* of the coil and is expressed in a quantity called *henrys* (H).

An inductor creates a self-induced EMF to oppose any changes in the current passing through it. This counter EMF has a value determined by the inductance L of the inductor and the rate of change of the current. If the current changes an amount Δi in the time interval Δt, then the rate of change of the current is $\Delta i/\Delta t$. The Δ means a change in the quantity that follows. The counter EMF is

$$v = L\,\Delta i/\Delta t$$

In the terminology of calculus, the counter EMF $v(t)$, as a function of time across an inductor with inductance L, is L times the time derivative of the current $i(t)$, or

$$v(t) = L\frac{\mathrm{d}}{\mathrm{d}t}i(t)$$

where $\frac{\mathrm{d}}{\mathrm{d}t}$ represents the derivative with respect to time.

Inductance

Electromagnetism:

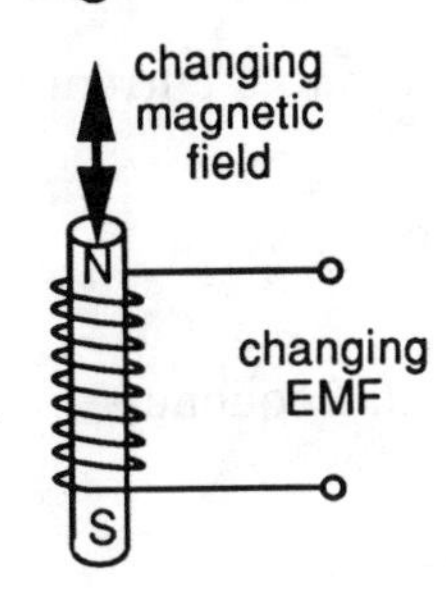

Self Induction:

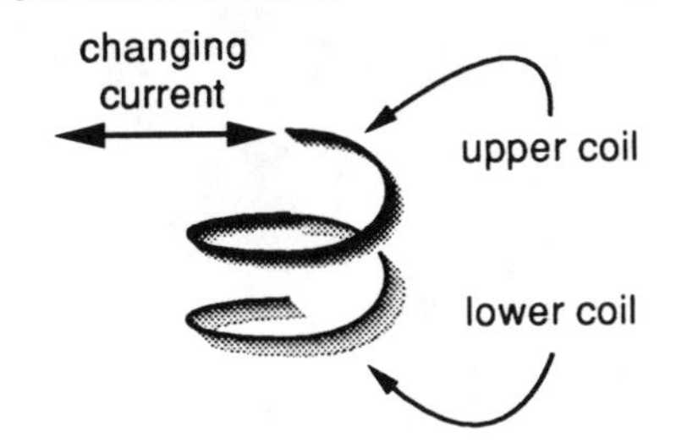

Changing current in upper coil creates a changing magnetic field that induces a counter EMF in the lower coil in a direction to oppose the change of current.

Inductance:

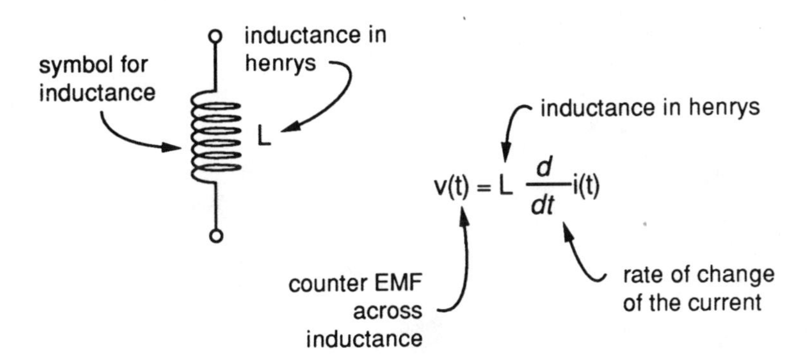

INDUCTANCE—FREQUENCY EFFECTS

An inductor appears as a closed circuit to dc current and offers very little opposition to very low frequencies. An inductor opposes high frequencies and does not pass them easily.

The reactance for an inductor with L henrys to a current of frequency f hertz is

$$X_L(f) = 2\pi fL$$

Thus, inductive reactance increases linearly with frequency.

Frequency Effects

Frequency Behavior:

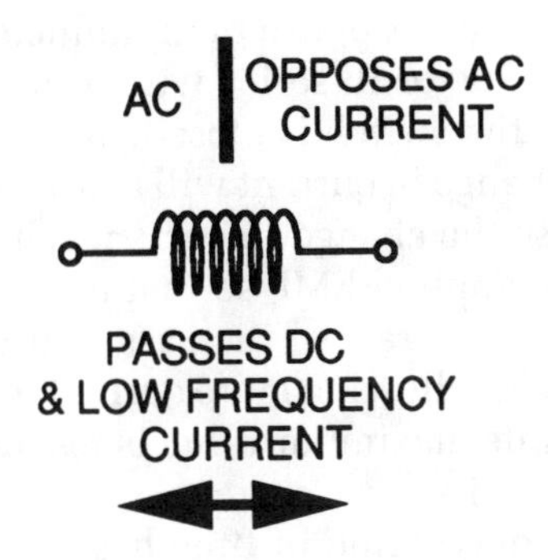

Reactance:

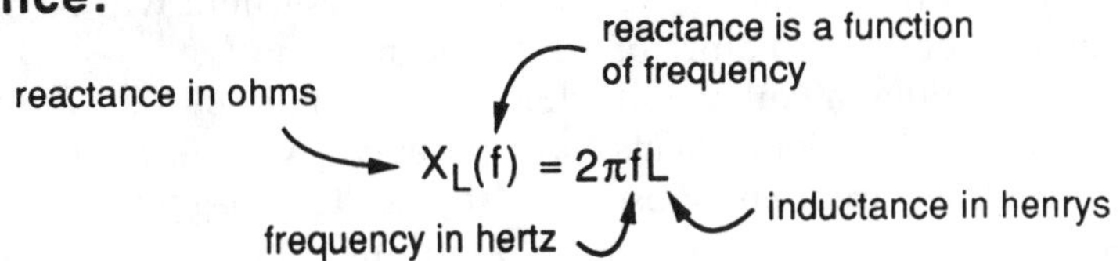

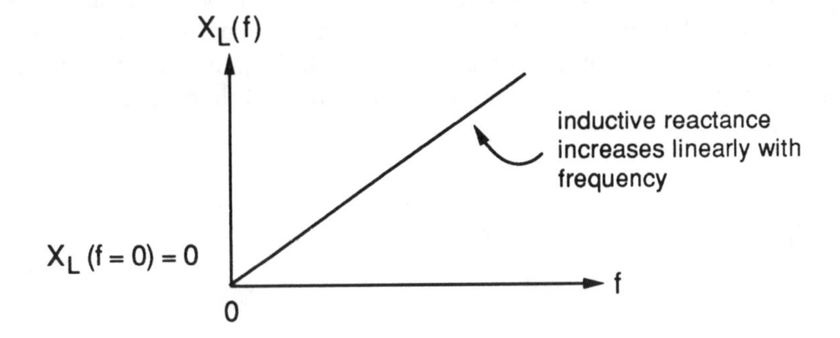

INDUCTANCE—TRANSIENT EFFECTS

A sudden change in voltage or current in an electrical circuit is called a *transient*. The charging of a capacitor caused by a sudden connection of a voltage across its terminals is called the *transient response* of the capacitor. Here, we will investigate the transient response of an inductor.

Consider an inductor in series with a resistor and a source of a constant EMF. At time $t = 0$, the switch is closed, and current will suddenly attempt to flow. The sudden change in current will cause a counter EMF to be self-induced in the coil to oppose the change in current. This counter EMF will initially be exactly equal to the applied EMF so that no current will initially flow. As the current starts to flow, its rate of increase will decrease, and the counter EMF will likewise decrease, allowing more current to flow. The current will increase exponentially until its maximum value is reached, as determined by the series resistance in the circuit.

After a fairly long period of time has passed, the initial transient effects will have died down and a steady dc current will be flowing through the inductor. This dc current will create a constant magnetic field within the inductor. Any attempt to change quickly this constant field will be opposed by the inductor. Such is the nature of electromagnetism.

Assume that a coil or inductor has a constant current flowing through it and all initial transient effects have decayed. At time $t = 0$, the coil will be instantly short circuited across a resistance. The magnetic field set up by the constant current cannot suddenly do nothing. The field will slowly collapse, eventually causing a current to flow until there is no more magnetic field and, correspondingly, no current is flowing. This decrease in current is exponential.

This effect is somewhat akin to pushing a car. The car has a lot of mass and is hard to get going. However, once it is in motion, if the force pushing it suddenly stops, the car will continue moving and then, slowly eventually come to a halt.

Inductance—Transients

Charging:

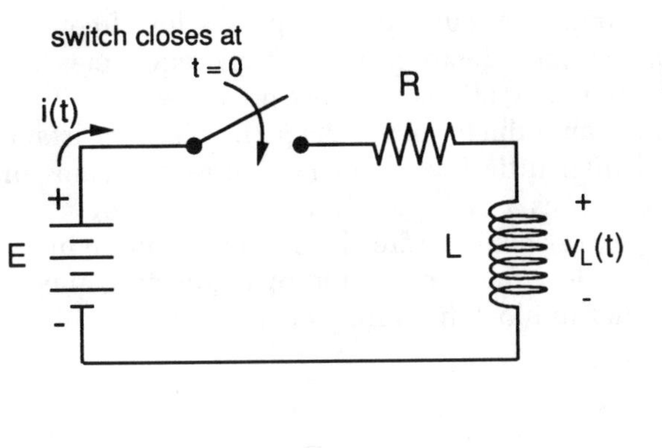

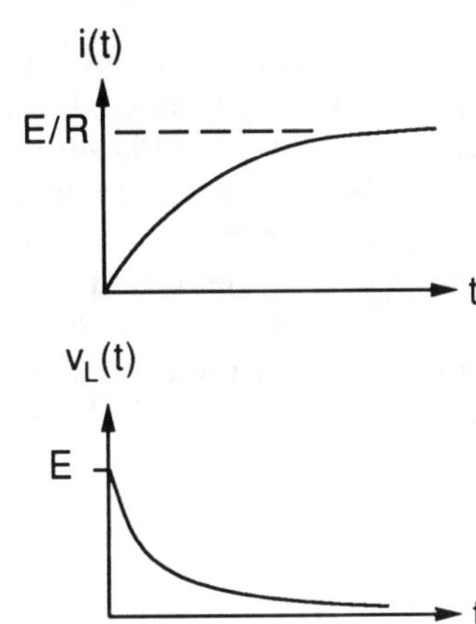

Discharging:

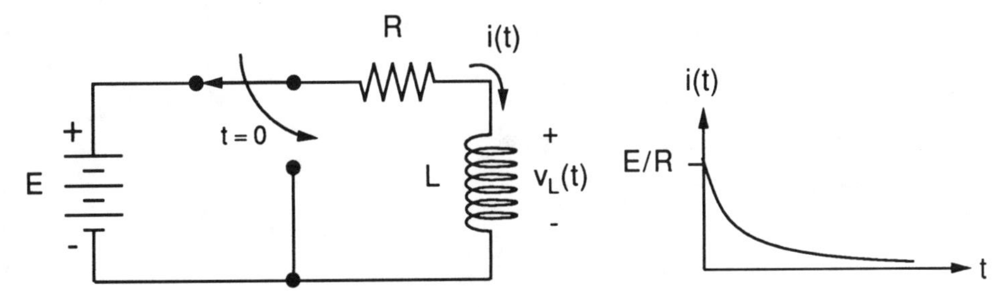

INDUCTANCE—FILTER EFFECTS

Depending on whether an inductor is connected in series or in parallel with a resistive load, the inductor can create either a *low-pass filter* or a *high-pass filter.*

An inductor blocks high frequency currents and passes low frequency currents. When connected in series, an inductor prevents high frequencies from reaching the load. This series circuit functions as a low-pass filter.

When connected in parallel, an inductor offers the path of least resistance to low frequency current, and then little low frequency current reaches the load. This parallel circuit functions as a high-pass filter.

A loudspeaker crossover network could be created by connecting an inductor across the tweeter to short out low frequencies and by connecting another inductor in series with the woofer to block high frequencies.

Filter Effects

Low-Pass Filter:

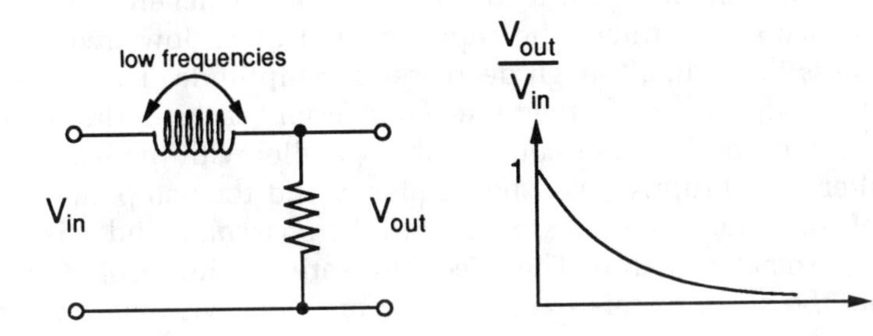

High-Pass Filter:

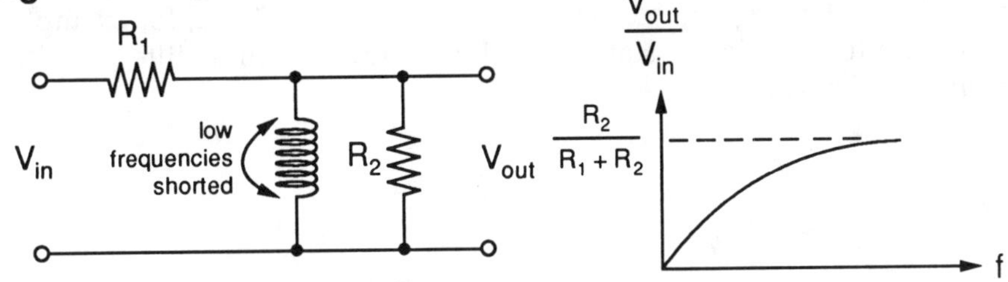

Loudspeaker Crossover Filter:

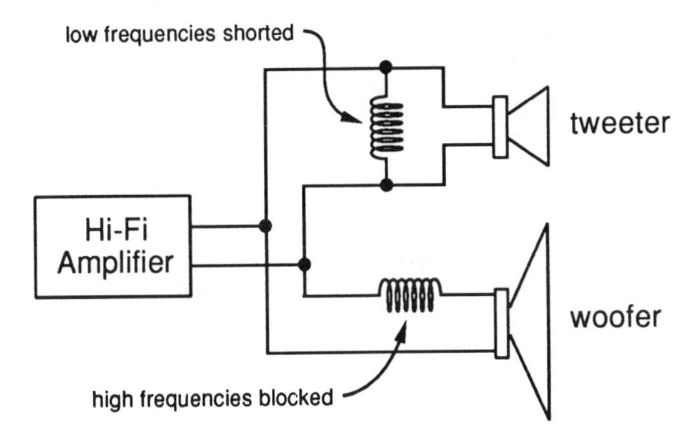

CAPACITANCE AND INDUCTANCE

A capacitor can be used together with an inductor to create a *high-pass filter.* The inductor is connected across the load, and the capacitor is connected in series with the parallel inductor-load combination. The net effect is to improve the performance of the filter. The capacitor will block low frequencies, but some will pass through, although decreased in amplitude. Those frequencies that pass through will be further prevented from reaching the load by the shorting action of the inductor connected in parallel with the load.

No filter can abruptly pass one frequency and then stop another. There is a gradual transition from the *stopband* to the *passband,* and this transition has a slope associated with it. The effect of creating a filter from two reactors is to double the slope of this transition as compared with what it would be were there only one reactor.

Capacitors and inductors can be fixed or variable. If variable, they are shown symbolically with an arrow through them. Variable capacitors are used to vary the center frequency of very narrow *bandpass filters,* for example, to tune a radio to a particular station. Such a narrow bandpass filter is called a *sharply tuned circuit.*

Capacitance and Inductance

High-Pass Filter:

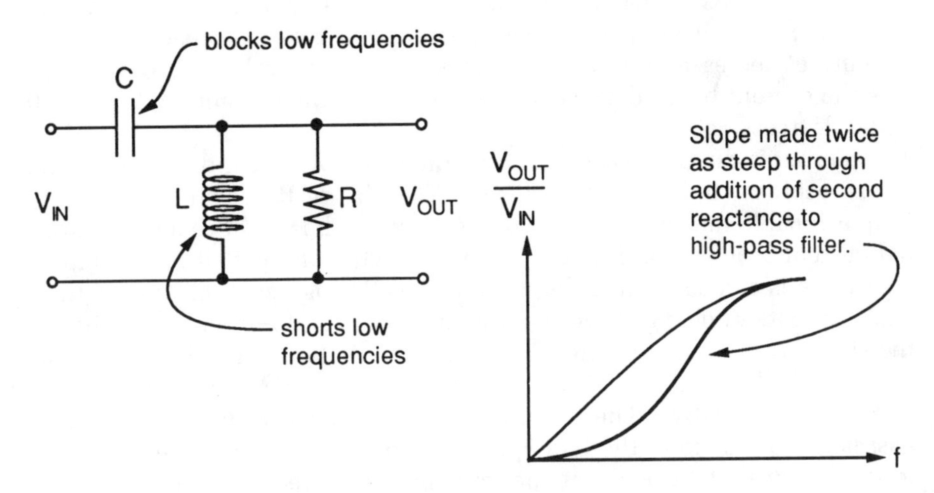

Variable Reactors:

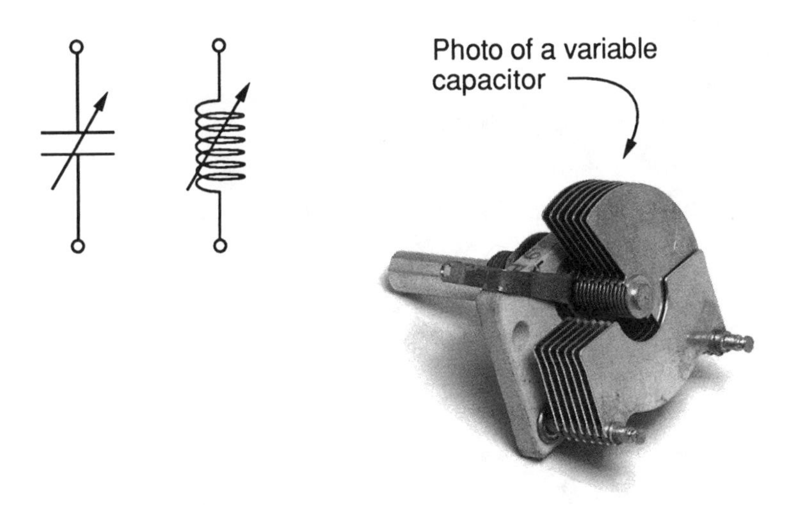

PHASE AND REACTANCE

The current passing through a reactance is not in phase with the applied voltage.

For a capacitive reactance, the maximum charging current flows when the voltage across the capacitor is at its lowest value. For a sinusoidal EMF at the source, the current peaks when the voltage across the capacitor is zero. As the current decreases, the voltage increases. Thus, the voltage across a capacitor lags the current by 90 degrees. Or, equivalently, the current leads the voltage by 90 degrees.

For an inductive reactance, the voltage across the inductor will oppose the sinusoidal changes in the current. The size of this counter EMF will be proportional to the rate of change of the current. The rate of change of a sinusoidal current is itself continuously changing. The rate of change, or slope, of a sine wave is steepest when the wave passes through zero, or crosses the zero axis. The rate of change is zero at the maximum and minimum amplitudes of the sine wave. So, the counter EMF is maximum when the sinusoidal current begins its full cycle, since at this time the rate of change, or slope, of the wave is maximally positive. The slope is maximally negative when the wave next passes through zero at the half cycle. The net effect is that the counter EMF leads the current through an inductor by 90 degrees. Or, equivalently, the current lags the voltage by 90 degrees.

Phase and Reactance

Capacitance:

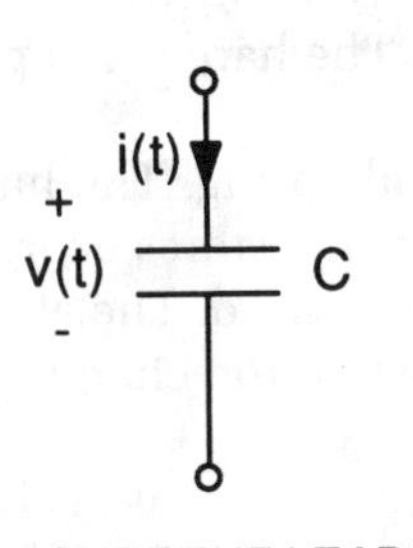

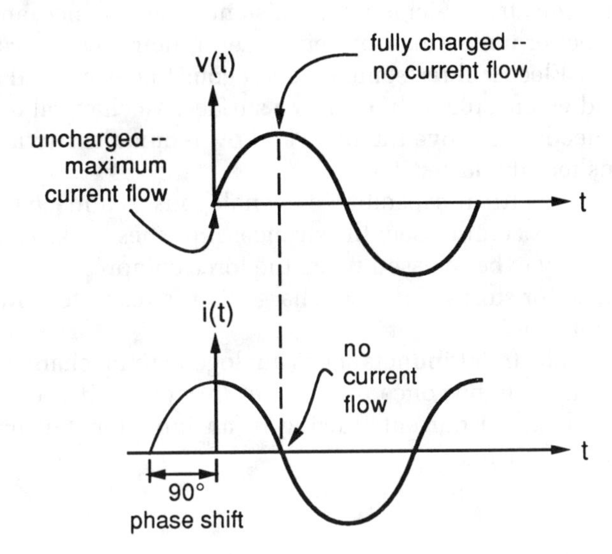

CURRENT LEADS
VOLTAGE BY 90°

Inductance:

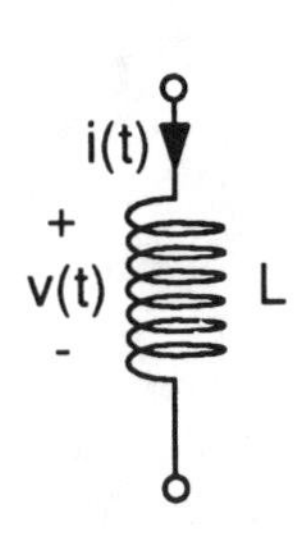

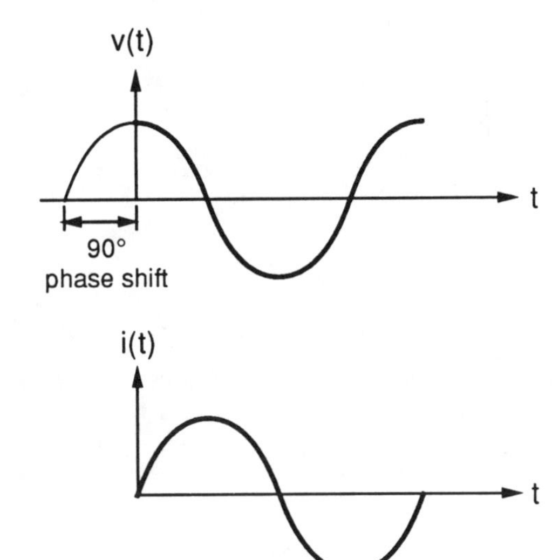

VOLTAGE LEADS
CURRENT BY 90°

MECHANICAL COUNTERPARTS

Such electrical elements as resistance, capacitance, and inductance can perhaps be better understood by reference to their mechanical counterparts.

Electrical resistance is analogous to mechanical friction. Friction is generated when a plate slides over a surface. Mechanical force, analogous to an EMF, is needed to move the plate and overcome the friction. The harder the plate is pushed, the faster it slides.

Electrical capacitance is analogous to a mechanical spring. The harder a spring is compressed, the stronger it pushes back. The spring stores energy that is ready to be released when the force compressing it is released. The electrical capacitor stores electrical charge that is ready to flow when the charging EMF is removed.

Electrical inductance is analogous to mechanical mass. A mass is hard to get moving, but once mass is set in motion, it keeps moving. It is difficult to get electrical current flowing in an inductor, but once current is flowing, it keeps flowing.

Mechanical Counterparts

Electrical | **Mechanical**

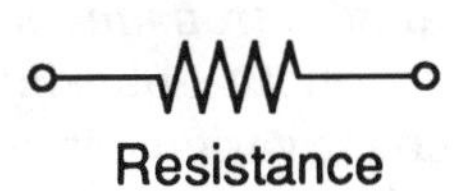
Resistance

Friction

Capacitance

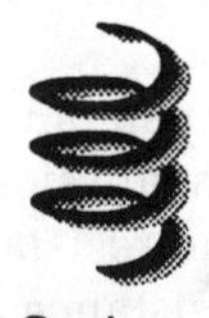
Spring

Inductance

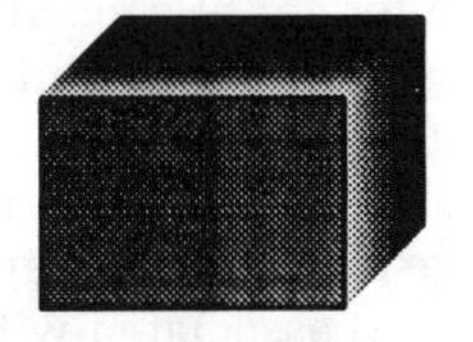
Mass

TRANSMISSION LINES

The flow of an electrical current between a source and a load requires two conductors so that a complete circuit is formed. If the source and the load are distant from each other, the two conductors will be quite long in length and will cover most of the distance to be quite close to each other. Two parallel conductors in close proximity to each other are called a *transmission line.*

Capacitance and inductance are associated with the physical devices called capacitors and inductors used in circuits to create various types of filters. Capacitance and inductance are also encountered whenever electricity is flowing in conductors. In this situation, they are effects rather than physical circuit elements.

Resistance, capacitance, and inductance are distributed uniformly along the length of a transmission line. Capacitance is a *parallel effect*, inductance is a *series effect*, and resistance is both a *series and parallel effect.* The parallel capacitance and resistance are called *shunt capacitance* and *shunt resistance.* All these characteristic effects are expressed on a per distance basis, for example, farads per mile for shunt capacitance.

The series inductance and shunt capacitance limit the high frequency performance of a transmission line, and in so doing determine the bandwidth of the line. A sharp pulse or sharp square wave at the input to a transmission line will emerge distorted in shape to the extent that the line does not pass the higher frequencies associated with sharpness.

Physical aspects of the transmission line, such as the spacing between the conductors, their diameter and conductivity, and the dielectric constant of the insulation, determine the actual values of the transmission characteristics. So, different types of transmission lines (for example, bare wire, twisted pair, and coaxial cable) will have quite different transmission characteristics and bandwidths.

Transmission Lines

Transmission Line:

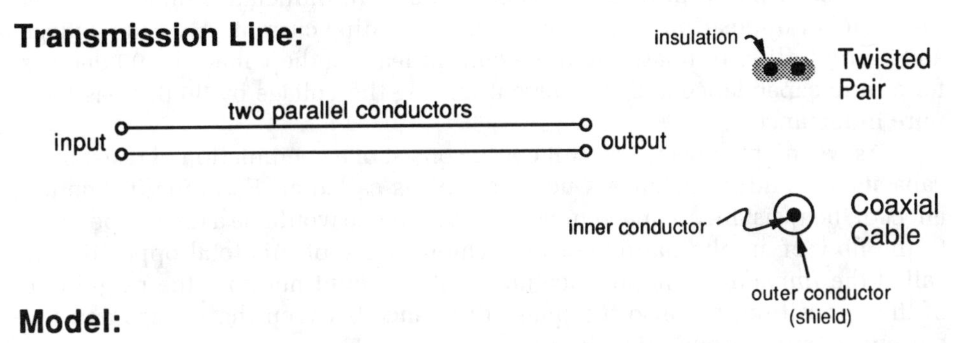

Model:

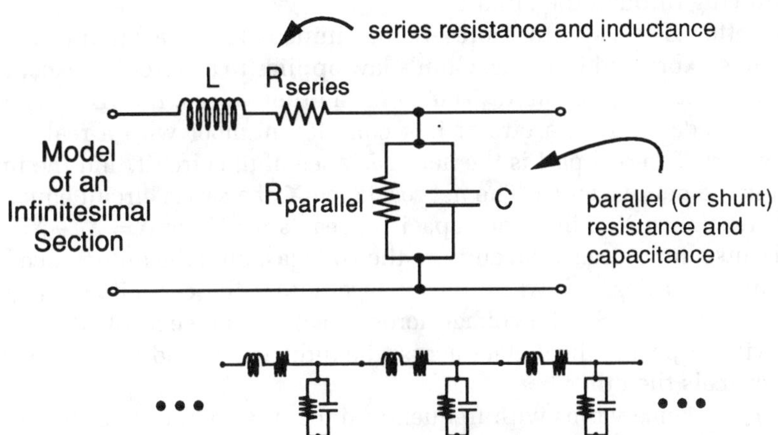

The overall transmission line is modeled as an infinite number of infinitesimal sections connected in series.

Bandwidth:

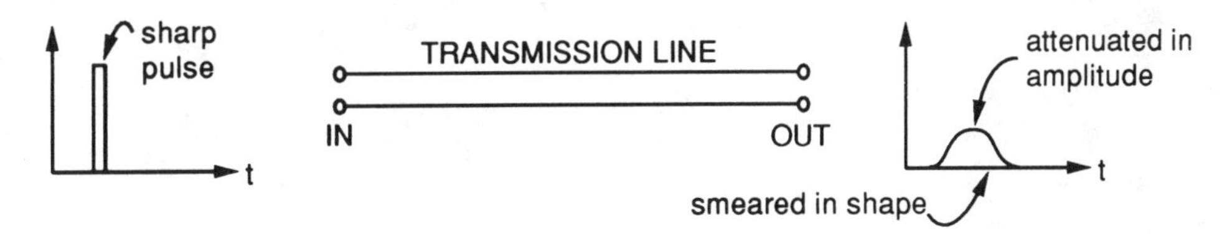

IMPEDANCE

We have seen that a pure capacitance and a pure inductance offer electrical "friction" or opposition to the flow of an alternating current. Also, the voltage and current differ in phase, with the current leading the voltage by 90 degrees for a pure capacitance and the current lagging the voltage by 90 degrees for a pure inductance.

As we might expect, a circuit could consist of a combination of resistance, capacitance, and inductance. Such a circuit is called an *RLC circuit*. The net effect of the resistance, capacitance, and inductance would be a total opposition to the flow of an alternating current. The measure of this total opposition is called the *impedance*. Impedance takes into account not only the magnitude of the opposition, but also the phase difference between the voltage and the current flowing through the circuit.

The letter Z is used to represent the impedance of a circuit, and the impedance is expressed in ohms. Ohm's law applies to ac circuits, except that the impedance Z replaces the resistance R. In other words, $E = IZ$.

The impedance Z of a circuit is a complex number with a real and an imaginary part. The real part is the net resistance of the circuit, and the imaginary part is the net reactance. The net reactance X of a series circuit equals the inductive reactance X_L minus the capacitive reactance X_C, or $X = X_L - X_C$. This occurs because, for a sinusoidal current, the voltage across the inductance leads the current by 90 degrees, while the voltage across the capacitance lags the current by 90 degrees. So, the voltage across the capacitance is 180 degrees out of phase with respect to the voltage across the inductance, and then one voltage partially cancels the other.

The impedance varies with frequency since it is composed of frequency-dependent reactances.

Impedance

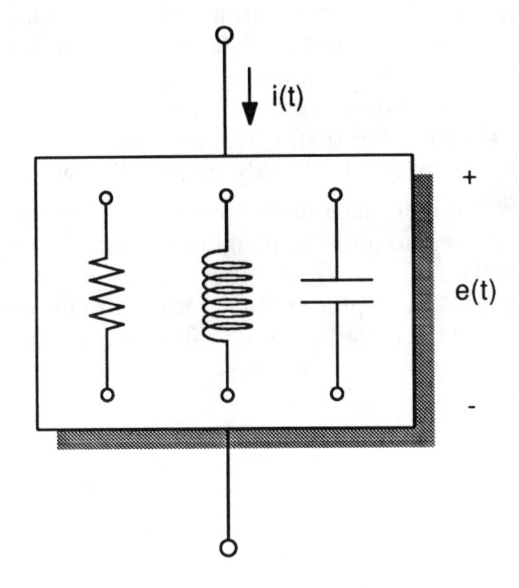

2

Z is the impedance in ohms.

$$E = I Z$$

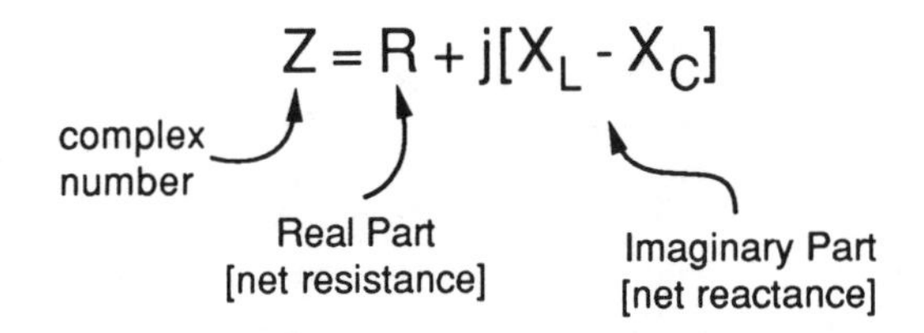

Z is a function of frequency:
Z(f)

COMPLEX NUMBERS

A complex number defines a point in a two-dimensional space. The real part of a complex number indicates the distance along the x-axis (or abscissa), and the imaginary part indicates the distance along the y-axis (the ordinate). The imaginary part is indicated by placing a **j** in front of the imaginary part; mathematicians use the letter i rather than **j**. The **j** symbolizes the square root of minus one, or $\mathbf{j} = \sqrt{-1}$, which clearly is aptly named "imaginary!"

A complex number has a magnitude equal to the square root of the sum of the squares of the real and the imaginary parts. The angle θ formed by the magnitude with the real axis is the inverse tangent of the real part divided by the imaginary part. The representation of a complex number in terms of its magnitude and angle is called polar coordinates.

Complex Numbers

Rectangular Coordinates:

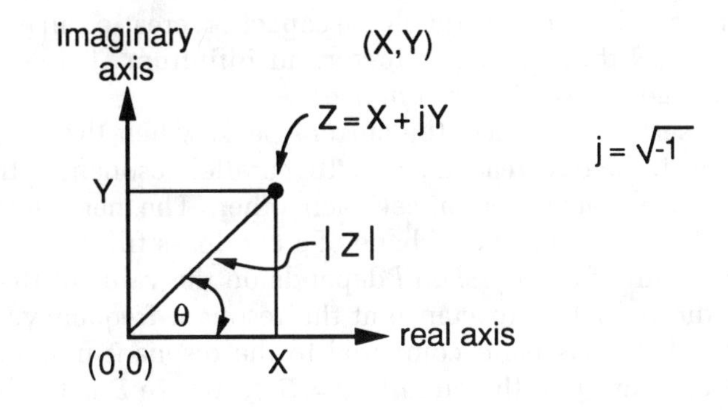

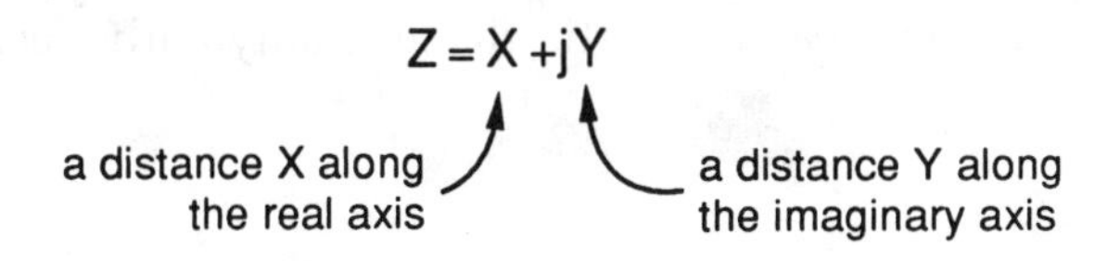

Polar Coordinates:

$$Z = |Z| \angle \theta$$

magnitude

$$|Z| = \sqrt{X^2 + Y^2}$$

$$\theta = \tan^{-1}(X/Y)$$

angle

RESONANCE

Interesting interactions occur when a capacitor and an inductor are connected in series or in parallel. These interactions, called resonance, occur as the energy stored in the capacitor discharges through the inductor, creating a magnetic field that then discharges through the capacitor, creating an electrostatic charge that discharges through the inductor, ad infinitum. This oscillatory flow of electric charge is *electrical resonance.*

With series resonance, the current peaks when the capacitive reactance cancels the inductive reactance. With parallel resonance, the voltage peaks when the two reactances cancel each other. The net effect is a frequency response that has the peaked shape of a bandpass filter.

The width of the *passband* depends on the value of the resistance compared to the inductive reactance at the resonant frequency, f_R. A measure of the width of the passband compared to the resonant frequency is called the quality factor, or Q, of the circuit. $Q = B/f_R$, where B is the bandwidth of the passband. The peakedness of a curve is called cortosis. A leptocortic shape is very sharply peaked while a platocortic shape is only flatly peaked.

Resonance

Series Resonance:

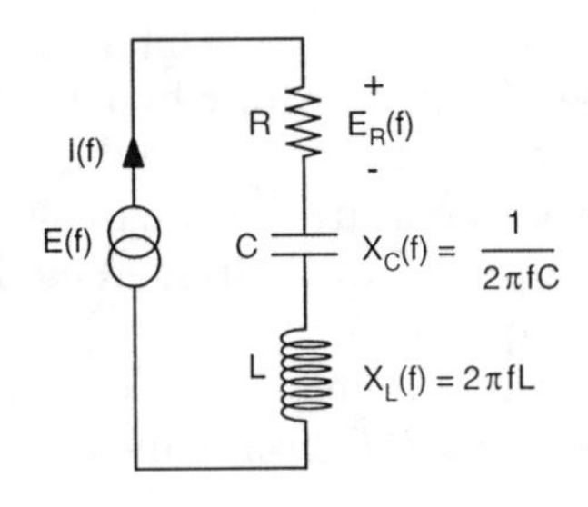

Ohm's Law:

$$E(f) = I(f)\,Z(f)$$

or

$$I(f) = E(f) / Z(f)$$

$$Z(f) = R + j\,[X_L(f) - X_C(f)]$$

$$= R + j\,[\,2\pi fL - \frac{1}{2\pi fC}\,]$$

$$|\,Z(f)| = \sqrt{R^2 + [\,2\pi fL - 1/(2\pi fC)\,]^2}$$

$$|\,Z(f)| = R \text{ when } 2\pi fL = 1/(2\pi fC)$$

and that occurs at the resonant frequency:

$$f_R = \frac{1}{2\pi\sqrt{LC}}$$

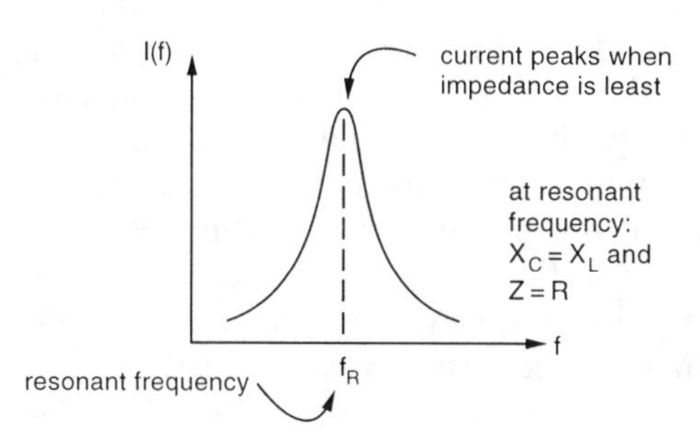

Parallel Resonance:

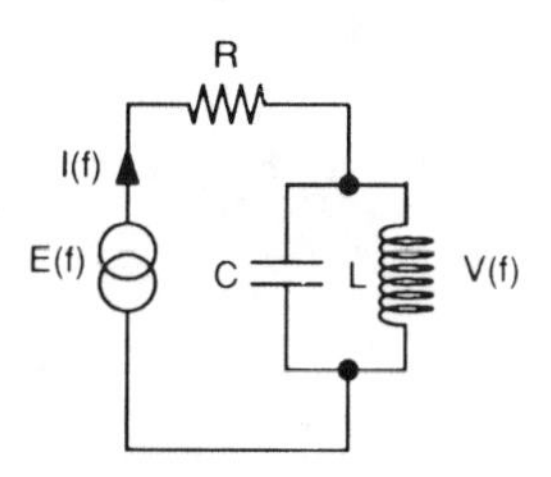

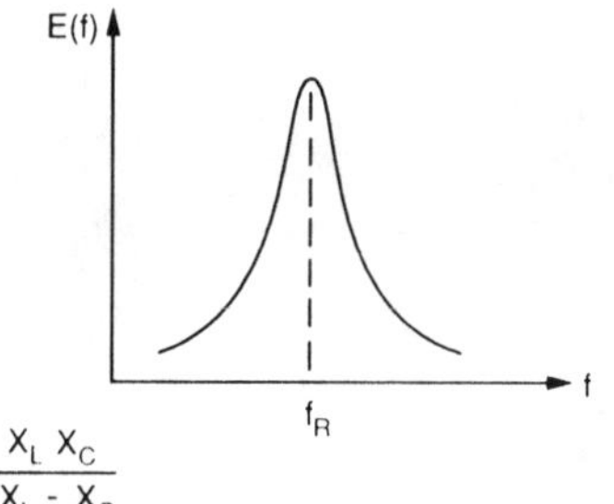

net reactance of parallel combination of L and C $= \frac{X_L\,X_C}{X_L - X_C}$

Quality:

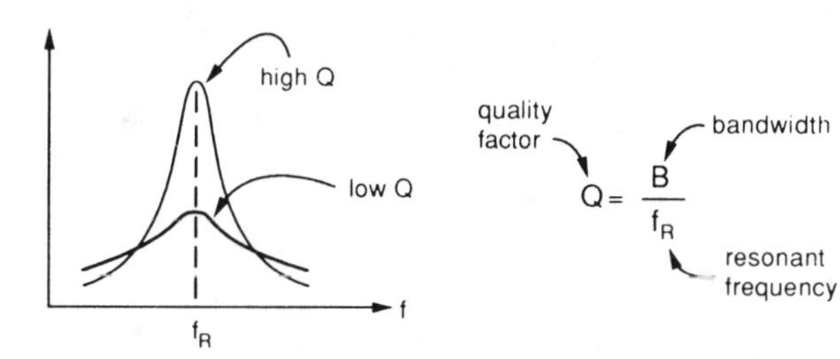

RESONANCE MATH

Consider the series connection of a resistor, a capacitor, and an inductor. The reactive portion of the impedance of this series connection is the inductive reactance minus the capacitive reactance. Since the inductive reactance, $X_L = 2\pi fL$, increases with frequency, and the capacitive reactance, $X_C = 1/(2\pi fC)$, decreases with frequency, there will be a frequency for which the two reactances cancel each other, leaving only the effects of the resistor in the circuit to oppose the flow of current. So, the magnitude of the impedance will show a dip at this frequency, and since the current is the applied EMF divided by the impedance, the current will show a peak at this frequency.

The frequency at which the current reaches its peak value is called the *resonant frequency* of the circuit. For a series *RLC* circuit, the resonant frequency f_R is

$$f_R = \frac{1}{2\pi\sqrt{LC}}$$

The second example of resonance occurs with the parallel connection of a capacitor and an inductor. The impedance of a capacitor in parallel with an inductor is the product of the reactances divided by the inductive reactance minus the capacitive reactance. There will be a frequency—the resonant frequency—at which the two reactances cancel each other, causing an infinite impedance of the parallel combination. When the impedance is infinite, no current flows in the circuit, and the voltage across the parallel combination equals the source EMF. At other frequencies, the impedance across the parallel *LC* combination decreases as the frequency increases beyond the resonant frequency, or as the frequency decreases below the resonant frequency.

Resonance Math

Series Resonance:

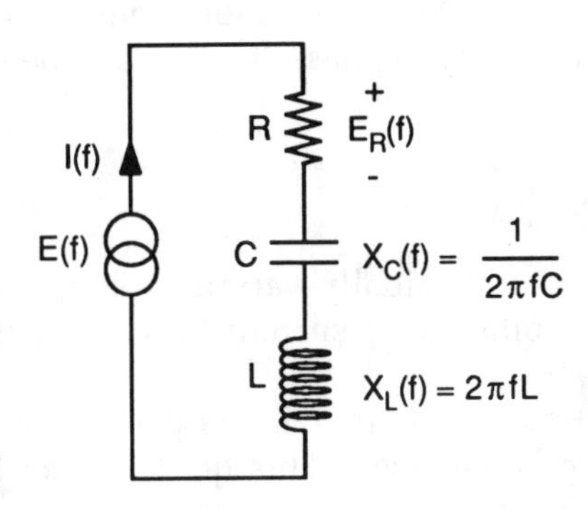

Ohm's Law:

$$E(f) = I(f)\,Z(f)$$

or

$$I(f) = E(f)\,/\,Z(f)$$

$$Z(f) = R + j\,[X_L(f) - X_C(f)]$$

$$= R + j\,[\,2\pi fL - \frac{1}{2\pi fC}\,]$$

$$|\,Z(f)| = \sqrt{R^2 + [\,2\pi fL - 1/(2\pi fC)\,]^2}$$

$$|\,Z(f)| = R \text{ when } 2\pi fL = 1(2\pi fC)$$

and that occurs at the resonant frequency:

$$f_R = \frac{1}{2\pi\sqrt{LC}}$$

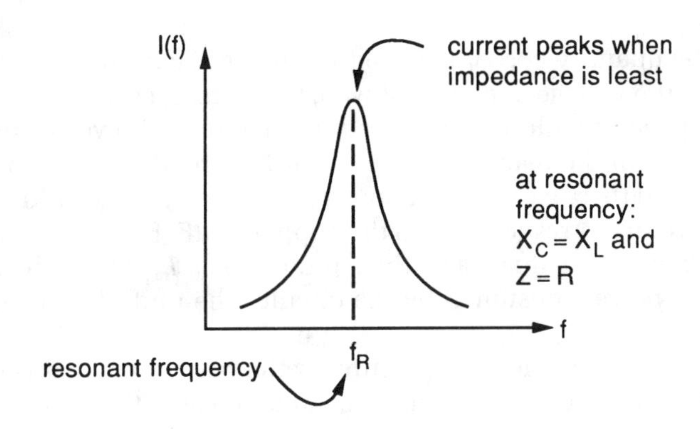

POWER IN REACTIVE CIRCUITS

The power consumed by a resistor in an ac circuit is continuously changing, leading to the use of the concept of a dc effective power. This dc effective power is the rms value of a sinusoidal signal.

Alternating current circuits containing reactances are more complicated because the voltage and current are not necessarily in phase. The instantaneous power $p(t)$ in a reactive circuit is

$$p(t) = E_p \sin(\omega t)\ I_p \sin(\omega t + \phi)$$

where E_p and I_p are the peak values of the sinusoidally varying voltage and current, and ϕ is the phase between the voltage and current flowing in the circuit.

The area under the $p(t)$ curve over time T, where T is the period corresponding to the frequency ω, is the energy consumed. This quantity can be obtained using calculus by integrating $p(t)$ over the interval 0 to T, and is

$$\text{Energy} = \frac{E_p I_p}{2} \cos(\phi) T$$

Since the peak value of a sine wave divided by the square root of two is the rms value of the wave, the time averaged value of the energy, which is the power consumed, becomes

$$\text{Power} = E_{\text{rms}} I_{\text{rms}} \cos(\phi)$$

The quantity $\cos(\phi)$ is called the power factor of the circuit. If the angle between the voltage and current were 90 degrees, $\cos(\phi)$ would equal zero, and no real power would be consumed in the circuit. Power would flow back and forth between the reactive elements in the circuit and the source.

Even if there were no real power consumed, there would be current flowing in the circuit, corresponding to the applied EMF. Both the current and the EMF would have rms values, and their product $E_{\text{rms}}\ I_{\text{rms}}$ would be a measure of the *apparent power* consumed by the circuit. The unit of apparent power is *volt-amperes*.

Power companies charge only for the real power that is consumed. However, the distribution wires must be able to carry the rms currents associated with the load. So, power companies attempt to keep the power factor as close to one as possible by balancing highly inductive loads, such as motors, with large capacitors, and balancing highly capacitive loads, such as fluorescent lights, with large inductors.

Power in Reactive Circuits

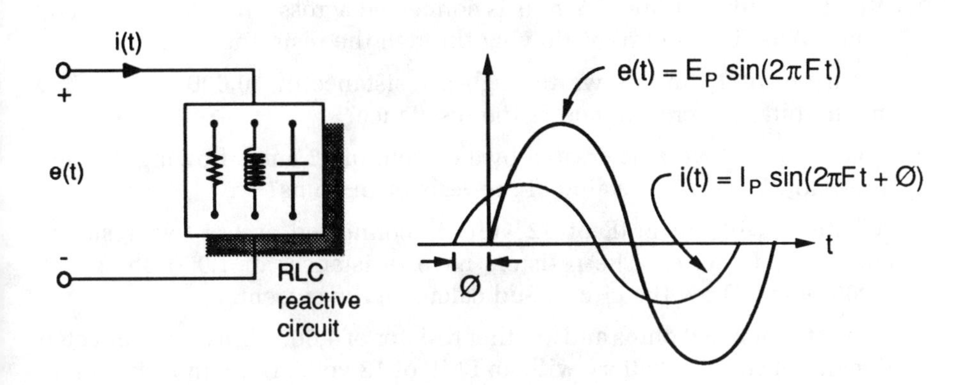

$$P_{effective} = \frac{E_P I_P}{2} \cos(\varnothing)$$

$$= E_{RMS} I_{RMS} \cos(\varnothing)$$

power factor

$$P_{apparent} = E_{RMS} I_{RMS} \quad \text{(Volt-Amperes)}$$

PROBLEMS

2.1. A battery with an EMF of 5 volts is connected across a resistance of 1,000 ohms. What is the current flowing through the resistance?

2.2. A current of 10 mA flows through a resistance of 10,000 ohms. How much voltage is created across the resistance?

2.3. A voltage of 12 volts is created by a current of 12 amps flowing through a resistor. What is the value of the resistor in ohms?

2.4. A battery with an EMF of 12 volts is connected across two resistors connected in series. The resistors have resistances of 1,000 ohms and 5,000 ohms. Draw the circuit and calculate the current.

2.5. A resistor of 2,000 ohms and another resistor of 4,000 ohms are connected in parallel across a battery with an EMF of 12 volts. Draw the circuit and calculate the current flowing in each resistor. What is the equivalent resistance of the two resistors connected in parallel?

2.6. A 100-watt light bulb is connected across an EMF of 100 volts. What is the current flowing in the bulb?

2.7. An electric hair dryer is rated at 1,200 watts. How much current does it draw when connected across an EMF of 100 volts?

2.8. An air conditioner is rated at 800 watts and runs 5 hours per day. In an average 30-day month, how many kilowatt-hours of electric energy are consumed? How much is the monthly electric bill to run this air conditioner if electricity costs 12 cents per kilowatt-hour?

2.9. The ac supplied by the power company has a nominal rms value of 110 volts. What is the peak value of the voltage?

2.10. An electric heater is connected to a dc source of 100 volts and draws 10 amps. How much power is consumed? The same heater is then connected to an ac source with a peak voltage of 100 volts. How much effective power is consumed?

2.11. A transformer with a turns ratio of 20 is connected to an ac voltage of 2 kV. What is the voltage at the secondary of the transformer?

Chapter 3: System Elements

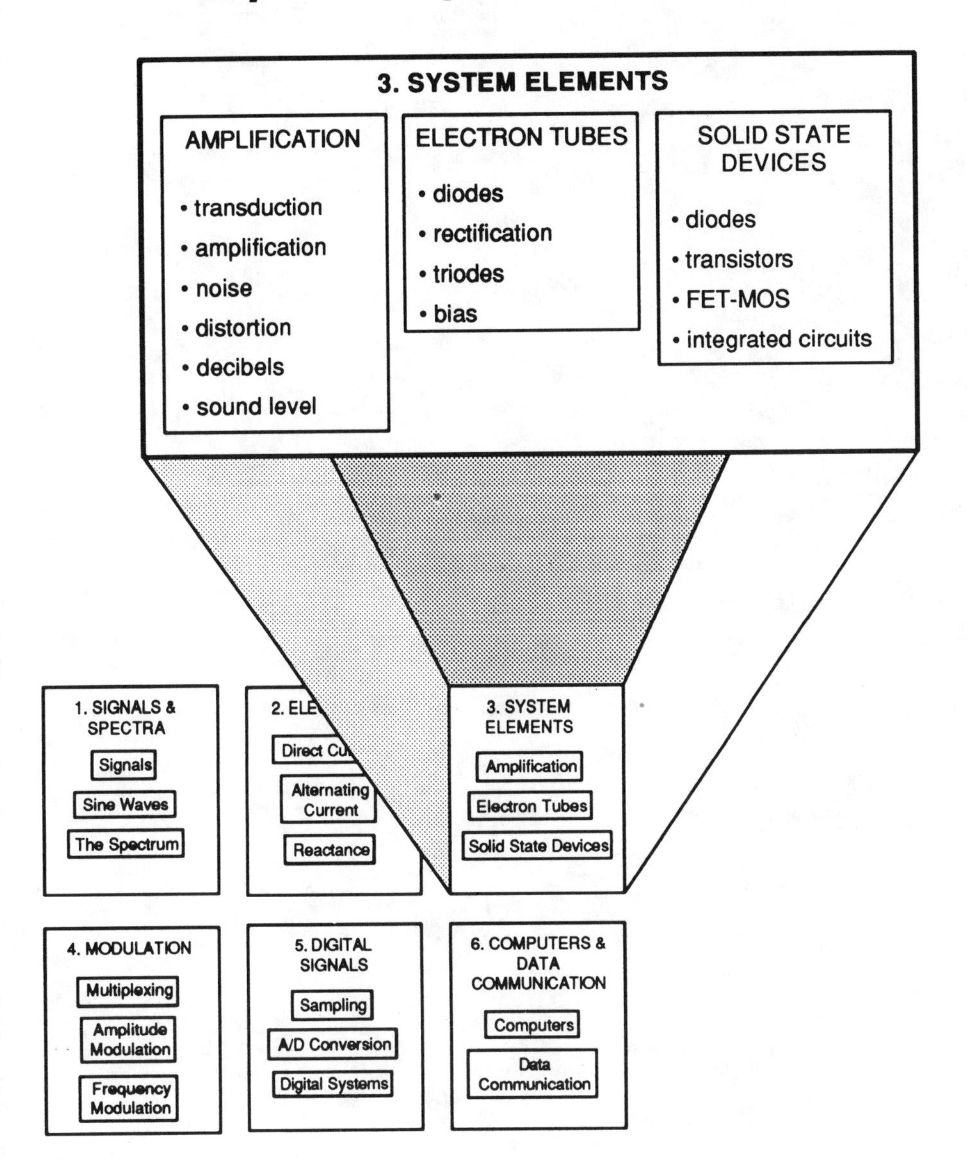

Introduction to System Elements

The preceding modules have treated the theoretical aspects of signals and electricity. This module describes many of the component parts of electronic systems. The components described in this module will be used in subsequent modules to assemble practical communication systems. So, this module is a bridge between theory and practical systems.

Most communication signals are processed as electric signals. However, since many signals also exist in forms other than electrical signals, means are necessary to convert signals from one form, or medium, into another. This conversion is accomplished by a class of devices called *transducers*. This module begins with a description of the workings of different types of transducers.

Signals are often very small and must be made larger through *amplification*. The process of amplification can introduce various distortions, noise, and limitations on bandwidth. The *decibel* is a measure of amplification that depends on ratios. Amplification and decibels are treated next in the module.

Amplification is accomplished by such active devices as *electron tubes* and *transistors*. Their workings are described last in this module, along with the marvels of *integrated circuits*.

Amplification

TRANSDUCERS

People do not generate electric signals, nor do people respond directly to electric signals, unless accidentally as a shocking experience! People generate *acoustic* speech signals and respond to speech sounds, but electric circuits respond to and generate *electric* signals. For people to use electric communication devices and systems, acoustic signals must be converted to electric signals and *vice versa*. This conversion is accomplished by a general class of devices called transducers.

Transducers that convert variations in sound pressure into electric signals are called *microphones* (or *transmitters*, if used in a telephone instrument). Transducers that convert electric signals into sound waves are called *loudspeakers* (or *receivers*, if used in a telephone instrument). The basic principles of operation are similar for both, although their physical size and electrical characteristics may be quite different, depending on the application.

The general class of devices called transducers is quite broad, and, in addition to devices that convert between electric signals and acoustic signals, includes devices that convert such signals as mechanical movement, variations in fluid pressure, and light into electric signals and vice versa. In general, a transducer converts a signal from one medium to another.

Transducers

Definition:

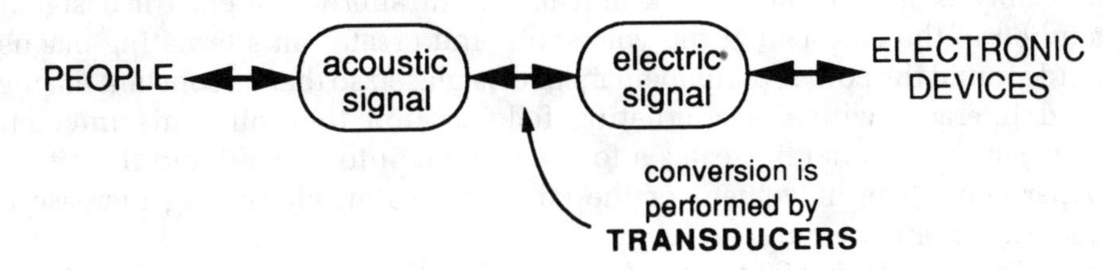

Microphones:

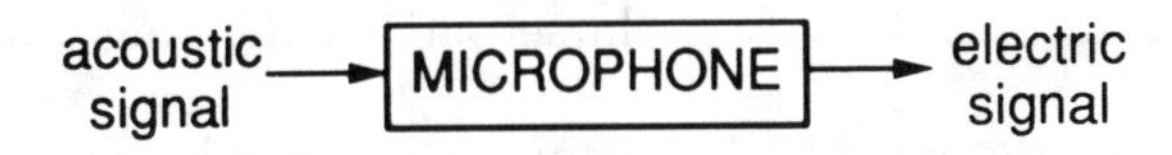

Loudspeakers:

3

TRANSDUCERS—LOUDSPEAKERS

Most loudspeakers operate using the principles of *electromagnetism*. A loudspeaker consists of a coil of wire to which a paper cone is attached. This whole assembly is free to move back and forth. An alternating electrical signal is applied to the coil, called the voice coil, and creates an alternating magnetic field around the coil. A permanent magnet is placed so that its constant magnetic field interacts with the alternating field around the coil. This interaction between the two fields creates a force on the coil to move it and the attached paper cone. The movement of the cone creates an alternating increase and decrease in air pressure.

The changes in air pressure result in sound waves that travel over distance, finally impinging on a person's ear. The sound waves are channeled down the ear canal until they meet the ear drum, which is set into vibration in synchrony with the varying sound waves. The motion of the ear drum ultimately stimulates nerve cells to create signals that are carried over nerve circuits to the brain, which processes the signals to create the sensation of hearing and the perception of sounds.

Loudspeakers designed for low frequencies, called woofers, need to move large volumes of air and so they have large cones. A sound wave emanates from both the front and the back of the cone. The wave from the back travels around to the front of the loudspeaker and is 180 degrees out of phase with the front wave. The two waves would then cancel. To prevent this from occurring, loudspeakers are placed in enclosures. Some of these enclosures are designed to absorb completely the back wave. Other designs allow the back wave to escape from the enclosure after delaying the wave enough so that it is in phase with the front wave.

High frequency sound waves have much smaller energy than low frequency waves, and so the loudspeakers designed for high frequencies, called tweeters, are quite small in size. Unlike low frequencies which are almost totally nondirectional, high frequencies are highly directional. So, tweeters must be designed for good dispersion of the high frequency sound waves generated.

Loudspeakers

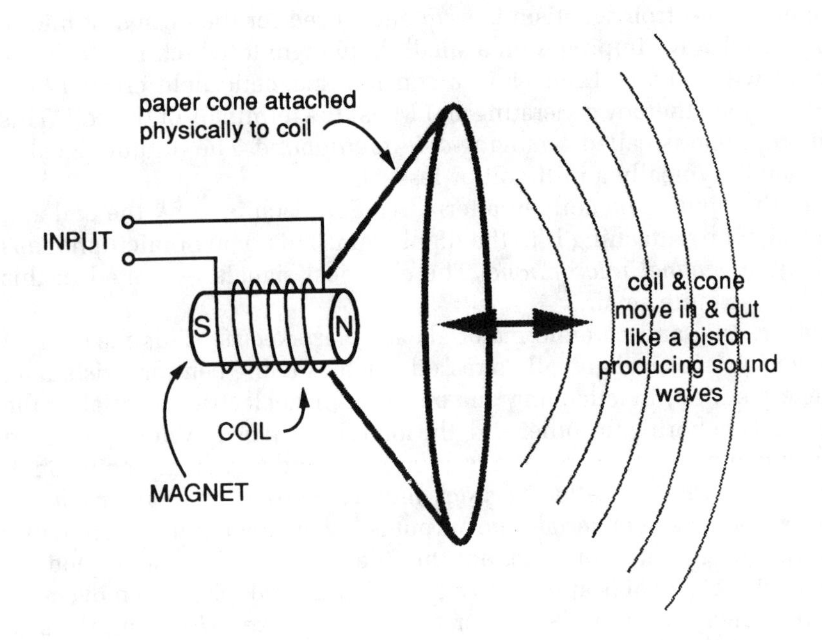

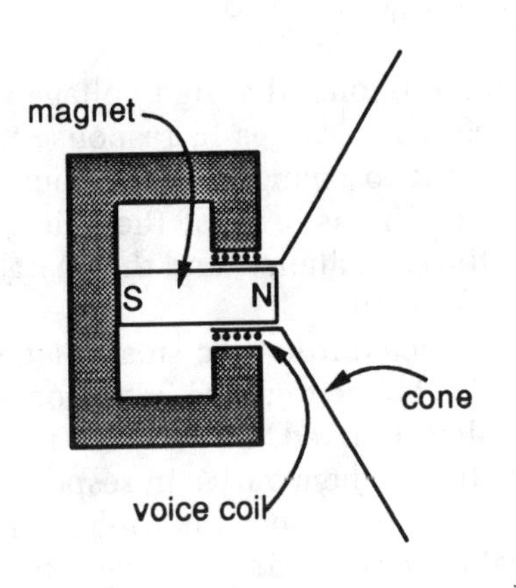

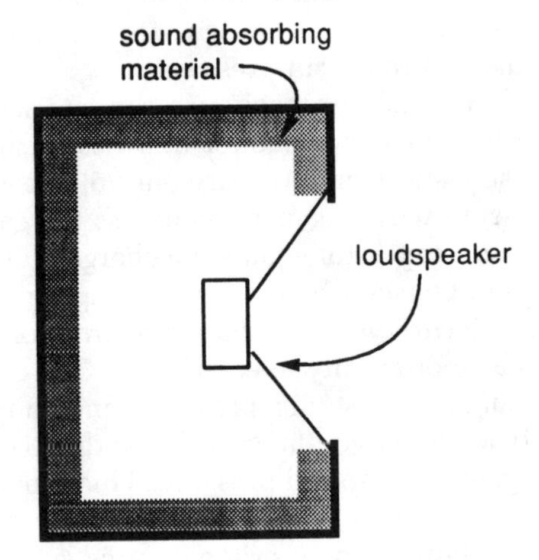

TRANSDUCERS—MICROPHONES

The principle of electromagnetism is frequently used for the design of microphones. A sound wave impinges on a small diaphragm to which is attached a small coil of wire. The coil moves in a constant magnetic field created by a permanent magnet, thereby generating an EMF at the terminals of the coil. This type of microphone is called a *moving-coil microphone*. The output signal is extremely small, typically a millivolt or less.

Rather than move the coil, an alternative approach is to fix the coil and move the magnet by attaching it to the diaphragm. This type of microphone is called a *moving-magnet microphone*. The electrical signals generated in this way are likewise quite small.

Certain crystal and ceramic materials are *piezoelectric*. This means that if these materials are mechanically stressed, such as being bent or twisted, an EMF is generated. So, by attaching one end of a piezoelectric material to the diaphragm and anchoring the other end, the material will be physically stressed by the movement of the diaphragm in response to sound waves impinging on it. This type of microphone is called a *crystal microphone* or *ceramic microphone*, depending on the type of material. The output is much larger than for a moving-coil or moving-magnet microphone, but the fidelity is usually not as good.

If an electrical signal is applied to a ceramic material, physical movement results. This principle can be used to create a loudspeaker. However, the size of movement is quite small, and so this type of loudspeaker is usually used for headphones that are worn close to the ear and do not need to produce large variations in sound pressure.

A variable capacitance can be used as a microphone. If a high voltage is placed across a capacitor that has a capacitance which varies in response to an acoustic signal, a variable voltage proportional to changes in the sound pressure will be generated across the capacitance. This is because the voltage across a capacitor equals the charge divided by the capacitance, and the charge is kept constant by the high voltage across the capacitor.

A *variable-capacitance microphone* is constructed from two small plates in close proximity to each other. One of the plates is free to move in response to changes in sound pressure, and the other plate is fixed. The capacitance between the two plates varies as the distance between them varies in response to changes in sound pressure. This type of variable-capacitance microphone is also called a *condenser microphone* and is capable of generating an output that is very faithful to the original sound signal. Such a microphone is also sometimes called an *electrostatic microphone*.

The high voltage is necessary for a condenser microphone to maintain a constant charge across its plates. Some materials can store their own charge, and so there is no need for an external source of high voltage. This type of condenser microphone is called an *electret microphone* and is extremely popular for both consumer-electronics and professional microphones.

Microphones

Moving Coil:

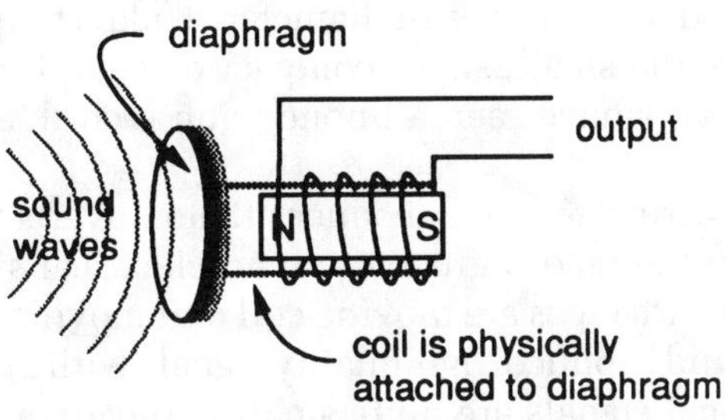

Moving Magnet:

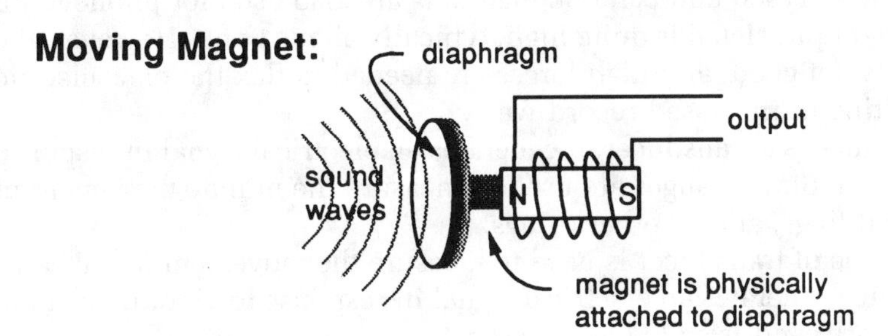

Piezoelectric:

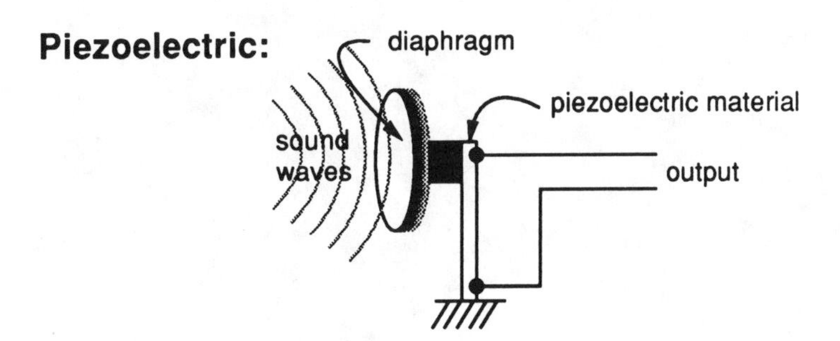

Variable Capacitance:

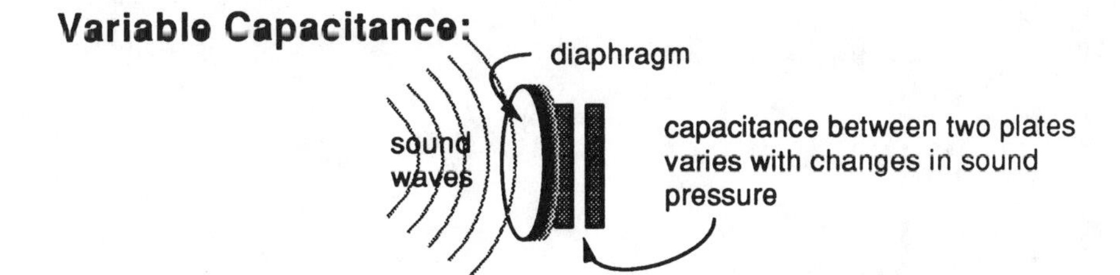

TRANSDUCERS—OTHER TYPES

Phonograph cartridges are used to play phonograph records made from black vinyl in the form of a disc 12 inches in diameter. Today, the phonograph record has been replaced by the small, silver compact disc, and in a few years, most young people will never have seen a phonograph record, except perhaps in a museum.

A phonograph cartridge is a transducer that converts the physical movement of the stylus in the record groove into an electrical signal. Two popular types of phonograph cartridges are moving coil and moving magnet. Both types are electromagnetic and produce high-quality signals with minimal record wear. However, the electrical signals are quite small, typically a millivolt or less.

Piezoelectric crystal and ceramic materials are also used for phonograph cartridges. The output signal is quite high, typically about 1 volt. However, the signal quality is not good, and high forces are needed to flex the piezoelectric material, resulting in increased record wear.

A strain gauge is a transducer. It generates an electrical signal in response to being stretched. Strain gauges are used to measure the minute movement of bridges and buildings, among other things.

Another type of transducer is used to measure the movement of the earth itself. This device generates an electrical signal in response to vibrations of the surface of the earth and is called a seismometer.

Other Transducers

Phonograph Cartridges:

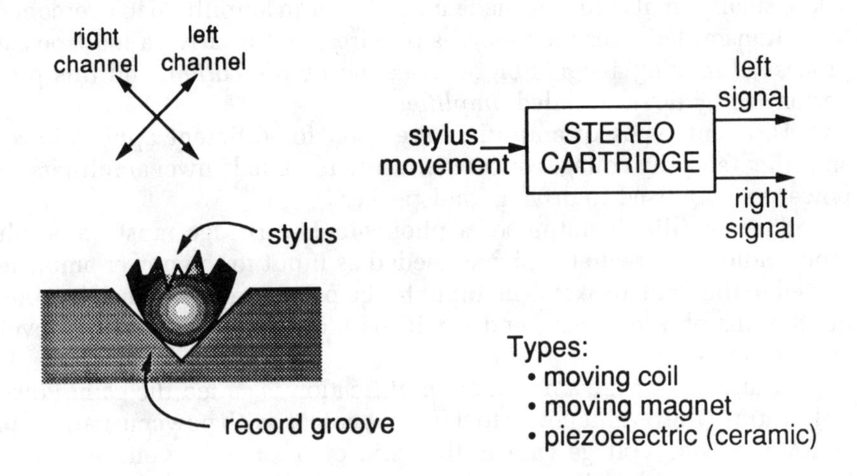

Types:
- moving coil
- moving magnet
- piezoelectric (ceramic)

Strain Gauges

Seismometers

AMPLIFICATION

The electrical signals generated by microphones, phonograph cartridges, and other transducers are usually quite small, typically a few millivolts or less. These small signals must be made much larger in amplitude to overcome noise, to be transmitted over long lengths of wires, and to drive a loudspeaker. The process of making a signal bigger is called *amplification*, and this process is performed by devices called *amplifiers.*

Different types of amplifiers are used for different applications. Some amplifiers amplify voltages or currents. Others, called power amplifiers, amplify power and are used to drive a loudspeaker.

The 1-millivolt output of a phonograph cartridge must be amplified a thousandfold to create the signal needed as input to the power amplifier connected to the loudspeaker. The input to the power amplifier may be about two-hundredths of a milliwatt, and it will be increased in power to a level of 10 watts or more.

A measure of the amount of amplification is called the gain. Power gain is the ratio of the output power to the input power, with power usually expressed as RMS power. Voltage gain is the ratio of the output voltage to the input voltage, again with voltage usually expressed as its RMS value.

Amplifiers need an external source of power to supply the energy needed to perform amplification. This power is provided by a power supply and is usually in the form of a constant dc signal.

Amplification

Process:

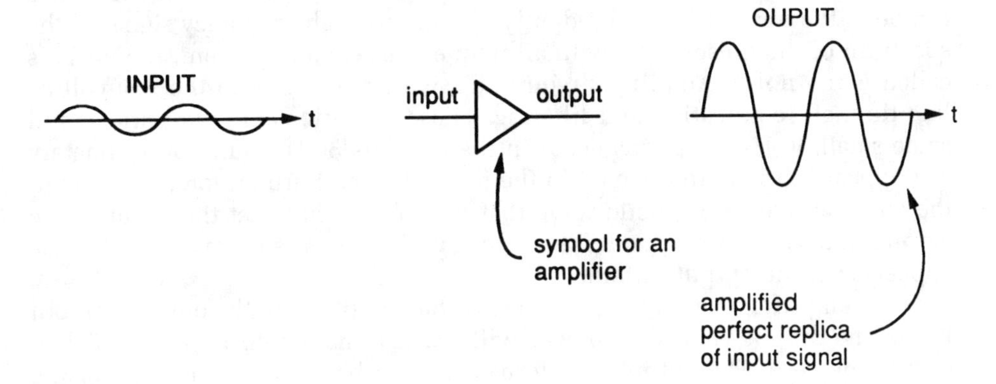

Gain:

$$\text{Power Gain} = \frac{P_{OUT}}{P_{IN}}$$

$$\text{Voltage Gain} = \frac{V_{OUT}}{V_{IN}}$$

Power Supply:

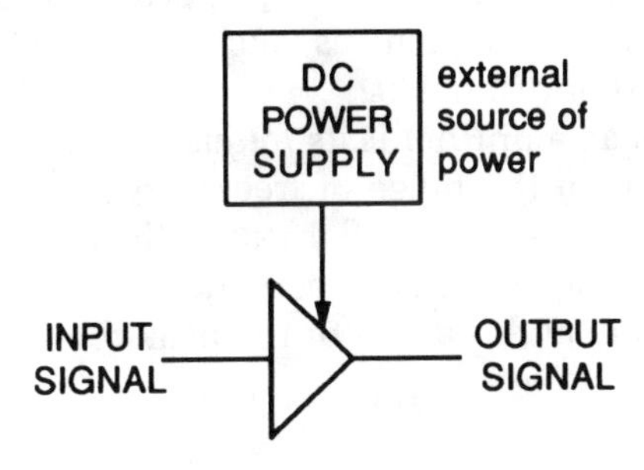

3

NOISE AND DISTORTION

In the ideal world, amplification is perfect in the sense that the output is an exact replica of the input, except for being larger in amplitude. In the real world, this does not occur. Noise and distortion creep into the process. Noise can be in the form of *hiss*, a randomly fluctuating high frequency signal. If the spectrum of the noise is flat with all frequencies equally present, the noise is called white noise. Amplifiers require dc, which comes from the 110-volt ac line through rectification and filtering. These operations are not perfect, and some small 60 Hz variations occur in the dc. This 60-Hz variation ultimately can appear in the output signal in the form of *hum*. Furthermore, 60 Hz is in the air as an electromagnetic wave that can be picked up at the input to the amplifier, ultimately being amplified along with the desired signal and likewise appearing at the output as hum.

An amplifier cannot keep increasing the amplitude of the output without limit. The output voltage or power will reach a maximum beyond which it cannot increase. The output waveform will then halt and be constant for the time that the output threshold is being exceeded. The waveform is said to be *clipped*, since it looks as if its top excursion and maximum negative excursion have both been clipped off with a pair of scissors. This type of clipping distortion does not sound appealing. It is what happens when the volume control is turned too high in a stereo unit.

A pure single-frequency tone can be used as input to an amplifier, but the output may consist not only of this tone, and will include additional tones at integer or harmonic multiples of the frequency of the input tone. This form of distortion is called *harmonic distortion*, since additional, unwanted harmonics have been added to the output.

Assume that two tones with frequencies f_1 and f_2 are used as input to an amplifier. If the amplifier performs ideally, the output will consist of only amplified replicas of the two tones at f_1 and f_2. However, sometimes a distortion occurs with actual amplifiers in which not only f_1 and f_2 appear, but also spurious tones at various sum and difference frequencies, such as $f_1 + f_2$. This type of distortion is called *intermodulation distortion*.

Another important characteristic of an amplifier is its *frequency response*. The bandwidth of an amplifier, in terms of the range of frequencies that are passed, must be specified along with a measure of the flatness of the response in the passband. Any peaks in the response are to be specified in such terms as the frequency at which the peak occurs and the size of the peak relative to some other portion of the response.

Noise and Distortion

Noise:

- **Hiss**

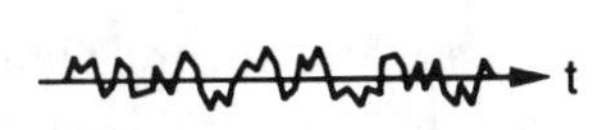

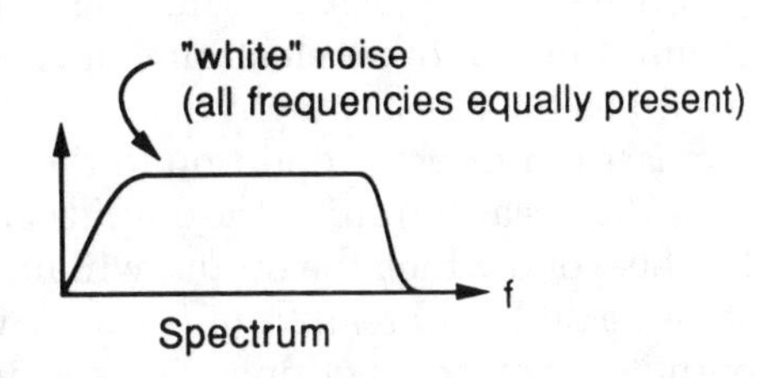

- **Hum**

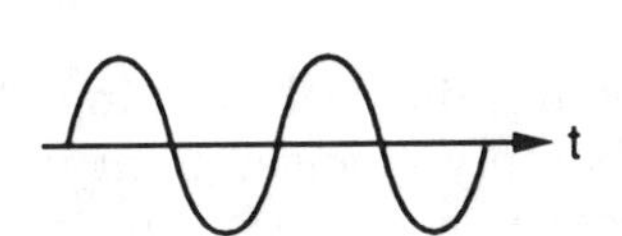

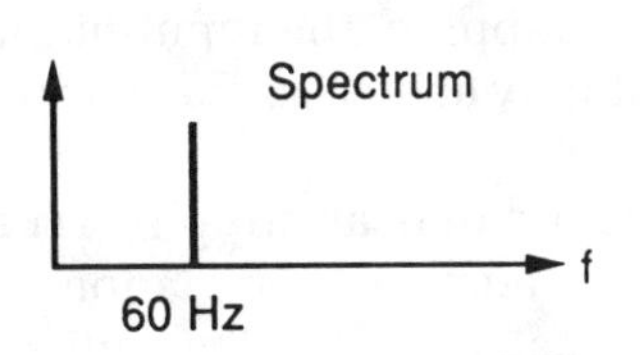

Distortion:

- **Clipping**

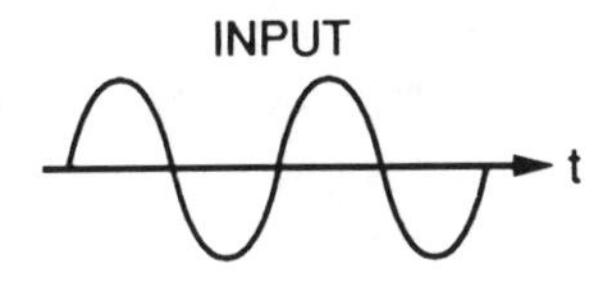

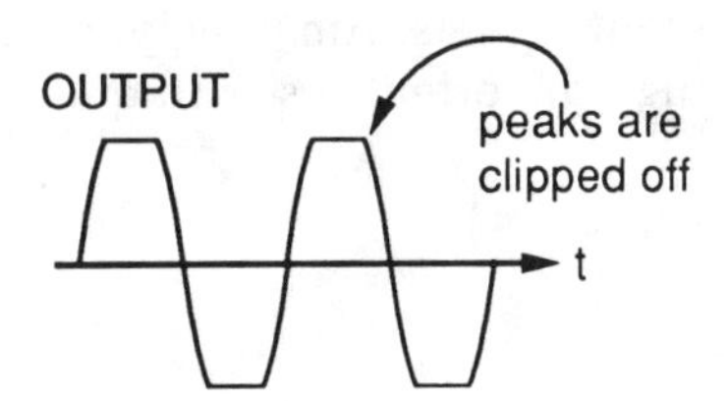

- **Harmonic**

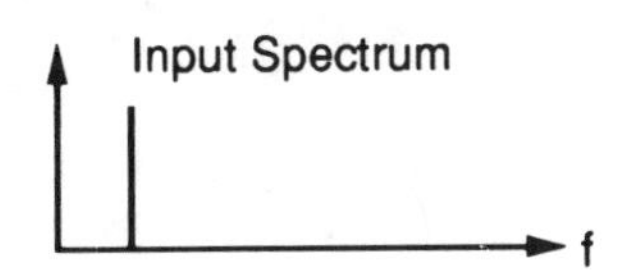

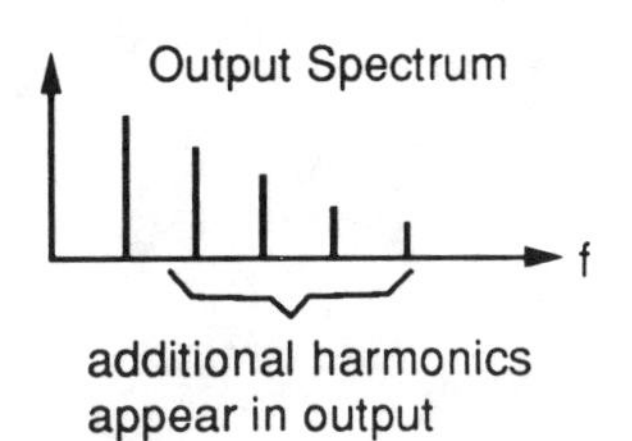

- **Intermodulation**

IN	OUT
f_1	f_1
f_2	f_2
	$f_1 + f_2$
	$f_1 - f_2$
	$f_1 + 2f_2$

3

LINEARITY

The output of an ideal amplifier is a perfect replica of the input. We could produce a graph that shows the relationship between the input and the output of an amplifier, or other electronic device. For an ideal amplifier, the relationship would be linear—a straight line. This linear relationship means that the output increases in direct proportion to the input.

For a real amplifier, the output cannot increase forever, and clamps at some level beyond which the output will not increase. The input-output relationship shows a sudden break from linearity where this occurs. The result is that the output waveform is clipped. This is an example of hard peak clipping.

With soft clipping, the input-output relationship does not have a sudden break but gradually decreases the output. The result is that the output waveform is squashed.

Another form of peak clipping is called infinite peak clipping. The output wave shows no amplitude variation and simply changes from one positive value to one negative value. The output looks like a square wave varying only in frequency. Only the zero-axis crossings are preserved with infinite peak clipping. One would expect such a processing to have horrible effects on speech and other signals. Amazingly, infinite peak clipped speech is still intelligible, although distorted in sound quality.

Linearity

Definition:

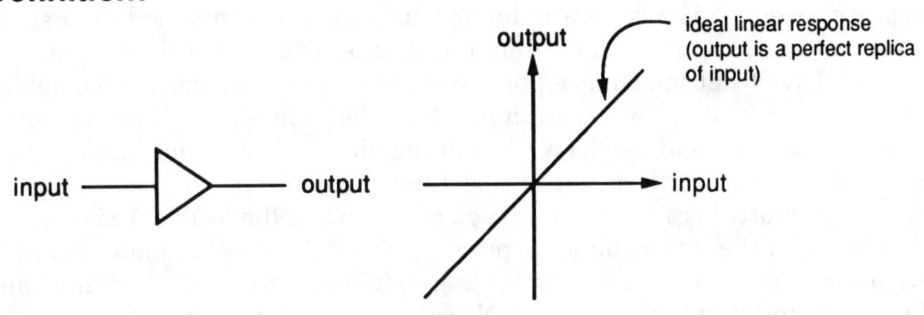

Clipping:

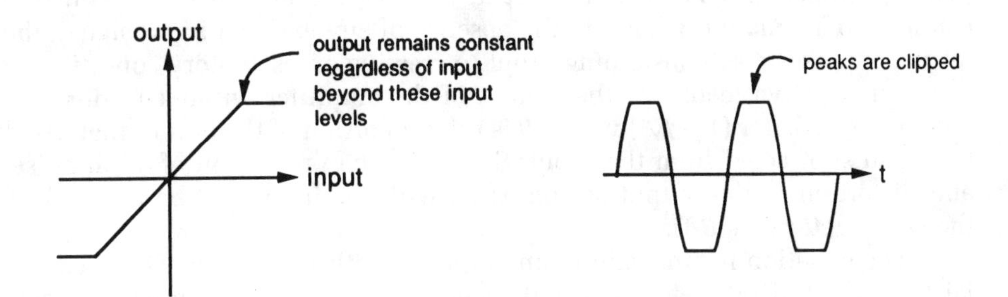

Soft Clipping:

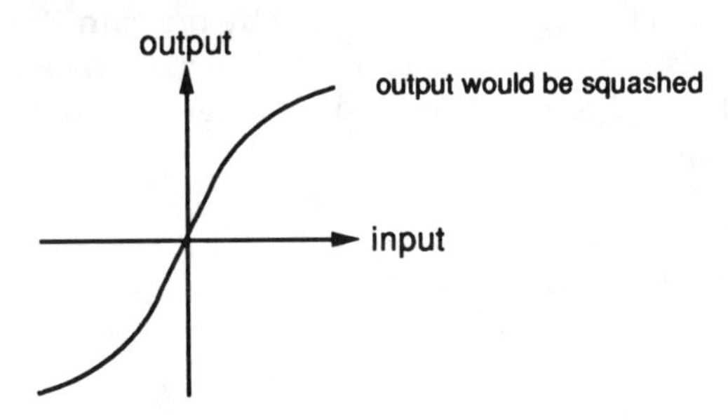

FEEDBACK

Actual amplifiers introduce hum, noise, and distortion into the amplification process. These deleterious effects can be reduced through the application of *negative feedback* in which a reduction in gain is accompanied by a similar reduction in the proportion of hum, noise, and distortion at the output.

We have feedback when a portion of the output is fed back and combined with the input to create a situation in which the input depends on the output. There are two types of feedback, depending on whether the feedback signal is in phase or is out of phase with respect to the input.

Degenerative or negative feedback occurs when the feedback signal is out of phase with the input and so opposes it. The final result is reduction in the overall gain along with a corresponding reduction in the proportion of hum, distortion, and noise at the output. Negative feedback was invented by Harold S. Black at Bell Labs in 1927 and made long-distance telephony practical.

Regenerative or *positive feedback* occurs when the feedback signal is in phase with the input and so reinforces it. The result is an increase in gain. This can lead to a situation where, in the absence of any external input signal, the feedback causes a self-sustaining input to creating an oscillatory condition.

For negative feedback, the gain A of the amplifier without feedback is reduced by a factor $(1 + \beta A)$ where β is the proportion of the output that is fed back and subtracted from the input. Similarly, the proportion of hum, noise, and distortion in the output as compared with the input will be reduced by the same factor $(1 + \beta A)$.

The equation for the gain of an amplifier with negative feedback can be fairly easily derived. Let A signify the gain of the amplifier without feedback. A proportion β (the Greek lower-case beta) of the output e_o is fed back to the input, where β is combined with the external input signal e_i. The feedback signal βe_o and the external input e_i are subtracted, giving the net signal $e_i - \beta e_o$, which is the input to the amplifier. This signal is amplified by the gain A and emerges as $(e_i - \beta e_o)A$, which equals the output signal e_o. This equation can be solved for the ratio of the output signal to the external input signal, and gives

$$\frac{e_o}{e_i} = \frac{A}{1 + \beta A}$$

Feedback

Types:

- DEGENERATIVE (negative)
- REGENERATIVE (positive)

Negative Feedback Circuit:

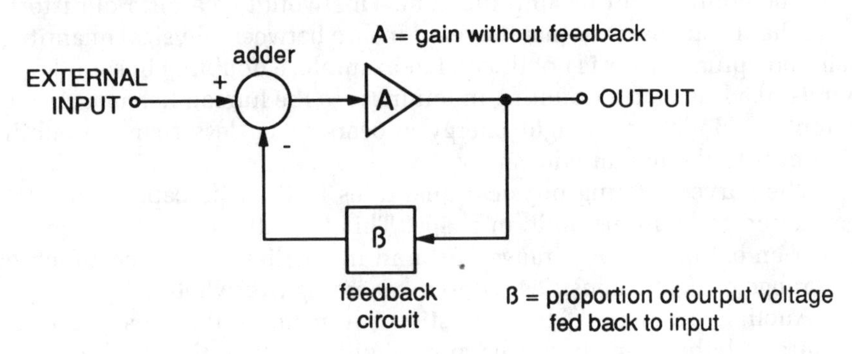

$$\frac{V_{OUT}}{V_{IN}} = \frac{A}{1 + \beta A}$$

Decibels

Some measure of the amount of amplification of an amplifier would be most useful. One possible measure would be the amplification factor. If a signal of 1 millivolt at the input were amplified 1,000 times to become 1 volt, the amplification factor would be 1,000.

If the amplification factor, or gain, were 1,000, then 2 millivolts at the input would produce 2 volts at the output, assuming that the amplifier were perfectly linear in its amplification characteristics. If the 2 millivolts resulted in something less than 2 volts at the output, perhaps 1.8 volts, then the amplifier would be nonlinear in its amplification. This would be a form of distortion.

It has been observed that the relationship between physical quantities and their perceptual effects is not linear. For example, a doubling in sound pressure sounds like less than a doubling in loudness to the human listener. As another example, a doubling in light energy appears to be less than a doubling in brightness to the human viewer.

The curves relating physical quantities to their perceptual effects have been found to be logarithmic in shape. This logarithmic response enables us to hear an extremely large range of sound intensities and to see an extremely large range of light intensities without becoming overwhelmed.

Another perceptual characteristic of humans is that changes in sound pressure or light intensity can be more significant than the absolute values of sound pressure or light intensity. So, the ratio between two quantities may be more important than their absolute values.

All of this suggests a measure that is the logarithm of the ratio of two quantities. This logarithmic measure is called the bel in honor of Alexander Graham Bell. A somewhat more useful measure, however, is the decibel, abbreviated dB. Ten decibels equal 1 bel.

The decibel is defined as ten times the logarithm of the ratio of two energies or measures of power. If electrical quantities are being measured, decibels are defined as

$$\text{decibels} = 10 \log_{10}(P_1/P_2)$$

where P_1 and P_2 are the two powers, in watts, that are being compared.

Usually, P_2 represents a reference power that can be the input power to some electrical circuit or device, and P_1 may be the output power. So, for this case,

$$\text{decibels} = 10 \log_{10}(P_{out}/P_{in})$$

Decibels

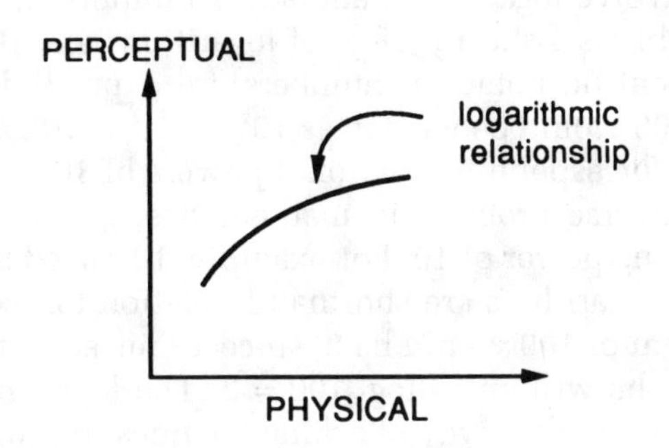
PERCEPTUAL
logarithmic
relationship
PHYSICAL

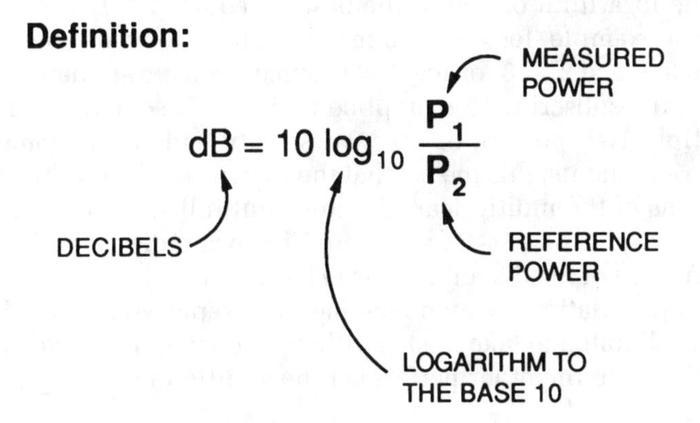
Definition:
MEASURED
POWER
$dB = 10 \log_{10} \frac{P_1}{P_2}$
DECIBELS
REFERENCE
POWER
LOGARITHM TO
THE BASE 10

3

LOGARITHMS

Decibels involve logarithms, and so it is important to know about logarithms. The following is a short review of logarithms and their use.

In scientific notation, numbers are expressed as powers of 10. So, for example, 100 would be written as 10^2; 1,000 as 10^3; and 1/100 as 10^{-2}. Negative exponents correspond to fractional powers of 10.

The inverse problem in mathematics, given the number, is to find the corresponding power of 10. For example, 10 raised to what power or exponent gives 100? Logarithms are shorthand notation for exponents. In this example, the logarithm of 100 would be 2, since 10 raised to the 2nd power equals 100. This would be written as $\log_{10}100 = 2$. The logarithm to the base 10 of some number N is simply a way of asking the question, what power of 10 equals N?

The subscript 10 indicates that 10 is being raised to the power. The subscript is called the base of the exponentiation. The equation $\log_{10}100 = 2$ would be read as "the logarithm of 100 to the base 10 equals 2." Logarithms can be to any base. For example, $\log_2 8$ would mean, 2 raised to what power equals 8? The answer is 3, since $2^3 = 8$, or $\log_2 8 = 3$. Since decibels are defined by using 10 as the base, the subscript 10 is implied and so it is sometimes deleted.

To multiply two powers of 10, you simply add the exponents. Since logarithms are exponents, this means that the logarithm of a product is the sum of the logarithms of the multiplicands. Thus, multiplication can be converted into the much simpler process of addition. This principle is used in the slide rule—an instrument of the distant past used by engineers to perform multiplications and other calculations, which now has been replaced by calculators. The slide rule was calibrated to take the logarithms of numbers and then the sliding of one rule relative to the other performed the addition necessary to add logarithms.

Because a fractional power of 10 has a negative exponent, the logarithm of a fraction is a negative quantity. Futhermore, it follows that the logarithm of one number divided by another equals the logarithm of the dividend minus the logarithm of the divisor.

A summary of all the preceding discussion might be helpful. If $10^n = N$, then $\log_{10} N = n$. The logarithm of a product equals the sum of the logarithms:

$$\log_{10} AB = \log_{10} A + \log_{10} B$$

The logarithm of a fraction is a negative quantity:

$$\log_{10} 1/N = -\log_{10} N$$

Logarithms

Definition:

exponent (power of 10)

If $10^n = N$

Then

$\log_{10} = n$

base of 10

Rules:

- MULTIPLICATION $\log (A \times B) = \log A + \log B$
- DIVISION $\log (A / B) = \log A - \log B$
- FRACTION $\log (1 / N) = - \log N$

LOGARITHMS (cont'd)

In practical use, logarithms of powers of 10, $\log_{10}10^n$, are simple to use, for example, since the logarithm of 10 raised to some exponent is the exponent itself. Another useful fact is the logarithm of 2: $\log_{10}2$ can be obtained from a calculator as about 0.3. An easy way to remember this is that the cube root of 8 is two. Since 8 is nearly 10, 2 is nearly the cube root or the one-third power of 10, and one-third is about 0.3. Thus, 10 to the 0.3 power is about 2, or $\log_{10}2 = 0.3$.

If two numbers expressed as exponents of the same base are multiplied, the result is the base raised to the sum of the exponents. For example,

$$10^3 \times 10^5 = 10^{3+5} = 10^8$$

So, the logarithm of the product of two numbers equals the sum of the logarithms of each individual number. For the preceding,

$$\log_{10}(10^3 10^5) = \log_{10}10^3 + \log_{10}10^5 = 3 + 5 = 8$$

As an example of division,

$$\log_{10}(10^3/10^5) = \log_{10}10^3 - \log_{10}10^5 = 3 - 5 = -2$$

Logarithms (cont'd)

Practical Facts:

$$\log_{10} 10^n = n$$

$$\log_{10} 2 \approx 0.3$$

Examples:

$$\log_{10} 100 = 2 \text{ since } 10^2 = 100$$

$$\log_{10} (10^3 \times 10^5) = \log_{10} (10^8) = 8$$

$$\text{or} = \log_{10} 10^3 + \log_{10} 10^5 = 3 + 5 = 8$$

$$\log_{10} (10^3 / 10^5) = \log_{10} 10^{-2} = -2$$

$$\text{or} = \log_{10} 10^3 - \log_{10} 10^5 = 3 - 5 = -2$$

DECIBELS (cont'd)

The definition of decibels is in terms of *power*. If the voltage or current is known, then power can be expressed in these terms in the equation. If the two powers are measured across the same resistance R, then,

$$\text{decibels} = 10 \log_{10}\frac{E_1^2/R}{E_2^2/R}$$

The R factors in the two denominators cancel and we have,

$$\text{decibels} = 10 \log_{10}\frac{E_1^2}{E_2^2} = 10 \log_{10}\left(\frac{E_1}{E_2}\right)\left(\frac{E_1}{E_2}\right)$$

Since the logarithm of a product equals the sum of the logarithms of the multiplicands,

$$\text{decibels} = 10 \log_{10}\frac{E_1}{E_2} + 10 \log_{10}\frac{E_1}{E_2}$$

$$= 20 \log_{10}\frac{E_1}{E_2}$$

If the currents are known, then, by similar reasoning, we have

$$\text{decibels} = 20 \log_{10}(I_1/I_2)$$

The above equations hold only if R is the same for both powers. If R is not the same, then the individual powers must be calculated and used in the decibel equation for the ratio of two powers. However, even if the R's are not the same, the $20 \log_{10} E_1/E_2$ equation can be used, but it should be stated then that the result is decibels for a voltage ratio.

It is essential to realize that decibels are a measure of the ratio between two quantities. Decibels are a measure of how one quantity compares to some reference quantity. Decibels are not an absolute measure. A statement that a voltmeter reads 10 dB is meaningless without knowing the reference voltage, for example.

Some reference quantities are quite prevalent, and decibel measures relative to these quantities have been given unique symbols. For powers, both 1 watt and 1 milliwatt are often encountered as reference powers. A decibel measure of a power relative to 1 watt is abbreviated dBW. A decibel measure relative to 1 milliwatt is abbreviated dBm. A decibel measure of a voltage or EMF relative to 1 volt as the reference is abbreviated dBV.

The addition of two signals to obtain their total power in decibels is performed by first adding the powers and then calculating the decibels. For example, if P_1 were the power of the first signal and P_2 were the power of the second signal, the total power (assuming that the signals added in phase, or coherently, and did not cancel each other) would be $P_1 + P_2$. The decibels relative to some reference power P_r would then be calculated from $10 \log_{10}(P_1 + P_2)/P_r$.

Decibels (cont'd)

Voltage-Current Ratios:

$$dB = 20 \log \frac{E_1}{E_2}$$

voltage ratio

$$dB = 20 \log \frac{I_1}{I_2}$$

current ratio

Standard References:

power in watts

$$dBw = 10 \log (P / 1)$$

power in milliwatts

$$dBm = 10 \log (P / 1)$$

voltage in volts

$$dBv = 20 \log (V / 1)$$

Power Addition:

for coherent addition:

$$dB = 10 \log \frac{P_1 + P_2}{P_{REF}}$$

DECIBELS—EXAMPLES

Some practical examples might be helpful in understanding decibels and their use.

An input voltage of 1 volt to an amplifier gives an output of 100 volts. The voltage gain of the amplifier in decibels is $20 \log_{10}100/1 = 20 \log_{10}10^2 = 20 \times 2 = 40$ dB. If an input voltage of 100 volts to some circuit gave an output of 1 volt, the gain would be $20 \log_{10} 1/100 = 20 \log_{10}10^{-2} = 20 \times (-2) = -40$ dB.

Clearly, if the output were less than the input, a loss occurred as the minus sign in the preceding example indicates. So, a loss is a negative gain. If a gain of 40 dB were followed by a loss of 40 dB, the net effect would be an overall gain of $40 - 40 = 0$ dB.

A doubling in voltage represents about a 6-dB change, and a doubling in power represents about a 3-dB change. This is because the logarithm of 2 is about 0.3. Remembering this can make decibels easier to calculate. Consider a voltage increase by a factor of 20. Twenty equals two times ten. A factor of 2 in voltage corresponds to 6 dB, and a factor of 10 corresponds to 20 dB. Simply adding the two numbers, remembering that the logarithm of a product is the sum of the individual logarithms, gives 26 dB as the decibel figure corresponding to a factor of 20.

Decibel Examples

#1:

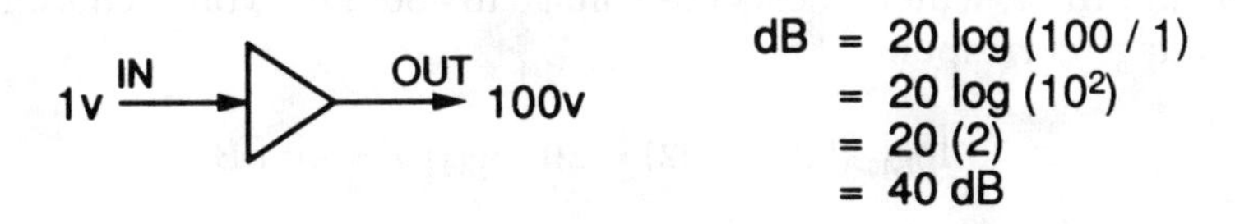

$$
\begin{aligned}
dB &= 20 \log (100 / 1) \\
&= 20 \log (10^{2}) \\
&= 20 (2) \\
&= 40 \text{ dB}
\end{aligned}
$$

#2:

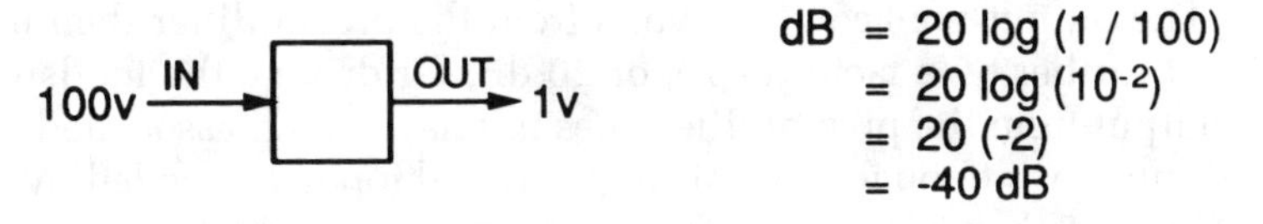

$$
\begin{aligned}
dB &= 20 \log (1 / 100) \\
&= 20 \log (10^{-2}) \\
&= 20 (-2) \\
&= -40 \text{ dB}
\end{aligned}
$$

#3:

IN OUT

+40 dB -40 dB

Net Overall Gain = +40 + (-40) = 0 dB

#4:

$$
\begin{aligned}
dB &= 20 \log 20 \\
&= 20 \log (2 \times 10) \\
&= 20 \log 2 + 20 \log 10 \\
&\approx 20 (0.3) + 20 \\
&= 6 + 20 \\
&= 26 \text{ dB}
\end{aligned}
$$

DECIBELS—STEREO EXAMPLE

Let us calculate a practical example for a stereo audio system. The output from a moving-magnet phonograph cartridge is about 2 mV. This signal is the input to the preamplifier, which boosts the voltage to about 0.2 volts. This amplification of voltage corresponds to

$$20 \log_{10} (0.2/0.002) = 20 \log_{10}10^2 = 40 \text{ dB}$$

The stereo loudspeaker has an impedance of 4 Ω, which is mostly pure resistance at audio frequencies. This loudspeaker consumes about 1 watt in producing a comfortable sound level. The voltage at the loudspeaker's terminals corresponding to 1 watt is obtained from the equation for power ($P = E^2/R$) and is 2 volts. The output voltage of 0.2 volts from the preamplifier then needs to be amplified further by a factor of 10, or 20 dB, for driving the loudspeaker.

The output from the preamplifier does not have much associated current, and so will not have enough current to generate the power needed by a loudspeaker. There is then a need for further amplification in both voltage and current to drive the low-impedance loudspeaker. This further amplification is accomplished by a power amplifier. The input impedance to the power amplifier is customarily about 50kΩ. The 0.2 volts at its input from the output of the preamplifier would then correspond to a power of $(0.2\text{V})^2/50\text{k}\Omega = 0.8 \times 10^{-6}$ watts. This extremely small amount of power is amplified to 1 watt, which is an increase in power by a factor of nearly one million to one, or a power gain of $10 \log_{10} 10^6 = 60$ dB.

The overall voltage gain and power gain from the output of the phonograph cartridge to the input of the loudspeaker can also be calculated. The phonograph cartridge output voltage of 2 mV is amplified to 2 volts, which is a gain by a factor of 1,000 or 60 dB. The power from the cartridge at the input to the preamplifier is $(0.002\text{V})^2/50\text{k}\Omega = 80 \times 10^{-12}$ watts or 80 picowatts. This extremely small power is amplified to 1 watt, which is a power gain by a factor of nearly 10^{10}, or 100 dB.

Examples (cont'd)

Stereo Hi-Fi:

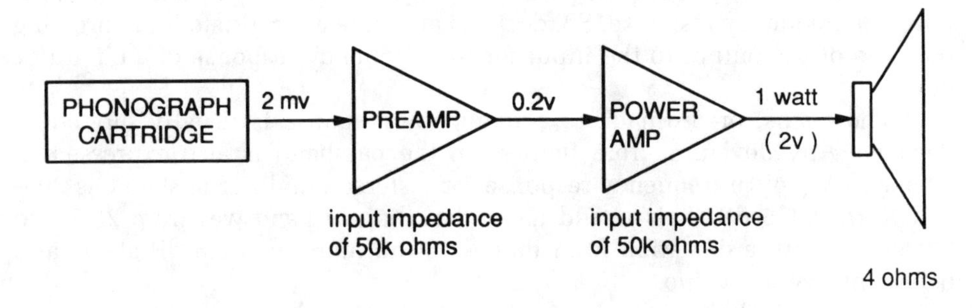

FREQUENCY RESPONSE

The frequency response of a circuit or device is its gain at a variety of frequencies in the passband. It is not the actual value of the output, but rather the ratio of the output to the input at each particular frequency that is important in specifying the frequency response. So, decibels are quite appropriate for expressing the ratio of the output to the input for the frequency response of a circuit or device.

The *y*-axis, or ordinate, of a frequency response is usually plotted in decibels. Any deviation from flatness in the passband is also expressed in decibels. A typical frequency response for a stereo amplifier is stated as 20 – 20,000 Hz ± 0.5 dB. This would mean that the passband was from 20 Hz to 20,000 Hz with a deviation from flatness of no more than 0.5 dB above and below the average value.

An electrical circuit or device does not abruptly begin passing a range of frequencies, then just as abruptly stop passing frequencies. There is instead a gradual transition between the passband and stopband, and this gradual transition has a *slope* associated with it. This slope is specified by comparing the change in decibels at a frequency in the passband just at the beginning of the transition and at a frequency twice (or half) this passband frequency. Since a doubling, or halving, in frequency is an octave, the change is a measure of the slope in decibels per octave.

A simple resistor-capacitor low-pass filter has a slope of 6 dB per octave. This means that the loss increases by 6 dB for each doubling in frequency outside the passband. If an inductor were added to this filter, the slope would be increased by an additional 6 dB, resulting in a final slope of 12 dB per octave.

Decibels can be used for many other ratio measures in addition to gain or loss. The *dynamic range* of a system is the decibel measure of the largest to the smallest signal handled by the system. In a stereo system, the *channel separation* is a decibel measure of how much of one stereo channel seeps, or leaks, into the other.

Frequency Response

Spectrum:

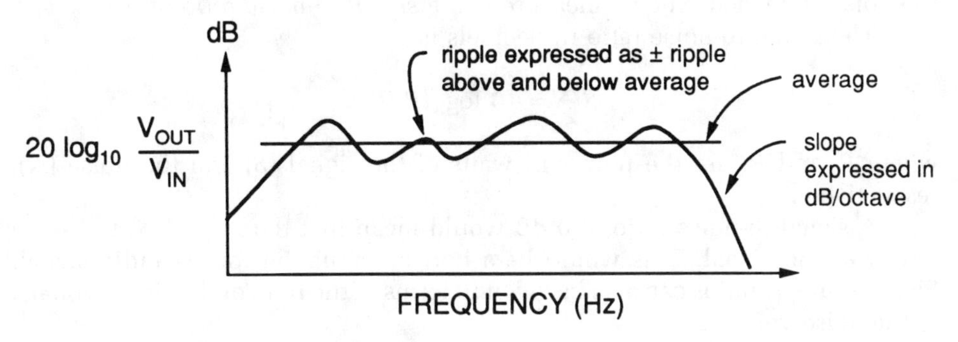

Low-Pass Filter:

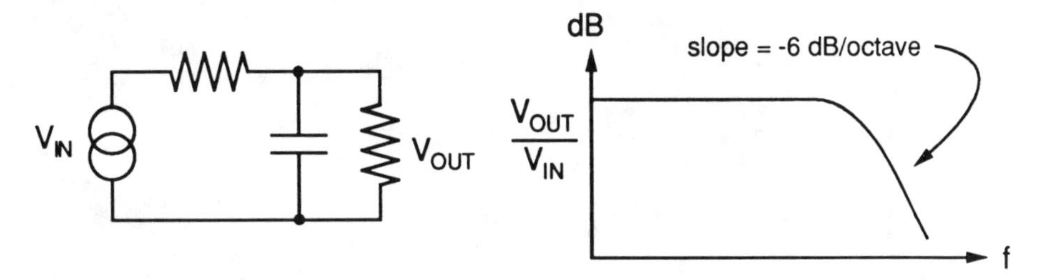

[octave = doubling in frequency]

SIGNAL-TO-NOISE RATIO

Noise is a problem that is always present in all electronic systems and that corrupts the desired signal. A measure of the strength of the signal relative to the noise is needed. Such a measure is the *signal-to-noise ratio* in decibels.

The signal-to-noise ratio in decibels is

$$S/N = 10\ \log_{10}(P_S/P_N)$$

where P_S and P_N are the power in watts of the signal (S) and the noise (N), respectively.

A signal-to-noise ratio of 0 dB would mean that the noise had the same power as the signal. This would be a horrible situation for an audio signal. Signal-to-noise ratios can also be given in terms of the ratio of the signal voltage to the noise voltage.

Signal-to-Noise Ratio

$$S/N = 10 \log_{10} \frac{P_S}{P_N}$$

$$S/N = 20 \log_{10} \frac{E_S}{E_N}$$

SOUND PRESSURE LEVEL

Sound intensity is a measure of the power in an acoustical signal. Sound intensity is proportional to the square of the sound pressure.

The range of human hearing from the softest to the loudest sound covers a range of nearly one million to one (10^6:1) in intensity. Decibels are used to measure sound intensity and sound pressure levels, since, in this way, the large ranges of intensities and pressures are more easily characterized.

Sound pressure is determined with an instrument called a sound-level meter. This meter has a calibrated microphone and an amplifier connected to a root-mean-square meter. Different frequency responses for the amplifier, called weightings, can be specified, depending on the type and level of the signal being measured. The sound-level meter gives sound pressure in decibels relative to a standard reference pressure of 0.0002 microbar, or equivalently 20 micro-newtons per square meter.

A sound level of 0 dB corresponds to the threshold of hearing, which is the smallest sound that can be heard. A soft whisper at 5 feet would measure 30 dB in sound level. A normal speech conversation would measure about 60 dB. A jet airplane takeoff at 200 feet would measure 120 dB. Initial pain would be felt for pure-tone sounds at levels of 140 dB.

There are three frequency weightings: A, B, and C. The C weighting is nearly flat or uniform over the frequency range from 32 Hz to 8 kHz, and sound-level readings made with this weighting give a good indication of overall sound pressure. The A weighting discriminates heavily against low frequencies, and sound-level readings made with this weighting give a good subjective measure of loudness and annoyance. The B weighting lies in between the other two.

Government regulations for permissible noise exposures are stated in terms of sound level with A weighting. A slow response of the meter is also specified to give an average reading that ignores short peaks. For example, a sound level of 90 dB (A weighting) should not be exceeded during a nine-hour exposure period, according to the regulations.

The reference sound intensity level is 10^{-12} watts per square meter. This reference corresponds approximately to the reference pressure of 0.0002 microbar.

Sound Pressure Level

Definition:

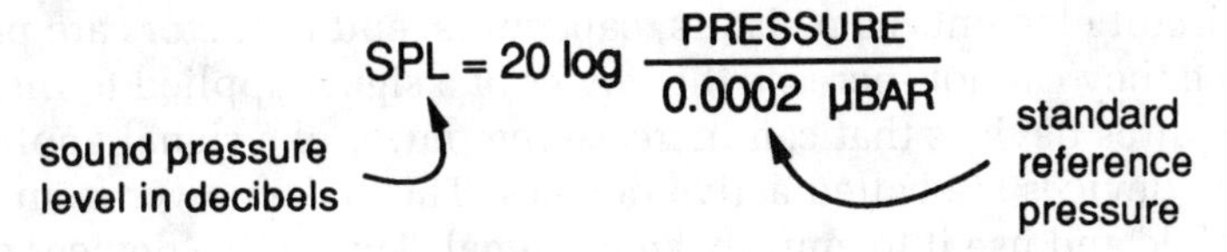

Table:

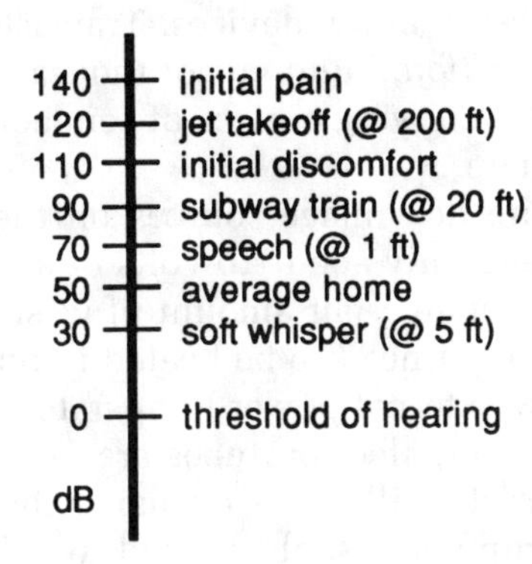

Sound Level Meter:

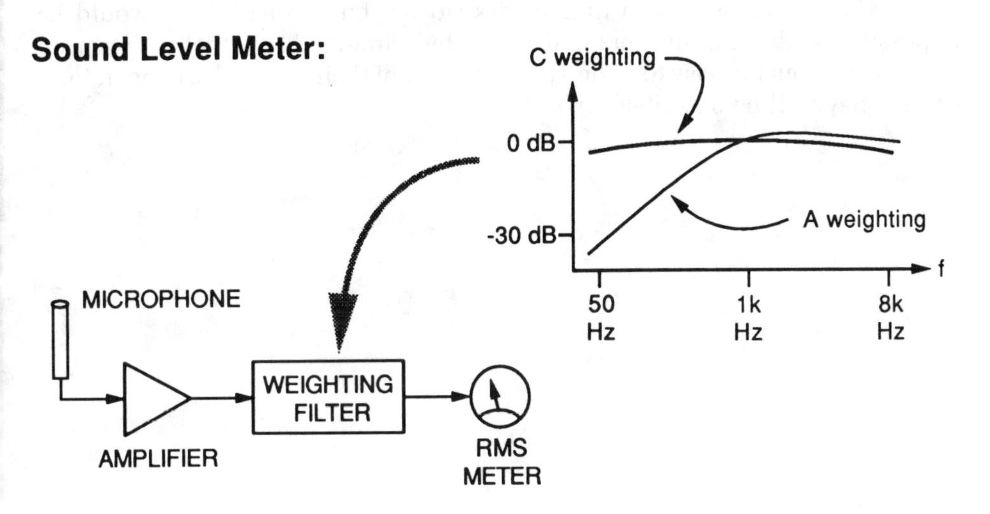

Electron Tubes

ACTIVE DEVICES

Such electrical circuit elements as resistors, capacitors, and inductors are passive devices, since they cannot increase the power of a signal applied to them. Amplification requires devices that can increase the power of a signal applied to them, and such devices are called active devices. They take power from an external source of dc and use it to amplify an ac signal. Two active devices are *electron* (or vacuum) *tubes* and *transistors.*

Transistors offer great improvements over electron tubes for most applications. Electron tubes are large, bulky devices. Transistors are very small, and with *very-large-scale integration*, hundreds of thousands of transistors can all be put on a single device occupying a piece of semiconductor material that is only one-quarter inch square.

Electron tubes require dc voltage sources on the order of hundreds of volts, while transistors need only about 10 volts or so. Electron tubes need to be heated, and so they generate a fair amount of wasted heat. Transistors are *solid-state devices*, and do not need to be heated to operate. As a result, they are cooler and require less external power to operate.

To a considerable extent, electron tubes are nearly obsolete, although a small number of hi-fi purists still insist on using them. The modern era of electronics—personal computers, sophisticated word-processing machines, robotics, the compact audio digital disk, to list but a small few—would be impossible without transistors and integrated circuits. Nevertheless, electron tubes are reasonably easy to understand in terms of their principles of operation, and so they will be described first.

Active Devices

Passive Devices:

- resistors
- capacitors
- inductors
- transformers

Active Devices:

- electron tubes
- transistors

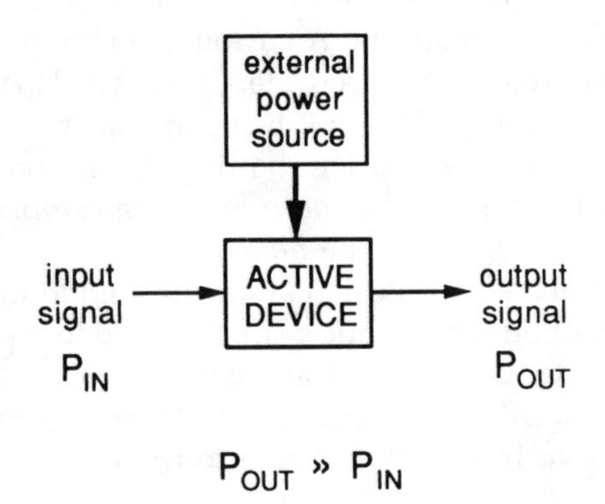

$P_{OUT} \gg P_{IN}$

Electron Tube vs Transistor

Electron Tube	Transistor
• large	• small
• large power	• small power
• hot	• cool
• large voltages	• small voltages

ELECTRON-TUBE DIODES

While investigating electric light bulbs, Thomas Alva Edison placed a second electrode in the bulb and discovered that a current could be caused to flow between the two electrodes. This was most surprising, since there was no physical contact between the two electrodes, and somehow electricity had to be flowing across the empty space between them. This phenomenon later was named the Edison effect in honor of its discoverer.

It is now known that if a material is heated, electrons will be freed from near the material's surface and will form a cloud near the surface. If a second electrode is placed nearby with a positive voltage applied to it, the electrons will be drawn away from the heated material and will flow between the two materials. This principle explains the operation of the vacuum (or electron) tube *diode*, invented in 1904 by John Ambrose Fleming in London.

An electron-tube diode consists of two electrodes located within a glass envelope from which all air has been evacuated, producing a vacuum. The outer electrode is a cylinder that surrounds the inner electrode. A wire runs within the inner electrode. A current flowing in this wire heats it up, and this heat is transferred to the electrode, causing it also to heat up. This inner electrode is called the *cathode*. When it heats up, electrons are "boiled" from its surface and form a cloud about it. This boiling of electrons from the surface of a material by heating it up is called *thermionic emission*. The wire within the cathode that heats it up is called the *heater*.

The outer electrode is called the *anode* or plate. If a voltage is placed across the cathode and anode so that the anode is positive relative to the cathode, the positive voltage will attract electrons from the cloud and these electrons will flow toward the anode. An electrical current then flows. If the anode is made negative in voltage relative to the cathode, the negative polarity repels electrons, and no current flows.

If the anode is positive relative to the cathode, the diode is said to be *forward biased*. Electrons are attracted to the anode, and a current flows. If the anode is negative relative to the cathode, the diode is said to be *reverse biased*. Electrons are repelled by the anode, and no current flows. A diode allows current to flow through it in only one direction. This property can be used to convert ac into dc, as we will show later.

Electron-Tube Diode

Edison Effect:

In a vacuum, electrons flow from heated negative electrode to positive electrode.

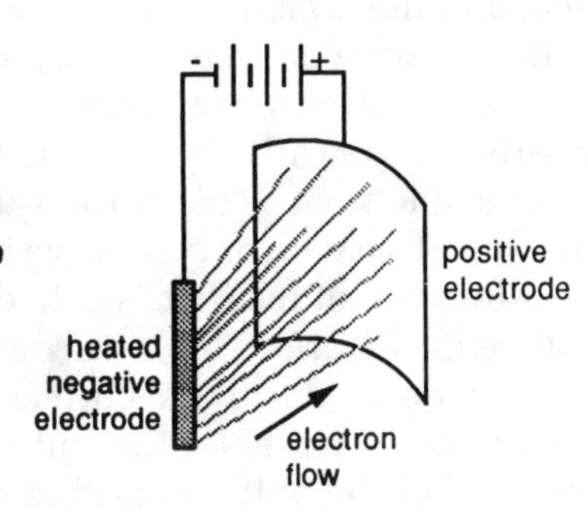

Diode:

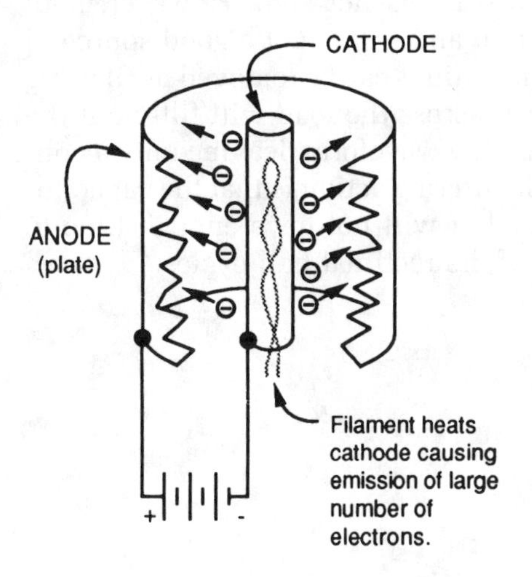

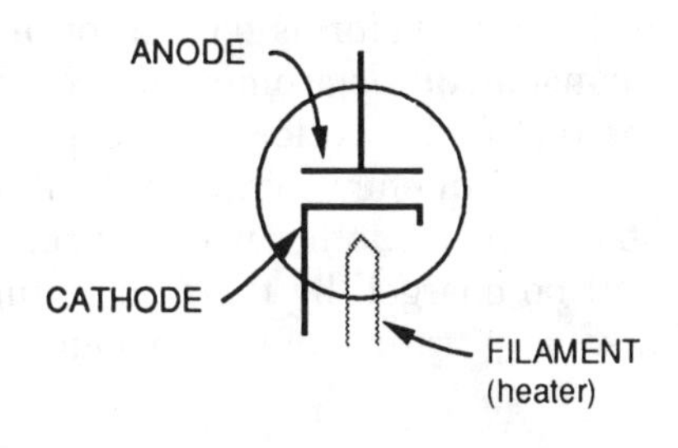

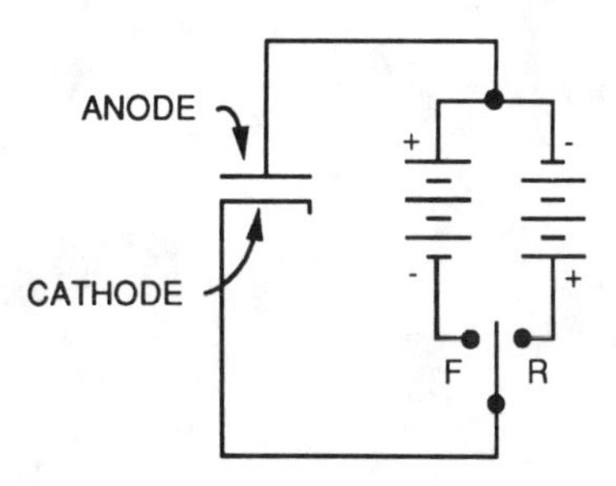

F = FORWARD BIASED (current flows)
R = REVERSE BIASED (no current flows)

RECTIFIERS

A vacuum-tube diode can be used to convert ac into dc. This process is called *rectification*, and the circuits using diodes in this way are called *rectifiers.*

Consider a diode connected in series with a source of a sinusoidal electromotive force. Current will flow through this circuit only when the anode is positive relative to the cathode. The current will then consist of only positive half cycles of a sine wave. This current flowing through the load resistor will also produce a voltage with only positive half cycles. Since only half the current wave flows, this type of rectifier is called a *half-wave rectifier.*

Full-wave rectification can be accomplished with two diodes. The source of ac power is connected to a transformer with a center-tapped secondary. The two diodes are connected to the secondary such that one of them conducts on the positive half of the voltage and the other conducts on the negative half. The net result is that the negative half of the voltage is "flipped," and both halves of the cycle flow in the same direction through the load resistor.

The average, or dc, value of a sine wave is zero. Clearly, the average value of a half-wave or full-wave rectified sine wave is not zero. However, both rectified waveforms have a lot of fluctuation, and so are not a good source of a constant dc electromotive force. These fluctuations can be removed by filtering the rectified waveforms. A capacitor placed across the load will filter out the higher frequency components of the fluctuating waveform, leaving only a constant value. Another way of visualizing this filtering action is that the capacitor will be charged by the fluctuating voltage, but will not have enough time to discharge appreciably between the peaks of the rectified half cycles.

Rectifiers

Half-Wave:

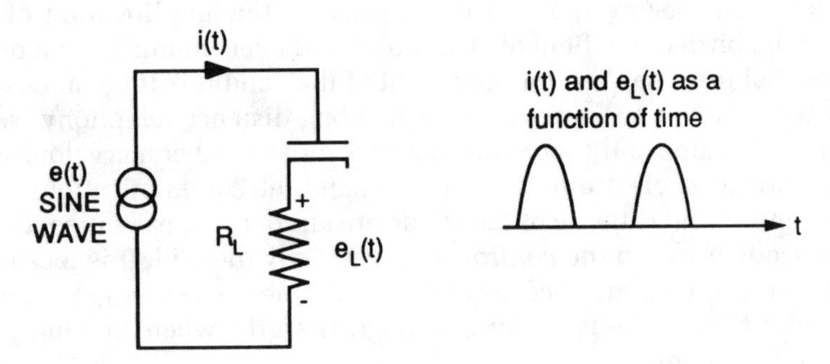

Full-Wave:

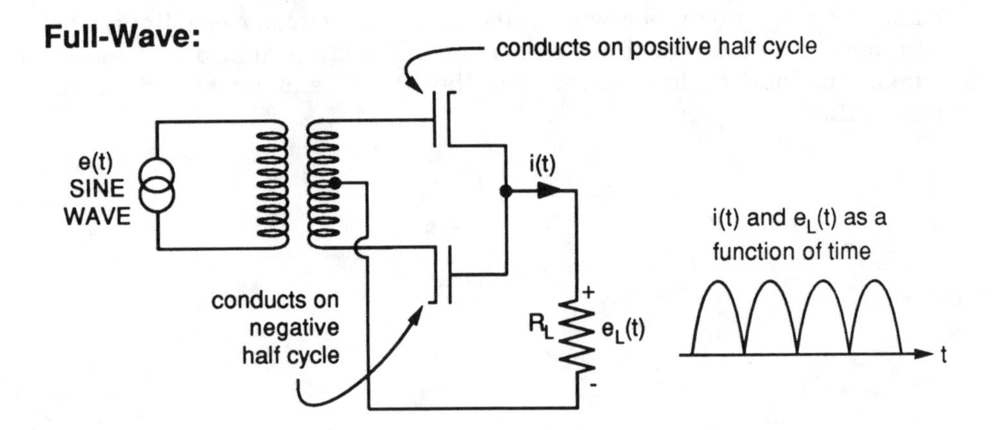

Filtering:

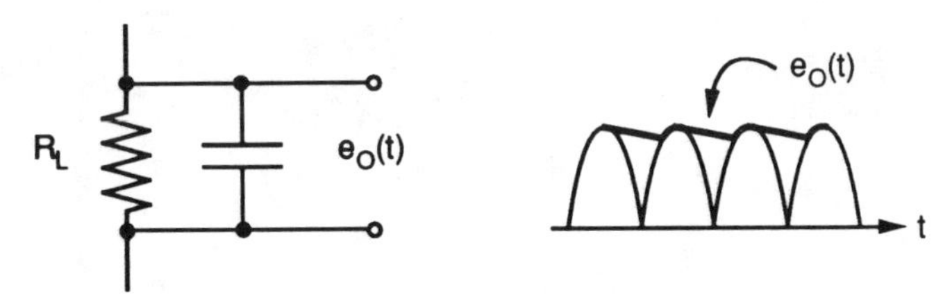

TRIODES

A third electrode can be placed between the cathode and the anode. The three-electrode electron tube produced in this way is called a *triode*. The triode was invented by Lee DeForest, and it made possible the amplification of signals, heralding the birth of the field of electronics and telecommunication over great distances. DeForest applied for the patent of the "audion" tube on October 25, 1906; it was granted on January 15, 1907. Long-distance telephony was made practical, and radio would soon follow along with the other marvelous wonders of today's world of electronics, computers, and telecommunications.

The anode and cathode of the triode are similar to those of the diode. The third electrode is called the *control grid*, or simply the grid. It is located much closer to the cathode, and because of this closeness exerts a great amount of control over the electrons passing through the grid when flowing from the cathode to the anode.

The grid is an open framework of wire through which electrons can easily pass. The grid usually is a wire in the shape of a circular or elliptical helix surrounding, but not touching, the cathode. The grid can also be a screen of wire in a cylindrical shape surrounding the cathode, and thus the grid is sometimes called the screen.

Triodes

Construction:

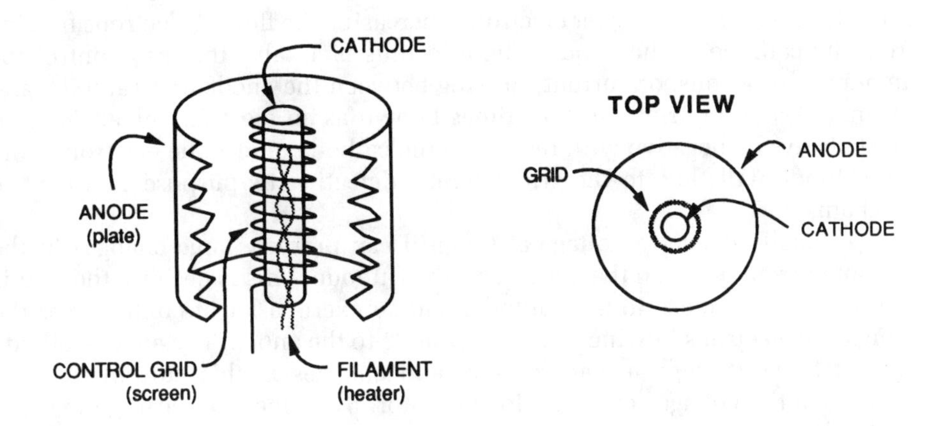

Symbol:

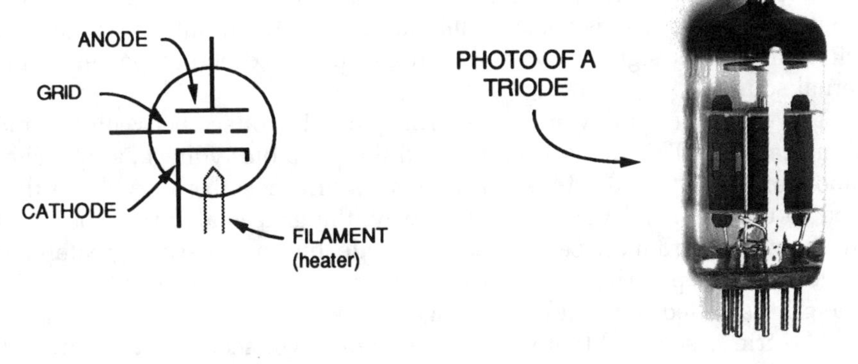

TRIODES (cont'd)

If the control grid is made negative in voltage relative to the cathode, the grid repels electrons and so impedes the flow of electrons from the cathode to the anode. If the control grid is made less negative in voltage relative to the cathode, the grid repels fewer electrons, increasing the flow of electrons flowing from the cathode to the anode. The grid thus is a valve that can control the amount of electrons, or current, flowing between the anode and cathode, and for this reason the triode is sometimes known as an electron valve. The grid must always be negative with respect to the cathode or else the electrons from the cathode will flow to the grid, thereby defeating its purpose as a control mechanism.

A small change in voltage at the grid can produce large changes in the flow of current between the anode and the cathode. This is because the grid is physically quite close to the cathode, and so exerts a lot of control over the number of electrons flowing from the cathode to the anode. The anode-cathode current flows through a load resistor, and changes in this current produce changes in the voltage across the load resistor. By properly choosing the value of this load resistor, small changes in voltage at the grid will produce large changes in voltage at the load resistor. So, a triode amplifies voltage.

Virtually no signal current ever flows in the grid circuit. However, substantial amplified signal current flows in the anode-cathode, or output, circuit. Remembering that signal power is the product of signal voltage and current, the output signal power is considerably larger than the input signal power. So, a triode amplifies signal power, and does so by extracting power from an external source.

The triode needs power in the form of the dc voltage between the anode and the cathode. This dc voltage is called the plate bias voltage, and is shown symbolically as B^+. The grid voltage consists of a negative dc voltage that is added to the ac signal voltage. In this way, the grid is always impeding the flow of electrons from the cathode to the anode, but to a lesser or greater extent depending on the polarity and value of the signal voltage. The dc voltage applied to the grid is called the grid bias voltage.

The grid is so small that very few electrons ever accumulate on it, and so there is virtually no current flowing in the grid circuit. The grid, or input, circuit is then isolated from the anode, or output, circuit in a triode.

Triodes (cont'd)

Process:

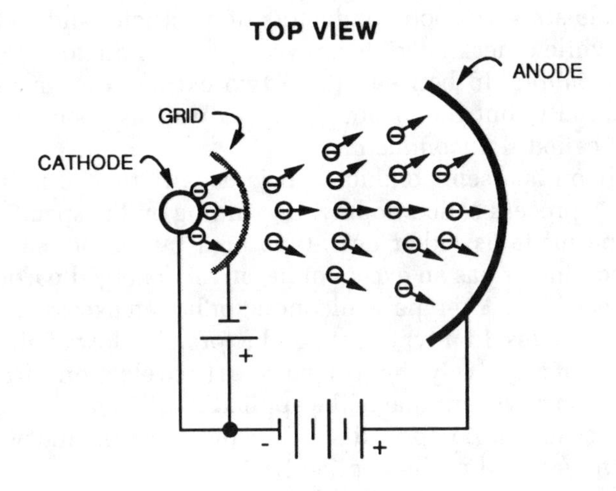

Circuit:

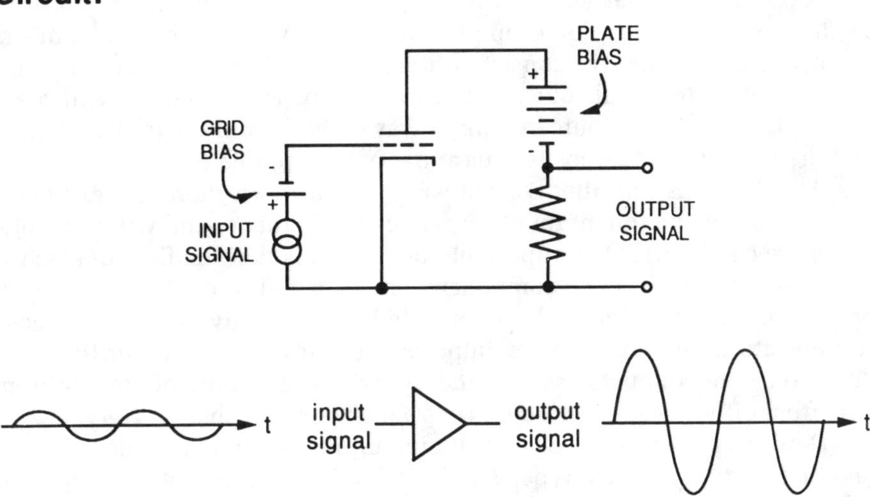

Solid-State Devices

SEMICONDUCTORS

Some materials are very good conductors of electricity and so they are called *conductors*, while other materials are very poor conductors of electricity and are called *insulators*. In between these two extremes are materials that can conduct electricity, but do so not nearly as well as good conductors. Such materials are called *semiconductors*.

Pure silicon is a semiconductor. It is usually treated by the addition of impurities in a process called *doping*. Depending on the specific impurity, the silicon can be made as either an *n*-type or a *p*-type of semiconductor. An *n*-type semiconductor has an excess of negatively charged particles (electrons) to conduct electricity. A *p*-type semiconductor has an excess of vacant electron particles, called holes. In *n*-type semiconductors, the flow of electricity is comprised of a flow of negatively charged particles (free electrons) from the negative terminal to the positive terminal of the applied EMF. In *p*-type semiconductors, the flow of electricity is comprised of a flow of holes from the positive terminal to the negative terminal of the applied EMF.

A semiconductor will usually have a predominance of either electrons or holes as the majority carriers of the flow of electricity. However, a small number of the opposite particles will also be present and contribute in a minor way to the flow of electricity. For example, the majority of the current flow in a *p*-type semiconductor will consist of holes flowing from the positive terminal to the negative terminal, but a small number of free electrons will also be present. They will contribute in a minor way to the flow of electricity by flowing from the negative to the positive terminal of the applied EMF.

All of this is admittedly confusing, but a more thorough explanation requires delving into quantum physics and atomic theory and will probably be even more confusing! One important conclusion is that the flow of electricity in a semiconductor has two components, namely, a flow of electrons (negative charges) in one direction and a flow of holes (seemingly positive charges) in the opposite direction. Another important finding is that impurities can be added to a semiconductor so that the majority of the flow of electricity may come from either negative charges (*n*-type) or positive charges (*p*-type).

Doping is the process of introducing impurities into a semiconductor to make it either *p*-type or *n*-type. The introduction of boron atoms into silicon will make it a *p*-type semiconductor. The introduction of arsenic, phosphorus, or antimony atoms into silicon will make it an *n*-type semiconductor.

Semiconductors

Types:

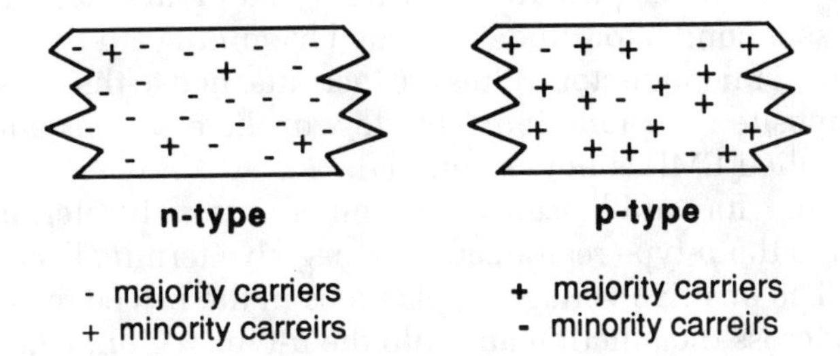

Conduction:

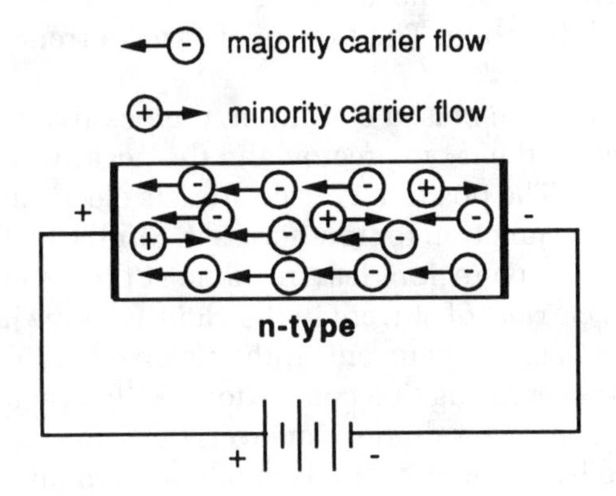

SEMICONDUCTOR DIODES

A semiconductor diode allows current to flow through it in only one direction, just like an electron-tube diode. However, the principle of operation of a semiconductor diode is quite different from that of an electron-tube diode.

A semiconductor diode is formed by joining a *p*-type semiconductor with an *n*-type semiconductor. Wires are then attached to the ends of the semiconductors, opposite their junction. The flow of the electrons and holes in response to an applied EMF is now examined.

In the forward-biased condition, the positive terminal of the EMF is applied to the *p*-type region, and the negative terminal is applied to the *n*-type region. The positive voltage repels holes in the *p*-type region, and causes them to flow across the junction and into the *n*-type region, where they are attracted by the negative voltage at the negative terminal. Similarly, the negative voltage repels electrons in the *n*-type region, and causes them to flow across the junction into the *p*-type region, where they are attracted by the positive voltage at the positive terminal. So, the semiconductor diode conducts electricity in the forward-biased direction. The flow of electric current is from the *p*-type region to the *n*-type region.

In the reverse-biased condition, the positive terminal of the applied EMF is connected to the *n*-type region and the negative terminal is connected to the *p*-type region. The positive voltage attracts electrons in the *n*-type region, leaving none at the junction to flow across it. Similarly, the negative voltage attracts holes in the *p*-type region, leaving none at the junction to flow across it. Since there are no carriers of current in the vicinity of the junction, the semiconductor diode cannot conduct current in the reverse-biased condition.

Nearly everything that can be done with a vacuum tube diode can be done with a semiconductor diode. However, the semiconductor diode does not need to be heated by a filament, and is much more robust and smaller than a vacuum tube diode.

Semiconductor Diodes

Forward Biased:

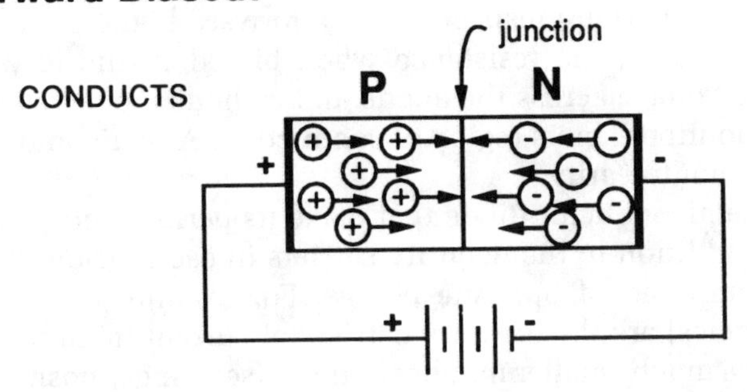

Reverse Biased:

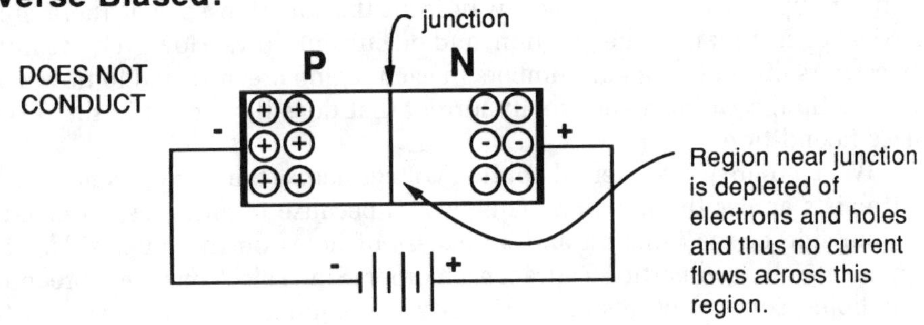

Symbol:

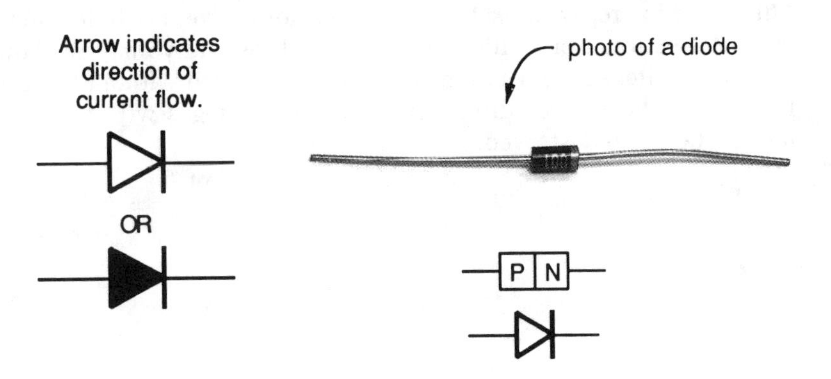

SEMICONDUCTOR DIODES (cont'd)

The current flowing through a diode can be plotted as a function of the applied voltage. For an ideal diode, this plot would be a horizontal line for reverse bias and a sloping line (pure resistance) for a forward bias.

For a vacuum tube diode, the resistance when biased in the forward direction varies with the voltage across the anode and cathode. So, a vacuum tube diode is said to be nonlinear in its region of conduction. A semiconductor diode also exhibits such nonlinearity.

Effects occur in a semiconductor diode that make its performance somewhat less than ideal. In addition to the majority carriers in each region, there are minority carriers composed of opposite charge. For example, electrons (negatively charged particles) are the majority carriers of current in an *n*-type region, but there are also a much smaller number of holes (seemingly positively charged particles) that behave as minority carriers of current.

In the reverse-biased condition, electrons in the *n*-type region and holes in the *p*-type region are pulled away from the junction so that there are no carriers of current near the junction, and no current flows. However, the minority carriers present in small numbers in each region are attracted to flow across the junction, creating a very small current that does flow, even in the reverse-biased condition.

When there is no external applied voltage across the diode, a small voltage still exists across the junction. This occurs because of an excess of electrons on one side of the junction and an excess of holes on the other side. These oppositely charged particles attract each other. Since electrons are more mobile than holes, some electrons in the vicinity of the junction flow across it, where they are neutralized by the holes near the other side of the junction. The final result is that electrons and holes are depleted in the vicinity of the junction. This depletion region must be overcome before current will flow in the forward-biased direction. For a silicon diode, this threshold voltage is about 0.6 volts.

The voltage across a diode cannot be increased without limit in either the forward or the reverse direction. Excessive voltages will cause the diode to burn out and be destroyed.

Diodes (cont'd)

Ideal Characteristics:

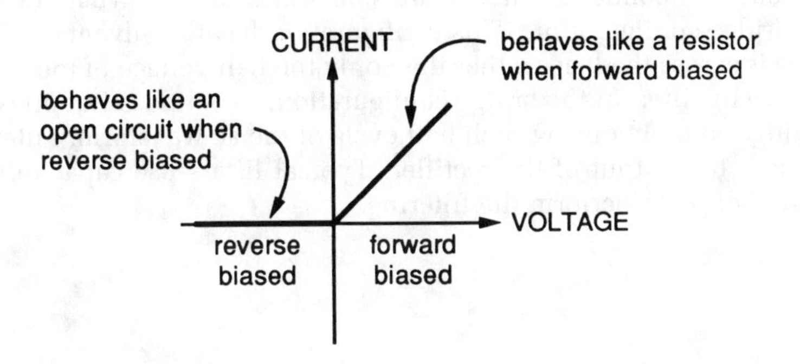

Real Characteristics:

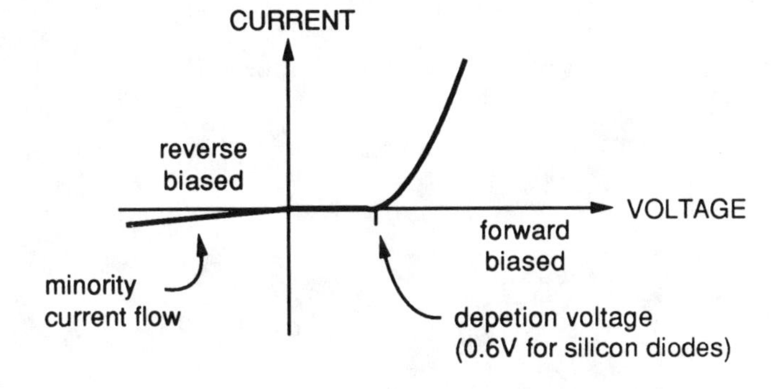

3

DIODE RECTIFIERS

Semiconductor diodes can be used to rectify ac into dc. They can be used for either half-wave or full-wave rectification in circuits similar to those used for electron-tube diodes.

Four semiconductor diodes are sometimes used in what is called a full-wave bridge configuration. This configuration has the advantage that the full voltage is across the load, rather than only the half voltage of the conventional full-wave rectifier. In the bridge configuration, two diodes in opposite arms of the bridge conduct during each half cycle of the ac waveform. Filters are used to smooth the output of the rectifier. Typical filters use capacitors and coils (called *chokes*) to perform the filtering.

Diode Rectifiers

Half-Wave:

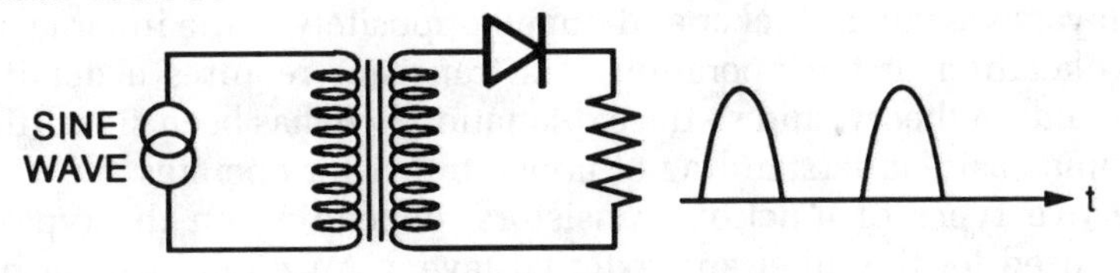

Full-Wave:

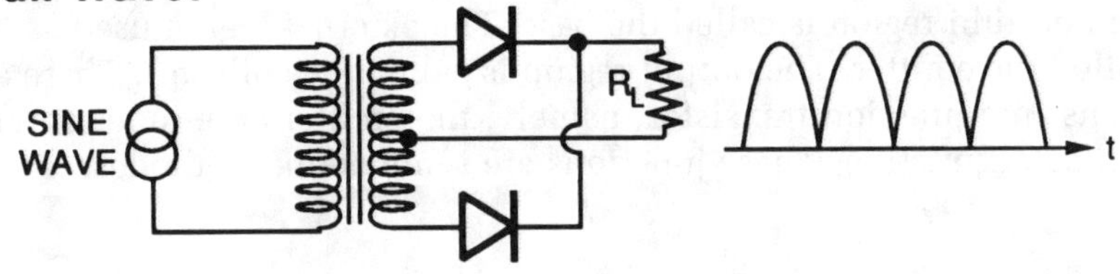

Bridge (Full-Wave):

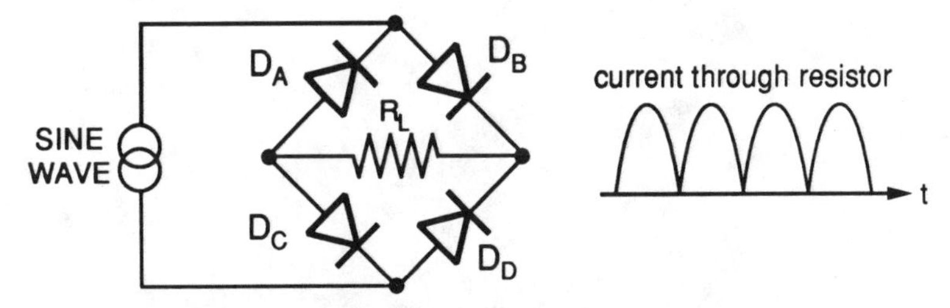

Diodes B and C conduct on positive half cycle.
Diodes D and A conduct on negative half cycle.

Filtering:

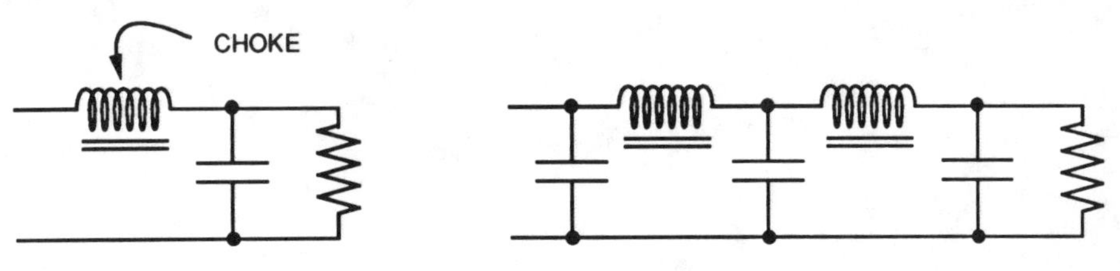

JUNCTION TRANSISTORS

The junction transisitor is able to amplify ac signals. It consists of a "sandwich" of three semiconductors. The inner layer is very thin and is only slightly doped. The outer two layers are much thicker and doped oppositely to the inner layer. A thorough explanation of the operation of a transistor requires a detailed knowledge of quantum theory, and so the explanation here has been simplified to try to give some basic understanding of how a transistor operates.

There are two types of junction transistors, depending on the type of semiconductors used for the inner and external layers. An *npn* transistor has *p*-type material for the inner and *n*-type material for the external layers. A *pnp* transistor has *n*-type material for the inner layer and *p*-type material for the external layers.

The inner, thin region is called the base. The external region used as the input is called the emitter. The output region is called the collector. There are two junctions in a junction transistor, namely, the emitter-base junction and the base-collector junction. These junctions are semiconductor diodes.

Junction Transistors

Types:

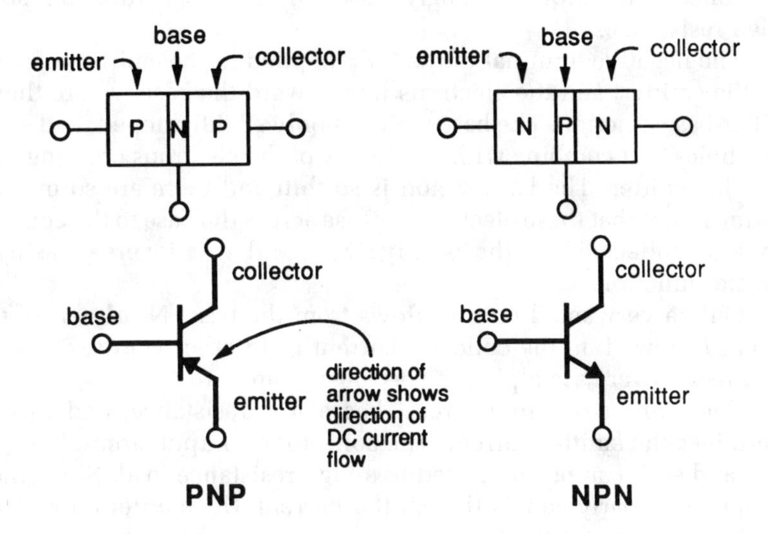

JUNCTION TRANSISTORS (cont'd)

The explanation that follows is for an *npn* transistor. The emitter-base junction is biased in the forward direction so that it has a low resistance. The base-collector junction is strongly biased in the reverse direction so that it has a high resistance.

The negative terminal of the EMF applied to the emitter injects electrons into the emitter, and the electrons flow toward the base, where they cross the emitter-base junction. The base region is only slightly doped, and so it has only a few holes that combine with only a few of the electrons crossing the junction from the emitter. The base region is so thin and there are so many electrons flowing into it that these electrons diffuse across the base to the collector, where they are "collected" by the positive voltage that is reverse biasing the base-collector junction.

Only a very small current flows from the base. Nearly all of the emitter current I_E flows into the collector current I_C. In other words, $I_C = \alpha I_E$, where α (the Greek lower-case alpha) is very nearly equal to 1.

The emitter or input circuit has a low resistance, and this resistance determines the emitter current. The collector or output circuit has a high resistance and so it can be connected to a high-resistance load. Since the collector current very nearly equals the emitter current, the emitter current is made to flow by the action of the transistor in a circuit with much higher than normal resistance. The transistor performs as a "transfer resistor," which takes its name *transistor.*

Acting in this way, changes in current in a low-resistance input circuit produce nearly equal changes in current in a high-resistance circuit. This is both a gain in voltage and a gain in power.

The junction transistor has a low input resistance and draws current from the input signal, which is very much unlike the triode which has a very high input resistance and draws very little current from the input signal. Furthermore, the input and output circuits of a junction transistor are interrelated, and do not offer the isolation between input and output of an electron tube. So, the junction transistor, though offering great advantages over the electron tube, is not without its problems. These problems are solved by the *field-effect transistor.*

Junction Transistors (cont'd)

Bias Voltages:

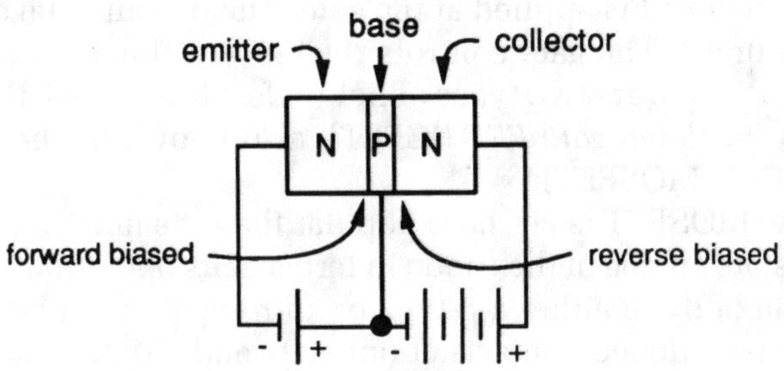

Signal Model:

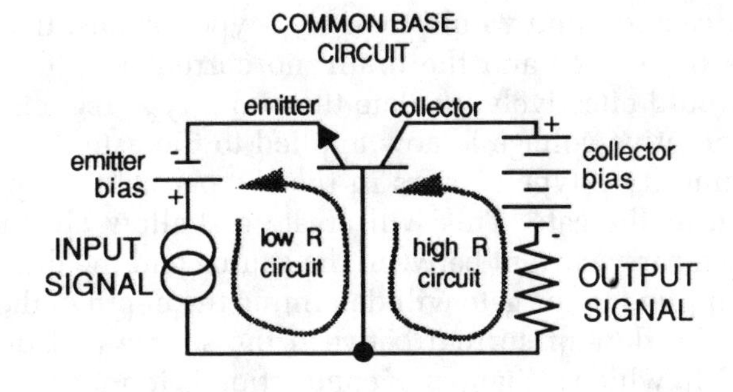

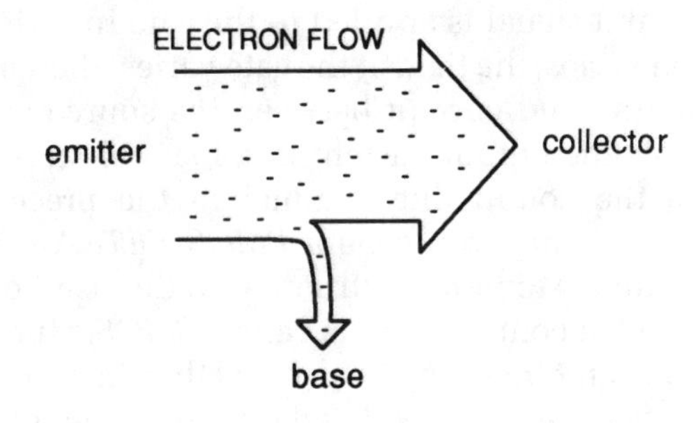

METAL-OXIDE-SEMICONDUCTOR FIELD-EFFECT TRANSISTORS

The field-effect transistor (FET) is a semiconductor device that is able to amplify ac signals. The FET has three terminals: the gate, the source, and the drain. The input signal is applied at the gate. The output is taken between the source and the drain. The gate controls the current flowing between the source and the drain. There are two types of field-effect transistors, the *junction FET* (JFET) and the *insulated-gate FET* (IGFET), also known as the *metal-oxide-semiconductor FET* (MOSFET).

The MOSFET is far more popular than the junction FET and is very much a solid-state analog of the triode in terms of its operation. The MOSFET consists of a main body of either a *p*-type or an *n*-type semiconductor. The source and the drain are doped contacts at opposite ends of this main body. The gate in a MOSFET is a metal contact that is insulated from the main body by a layer of silicon dioxide. The MOSFET then has a very high input resistance, very much like an electron tube.

Assume that the main body of a MOSFET is *n*-type semiconductor material. The source and drain would then be *p*-type regions. If a voltage were applied between the source and the drain, no current would flow, since the *n*-type region would effectively insulate the two *p*-type regions.

A negative voltage is now applied to the gate. Its electrostatic field will cause minority *p*-type carriers in the *n*-type main body to accumulate at the surface near the gate. This will create a shallow channel of *p*-type carriers, which can carry current between the source and the drain. The strength of the voltage applied to the gate will determine the depth of the channel, and so will regulate the flow of current between the source and the drain. This type of MOSFET in which a channel of conduction is induced by the electrostatic field at the gate is called an *enhancement MOSFET*.

The input signal is applied to the gate in series with the gate bias. As the input signal varies, the field at the gate varies, thereby increasing and decreasing the conduction and current between the source and drain. The input voltage thus controls the output current in a manner similar to the triode.

Since the conducting channel in the preceding example consisted of *p*-type carriers, this would be a *PMOS FET*. An *NMOS FET* would have an *n*-type channel. Modern electronic circuits use both NMOS and PMOS FETs together. Such a combination is called *CMOS*, since the two types are complementary to each other when used together in an electronic circuit.

Because the gate is insulated from the other elements, the input resistance is very high, typically from 10^9 to 10^{15} Ω. The MOSFET is somewhat akin to an electron tube, but unlike electron tubes, the MOSFET can be fabricated in microscopic sizes in integrated circuits.

MOSFET

Operation:

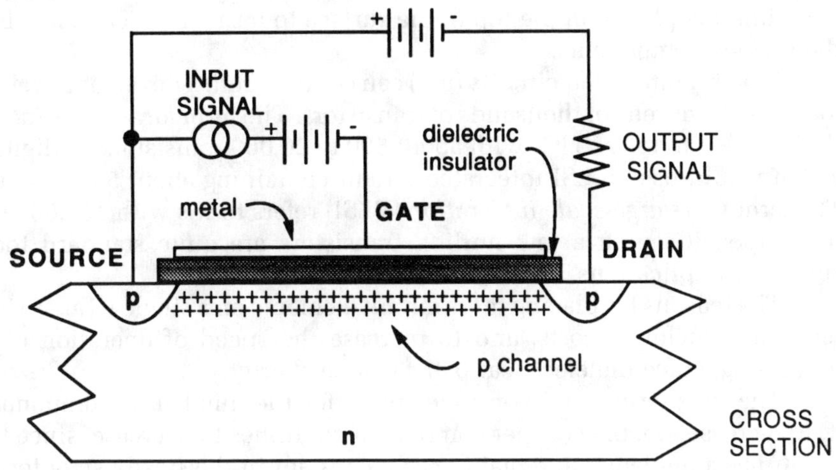

Symbols:

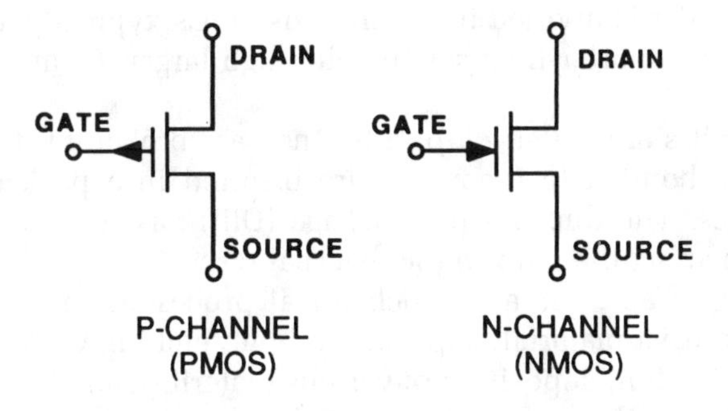

INTEGRATED CIRCUITS

An *integrated circuit* consists of many individual transistors, diodes, resistors, and capacitors that are all formed together on a single, small piece of silicon. Aluminum is plated on the top of the surface to form electrical paths between the various components.

Modern integrated circuits (ICs) can contain from a dozen or fewer transistors up to hundreds of thousands of transistors. The term *large-scale integration* (LSI) is used to refer to ICs with about 500 to 20,000 transistors. A digital wrist watch would use an LSI integrated circuit containing about 5,000 transistors. The term *very-large-scale integration* (VLSI) refers to ICs with 100,000 or more transistors. ICs with a few million transistors are quite standard today for computer applications.

The reasons for placing more circuit elements on a single IC are to increase reliability, to lower costs, and to increase the speed of operation of digital processing, since the electrical paths can be shorter.

The long-term trend continues to be for the number of components per IC to increase. So, the cost per component continues to decrease, since the cost per IC has a tendency to remain nearly constant in the steady state for a fixed degree of integration.

Integrated circuits are mass manufactured on a single wafer of silicon that is five inches in diameter, although larger wafers are being currently used. The silicon wafer is grown as a single crystal of pure silicon that is sliced into a thin wafer. The wafer is divided into hundreds of ICs, typically, each is a one-quarter-inch by one-quarter-inch square, although larger ICs are also standard today.

These little ICs are called *chips* after they are broken off the wafer. The chips have leads bonded to them and are inserted in a package for use in electronic devices. The dual in-line package (DIP) has two rows of pins for electrically connecting the chip to the external circuit.

The making of an IC is a photochemical process involving coating the wafer with photoresist material, exposure to light, etching with acids and solvents, gaseous diffusion, deposition of various materials, and heating in ovens.

Masks determine the various patterns to be placed on the wafer at different stages in its processing. Since an individual wafer contains hundreds of ICs and an individual IC can contain hundreds of thousands of components, the masks have extremely fine detail. The masks have become so detailed that the fineness of the patterns borders on the wavelengths of visible light.

The quest for greater component densities has led to the use of such esoteric techniques as x-ray and electron-beam lithography. These esoteric techniques can result in features that are only 10 times the space between atoms!

Clearly, there is a limit to how densely packed an IC can be. The quest for greater densities might then move to larger ICs or to a three-dimensional packing of components. It is intriguing to speculate what the results will be of such great densities in terms of cost and features for the consumer.

Integrated Circuits

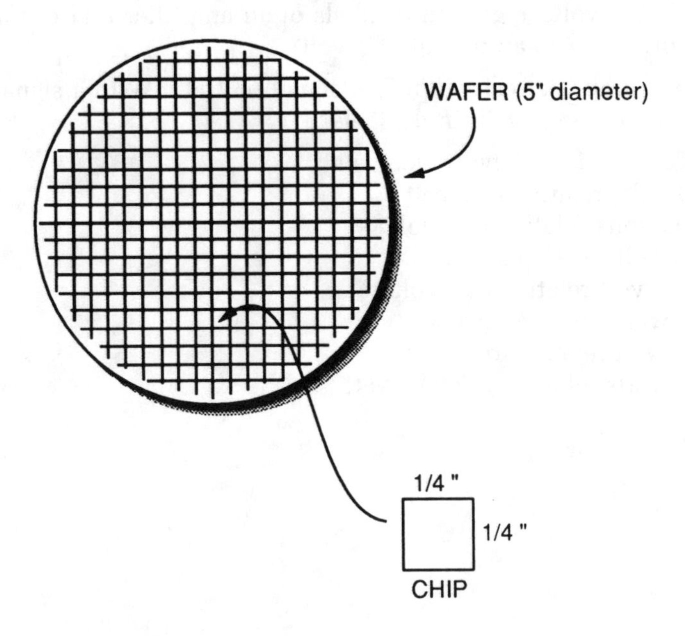

LSI (Large Scale Integration)

VLSI (Very Large Scale Integration)

PROBLEMS

3.1. What is the voltage gain in decibels of an amplifier that can amplify an input of 10 mV to an output of 1 volt?

3.2. A power amplifier can amplify a signal of 100 mW to a signal of 10 W. What is the power gain in decibels?

3.3. Calculate the following in decibels:
(a) 10 volts relative to 1 volt
(b) 100 volts relative to 1 volt
(c) 0.1 volt relative to 1 volt
(d) 0.01 volt relative to 1 volt
(e) 10 watts relative to 1 watt
(f) 0.1 watt relative to 1 watt
(g) 10 watts relative to 0.001 watt

Chapter 4: Modulation

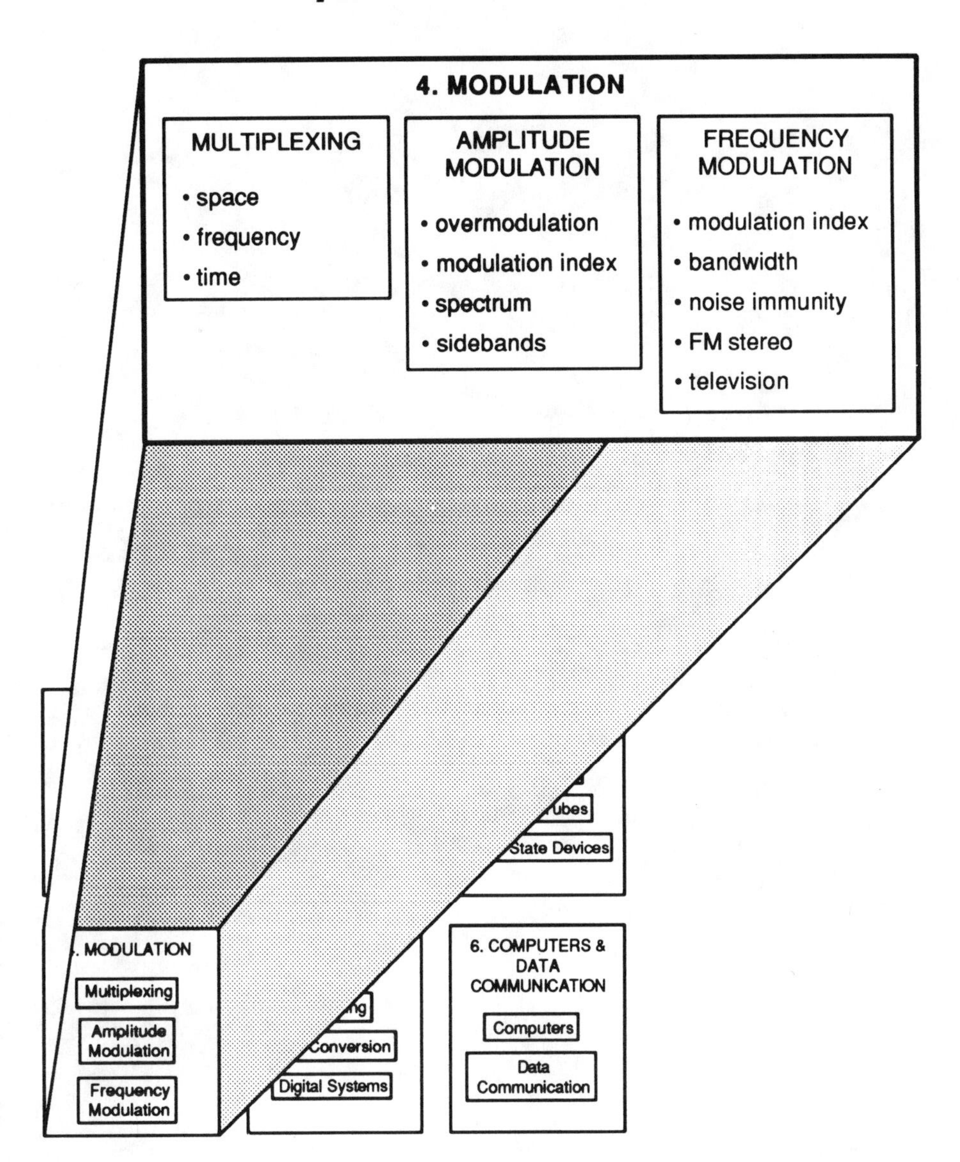

Introduction to Modulation

This module treats two types of modulation: *amplitude modulation* and *frequency modulation*. The concept of modulation is essential to communication systems since it enables a number of different signals to all share the same medium. One way in which this sharing can be accomplished is to place each signal in its own band of frequencies in the medium. Amplitude modulation and frequency modulation are two ways in which signals can be moved within the frequency domain to accomplish this placement and sharing. The combining of a number of signals to share a communication medium by dividing it into different frequency bands for each signal is called *frequency-division multiplexing.*

With amplitude modulation or frequency modulation, the amplitude or the frequency of a sine wave is varied in synchrony with the information-bearing signal. The modulated sine wave conveys, or carries, the information present in the information-bearing or modulating signal. So, the sine wave is called the *carrier*. Amplitude modulation is technologically quite simple, and the bandwidth of the amplitude-modulated carrier is at most twice the bandwidth of the modulating signal. However, an amplitude-modulated carrier is very prone to the deleterious effects of additive noise.

Frequency modulation is more complicated than amplitude modulation, and the bandwidth of the frequency-modulated carrier can be many times that of the modulating signal. However, the process of demodulating a frequency-modulated carrier eliminates much of the deleterious effects of additive noise. This trade-off between bandwidth and noise reduction characterizes most communication situations.

Multiplexing

TYPES

Consider a small room with two pairs of people who want to have two simultaneous conversations without interfering with each other. A precious medium—the air in the room that carries sound waves—needs to be shared somehow so that simultaneous conversation can occur. This sharing of a communication medium, circuit, or channel by two or more signals or messages is called *multiplexing*.

For the example of two pairs of people in the same room, one way that simultaneous conversation could occur would be for each pair to get close together at an opposite end of the room from the other pair. In this way, the physical space in the room would be divided for each conversation. This is an example of *space-division multiplexing*.

Another way that the two conversations could take place would be for each pair to alternate in conversing. The first pair might converse for a minute, then the second pair would converse for the next minute, and this sharing of time would continue, alternating back and forth. This is an example of *time-division multiplexing*.

One pair could speak in high frequency tones, while the other pair spoke in low frequency tones. The two people in a pair would then use filters to filter out the high frequency or low frequency speech of the other pair of conversants. This sharing of the frequency spectrum is an example of *frequency-division multiplexing*.

As just exemplified, there are three types of multiplexing, namely, space-division, time-division, and frequency-division multiplexing. These three types of multiplexing allow signals or messages to share a medium, circuit, or channel by dividing it in physical space, by allocating time slots, or by assigning different frequency ranges.

Frequency-division multiplexing is discussed in this module. This type of multiplexing requires a method to move signals into different frequency ranges. This is accomplished through modulation of a sine wave by varying its amplitude, frequency, or phase in synchrony with the instantaneous amplitude fluctuations of the signal. Amplitude modulation is described first and then frequency modulation. Phase modulation will be described elsewhere as a part of the discussion of modulation techniques used for data transmission.

Multiplexing

Types:

- Space-Division Multiplexing
- Frequency-Division Multiplexing
- Time-Division Multiplexing

MODULATION

Signals need to be moved about in the frequency spectrum so that they can be multiplexed together in frequency and also transmitted over various transmission media. Frequency shifting is accomplished by using another waveform to carry the information in the baseband signal into another range of frequencies. The waveform that carries the baseband signal to another range of frequencies is called the carrier waveform, or simply the *carrier.* Some property of the carrier is varied in synchrony with the information-bearing signal. The process by which this is accomplished is called modulation.

The information-bearing signal is used to modulate the carrier, and so is sometimes called the *modulating signal.* Since its frequency spectrum represents the basic information in the signal before any frequency shifting has occurred through modulation, the information-bearing signal is sometimes called the *baseband signal.*

The waveform that results from the modulation of the carrier by the baseband signal is called the *modulated carrier* or the *modulated signal.* The process of modulation then has two inputs: the baseband (or modulating) signal and the carrier. The process of modulation is performed by an electronic device called a *modulator.*

Although the carrier can be any waveshape, it usually is a sine wave that has its amplitude, frequency, or phase varied in synchrony with the baseband signal.

Modulation

Frequency Shifting:

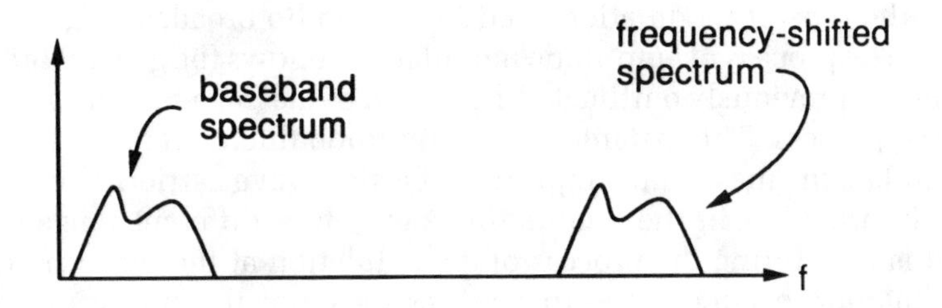

Process:

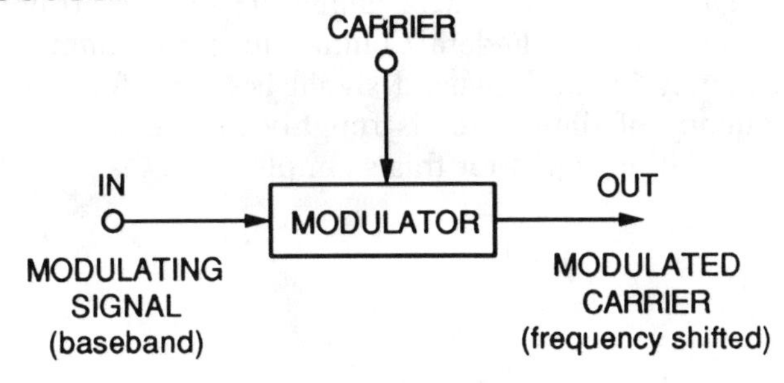

Sine-Wave Carrier:

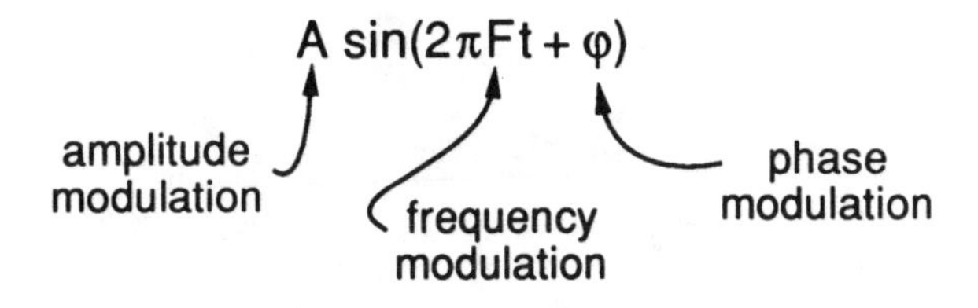

Amplitude Modulation

PROCESS

Amplitude modulation is conceptually the simplest type of modulation. It is also the type of modulation used for AM radio broadcasting.

The process of amplitude modulation follows the general process of modulation as previously outlined. A baseband or message-bearing signal is the input to the process. The instantaneous amplitude fluctuations of this signal vary or modulate the maximum amplitude of a sine-wave carrier. The carrier "carries" the information in the modulating signal to a different range of frequencies, and is lost during the process of demodulation at the receiver. The process of modulation occurs at the transmitter location; the process of demodulation occurs at the receiver location.

The frequency of the carrier is always much higher than the frequency of the modulating waveform. For AM radio broadcasting, the carrier frequency can be in the range of 550 kHz to 1,600 kHz, depending on the carrier frequency assigned to the radio station by the Federal Communications Commission (FCC). The maximum frequency in the baseband signal is 5 kHz for broadcast AM radio. So, the frequency of the carrier is roughly 200 times the maximum frequency in the modulating signal for this example.

AM Process

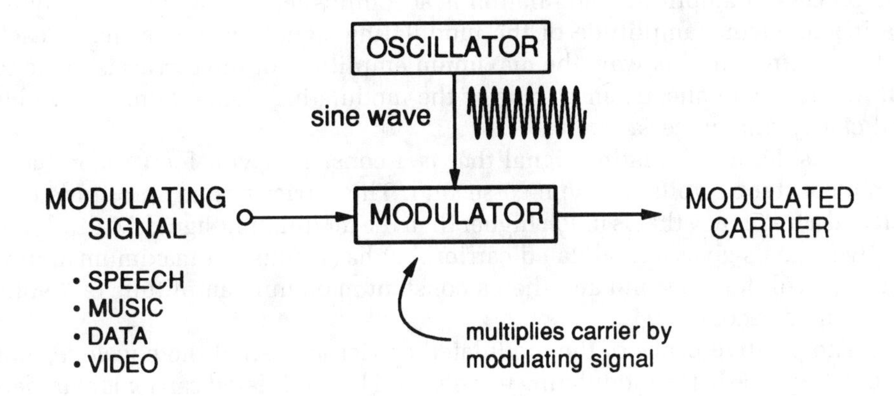
OSCILLATOR
sine wave
MODULATING SIGNAL
MODULATOR
MODULATED CARRIER
• SPEECH
• MUSIC
• DATA
• VIDEO
multiplies carrier by modulating signal

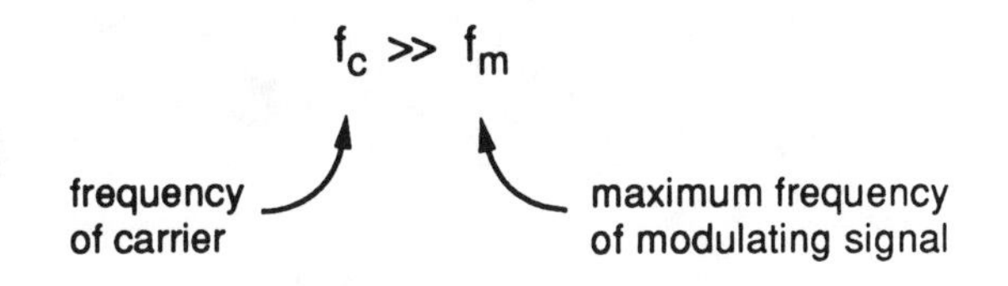
$f_c >> f_m$
frequency of carrier
maximum frequency of modulating signal

4

EXAMPLES

The process of amplitude modulation is accomplished simply by multiplying the instantaneous amplitude of the modulating signal by the carrier at each instant in time. In this way, the maximum amplitude of the carrier is made to follow the instantaneous amplitude of the modulating signal. Some examples will clarify this process.

Consider a modulating signal that is a constant 1 volt for 1 second and then a constant 2 volts for the next second. The carrier has an amplitude of 1 volt and a frequency that is much higher than the modulating signal. Multiplying the two signals gives a modulated carrier that has a constant maximum amplitude of 1 volt for 1 second and then a constant maximum amplitude of 2 volts for the next second.

The positive peaks of the modulated carrier are called the *envelope,* and follow the shape of the modulating waveform. The modulated carrier is symmetric, and its negative peaks are an inverse replica of the modulating waveshape.

Any waveshape can be used for the modulating waveform. Other examples shown are for a ramp waveshape and for a sine wave.

AM Examples

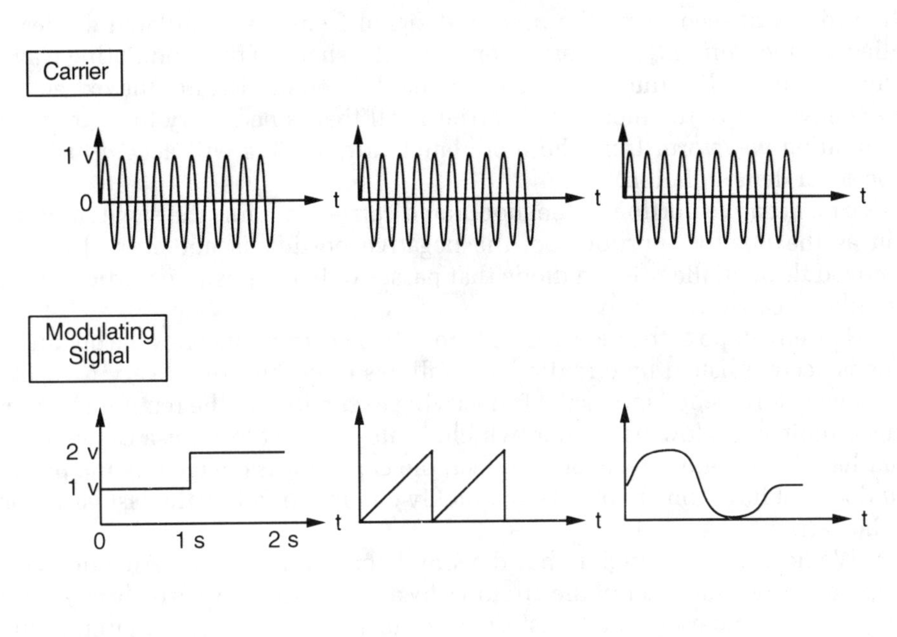
Carrier
1 v
0
t
t
t
Modulating Signal
2 v
1 v
0
1 s
2 s
t
t
t

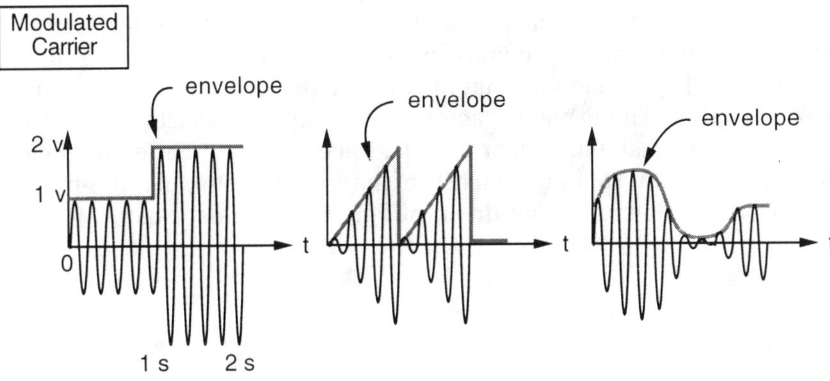
Modulated Carrier
envelope
envelope
envelope
2 v
1 v
0
t
t
t
1 s
2 s

DEMODULATION

The process of recovering the baseband signal from the modulated carrier is called *demodulation*. The information about the shape of the modulating waveform is contained in the envelope of the modulated carrier. So, the extraction of the envelope of the modulated carrier is all that is necessary to recover the modulating waveform from the modulated carrier. This will accomplish the process of demodulation.

The negative portion of the modulated carrier contains the same information as the positive portion. So, the negative portion is not needed and is removed through the use of a diode that passes only the positive portion of the modulated carrier.

The envelope of the rectified and modulated carrier must now be obtained. This is accomplished by circuitry that follows only the slowly varying peaks of the waveform, and ignores the fast varying excursions of the rectified carrier. This circuitry is a low-pass filter, which, in its simplest form, is a capacitor in parallel with an ouput resistor. In effect, the capacitor is charged by the peaks, but does not have time to discharge quickly enough to follow the fast variation of the carrier.

We now know enough to build a simple crystal AM radio. An antenna is coupled to the remainder of the circuitry by a transformer. A variable capacitor across the secondary of the transformer forms a parallel resonant circuit with the inductance of the secondary. This tuned circuit passes a narrow band of frequencies about a center frequency that can be changed by varying the value of the capacitor. This capacitor tunes the radio to the desired station. The crystal is a diode rectifier. The capacitor across the headphone tracks the envelope.

The circuitry that extracts, or detects, the envelope is called the detector. A block diagram of an AM radio consists of the antenna, a tuned radio-frequency (RF) amplifier, a detector, an audio amplifier, and a loudspeaker.

Demodulation

Process:

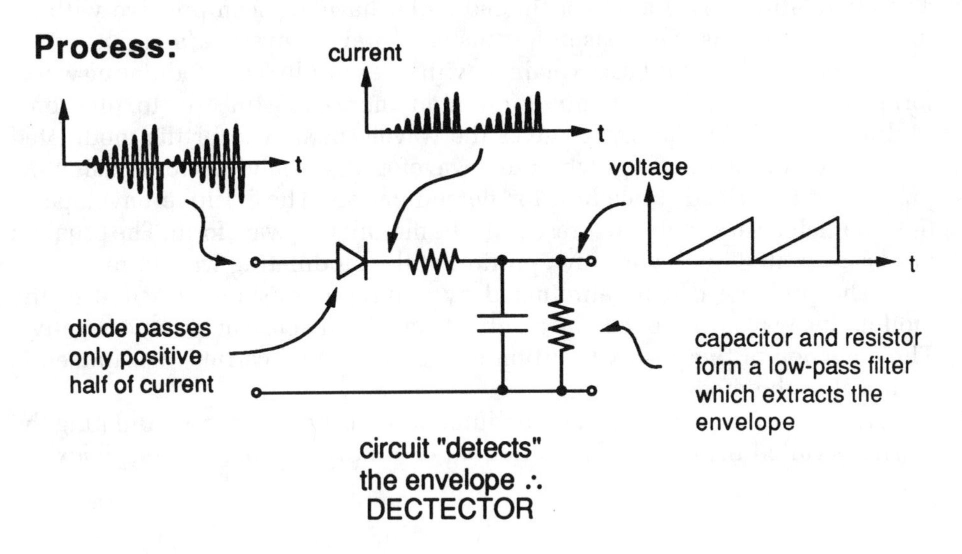

circuit "detects"
the envelope ∴
DECTECTOR

AM Radio Receiver:

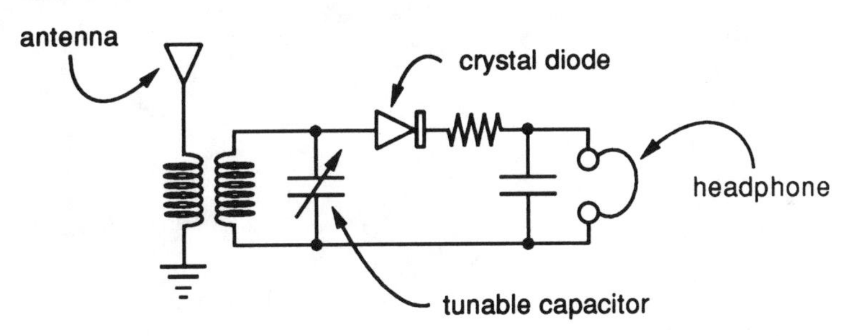

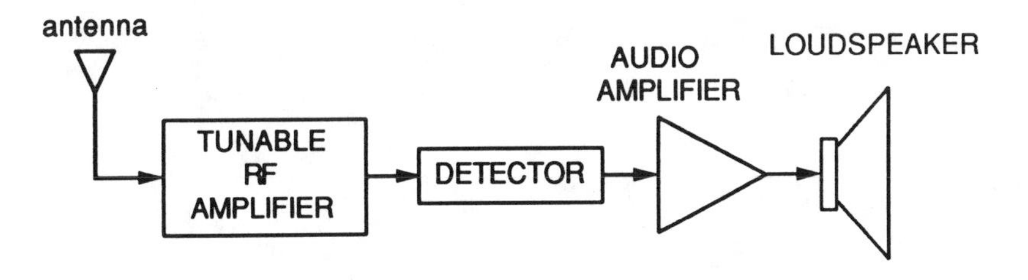

OVERMODULATION

The modulating waveforms considered so far have all been positive with no negative excursions. The reason for this is to prevent *overmodulation*.

Overmodulation is best explained with an example. The modulating waveform is a ramp starting at minus one and increasing linearly to plus one. Multiplying this by the carrier gives the waveform shown for the modulated carrier. The problem occurs when this waveform is demodulated by using the previously described procedure for demodulation. The positive envelope of the modulated carrier does not resemble the modulating waveform. The problem has been created by the negative portion of the modulating waveform.

The problem can be eliminated by adding a constant dc value to the modulating waveform so that the resultant waveform does not become negative. The envelope of the positive portion of the modulated carrier then indeed is identical to the modulating signal.

A measure of how close the modulation process is to overmodulating the carrier is called the index of modulation, or simply the *modulation index*.

Overmodulation

The Problem:

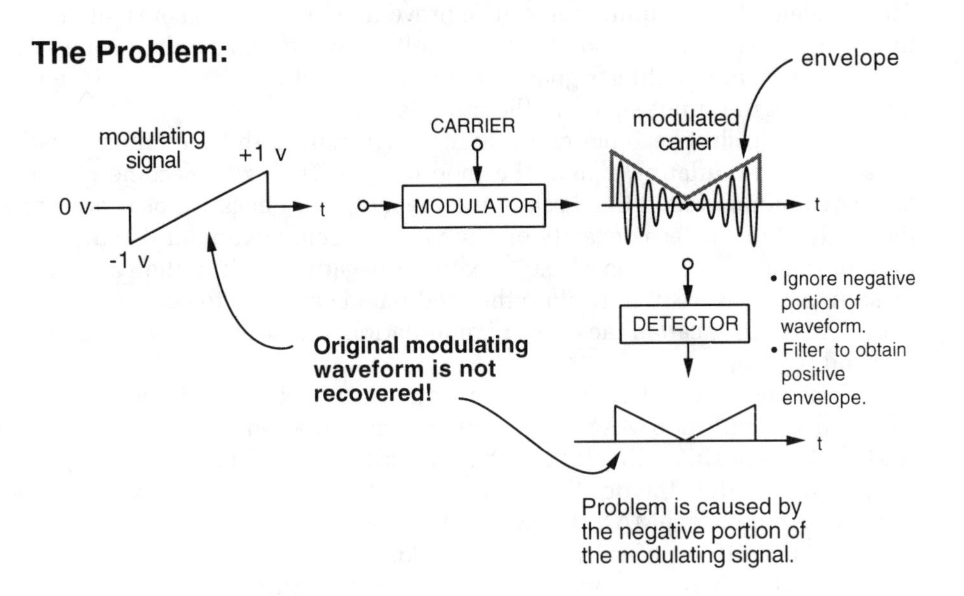

The Solution:

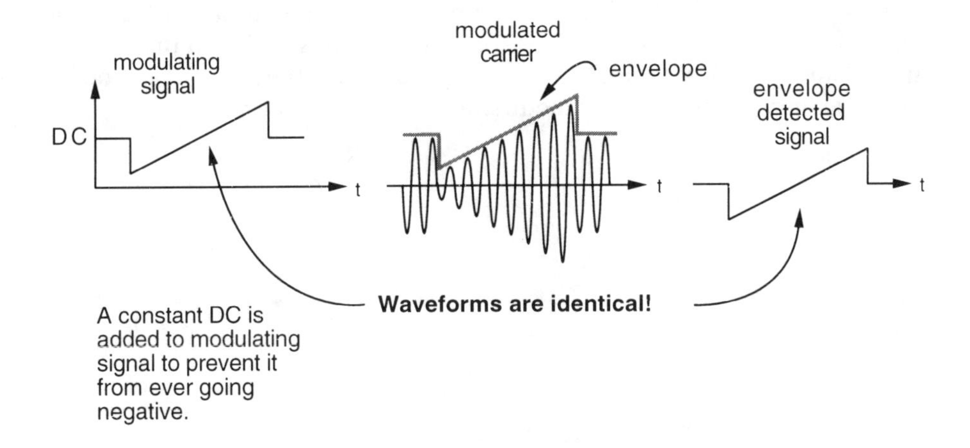

MODULATION INDEX

The problem of overmodulation can be prevented by adding a constant dc term to the modulating signal so that the resultant waveform is always positive. If $m(t)$ is the modulating signal, a constant dc value is added to it, giving $m(t)$ + DC as the input signal to the modulator.

As the dc shift becomes closer to the magnitude of the maximum negative peak in the modulating signal, the modulated carrier will become closer to being overmodulated. If the dc shift and the maximum negative peak are equal, the modulated carrier is exactly on the verge of being overmodulated. So, the ratio of the absolute value of the maximum negative peak to the value of the dc shift is a measure of how close the modulated carrier is to being overmodulated. This ratio is called the index of modulation, or, more simply, the modulation index.

Consider a modulating signal $m(t)$ that has positive and negative peaks less than an amplitude MAX. To prevent overmodulation, the dc shift DC must shift $m(t)$ in a positive direction so that the shifted waveform is never negative. To accomplish this, the dc shift DC must be at least equal to MAX. The index of modulation equals MAX divided by DC.

The magnitude of $m(t)$ in the preceding is MAX. Stated symbolically $|m(t)|$ = MAX, where the vertical lines indicate magnitude.

Assume that the modulating signal is normalized so that its magnitude is less than or equal to one, or $|m(t)| \leq 1$. In this case, a dc shift of one will ensure that overmodulation does not occur. The index of modulation then would be the maximum value of the magnitude of $m(t)$. If the modulating signal were a sinusoid with maximum amplitude of M, where $|M| \leq 1$, then if the input to the modulator were $[M \sin(2\pi ft) + 1]$, the amplitude M of the sinusoidal modulating signal would be the index of modulation.

Index of Modulation

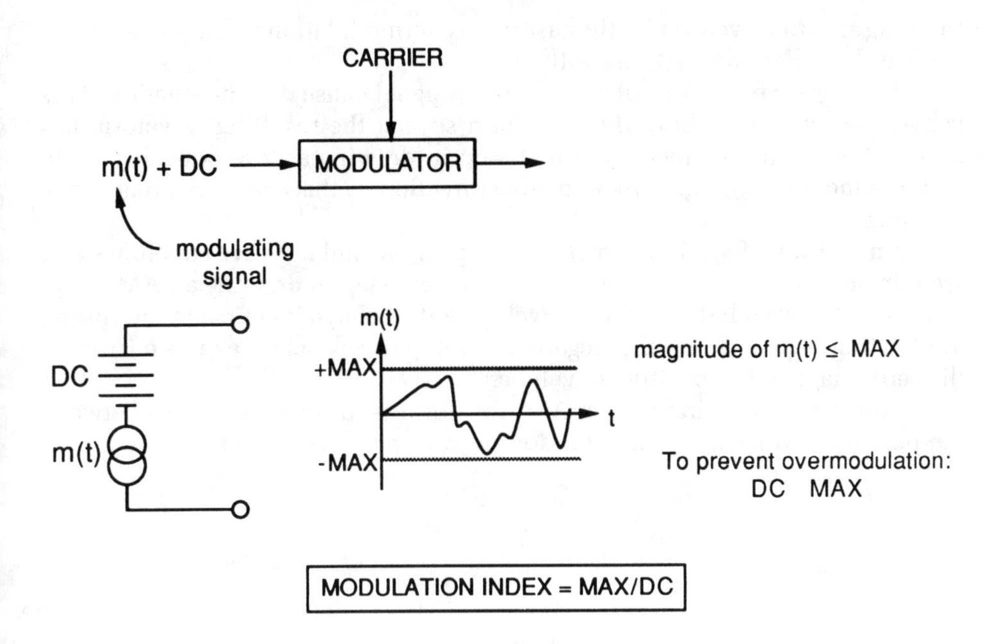

Examples:

for a
sinewave
modulating
signal

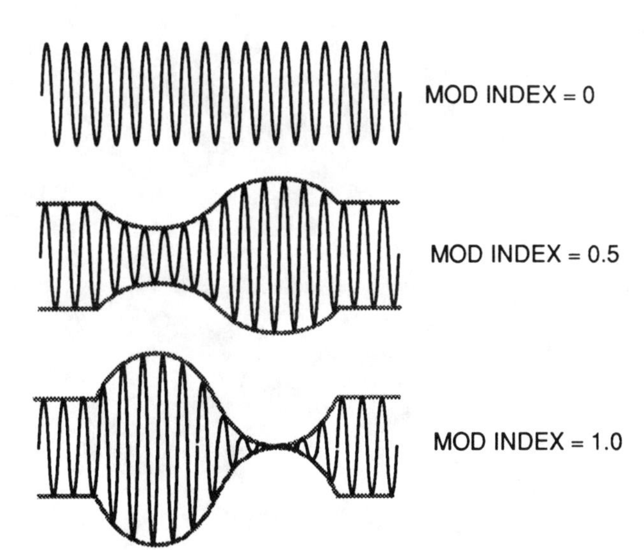

4

AM WAVEFORM

Drawing an AM waveform for the case of no overmodulation is fairly straightforward if the following steps are followed.

The negative portions of a modulating signal cause overmodulation. This is avoided by adding dc to the waveform so that the resulting waveform has no negative portions. Hence, the first step in drawing an AM waveform is to shift the modulating signal in a positive direction so that there are no negative portions.

An AM waveform is symmetric with positive and negative envelopes that are mirror images of each other. Hence, the next step in drawing an AM waveform is first to sketch the positive envelope with a shape identical to the shifted modulating signal. Then, the negative envelope is sketched as the mirror (or flipped) image of the positive envelope.

The last step in drawing an AM waveform is to fill in the space between the positive and negative envelopes with the sine wave carrier.

AM Waveform

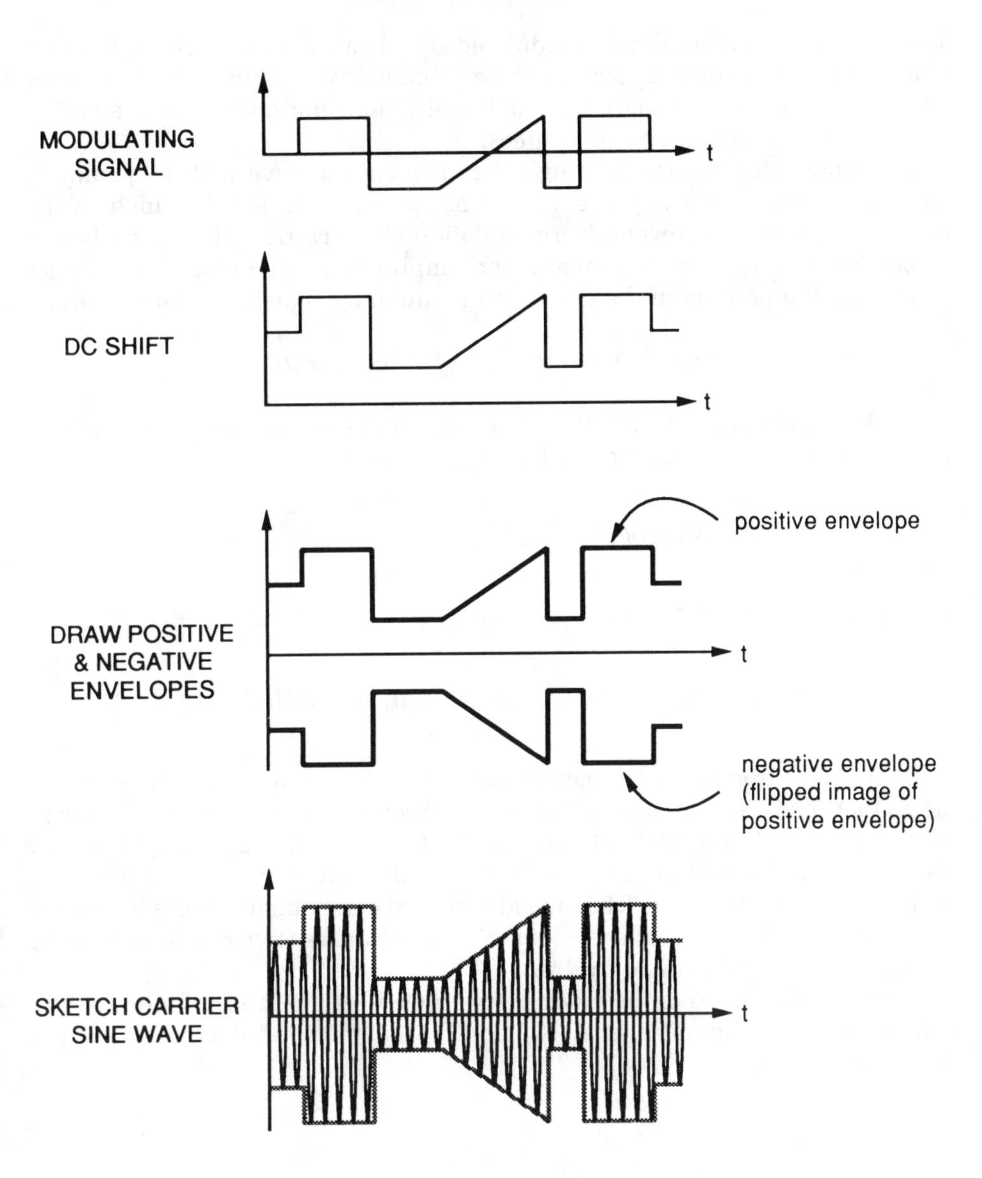

4

SPECTRUM

The spectrum of an amplitude-modulated signal can be calculated for the case of a sinusoidal modulating signal. The algebra is easier if cosine functions are used rather than sine functions, which really does not matter since a cosine wave is a sine wave shifted 90 degrees.

Suppose that the modulating signal is a cosine wave with frequency f_m and amplitude M, where $|M| \leq 1$. A constant dc level of 1 is added to the modulating signal to prevent overmodulation of the carrier. The carrier is also a cosine wave, but at a frequency f_c and amplitude A. The modulated carrier $x(t)$ equals the product of the dc-shifted modulating signal and the carrier, or

$$x(t) = A[1 + M\cos(2\pi f_m t)]\cos(2\pi f_c t)$$

This equation can be expressed in a different form by using the following trigonometric identity for the product of two cosines:

$$\cos(X)\cos(Y) = \frac{1}{2}\cos(X + Y) + \frac{1}{2}\cos(Y - X)$$

Letting $X = 2\pi f_m t$ and $Y = 2\pi f_c t$ gives, after a little algebra,

$$x(t) = A\cos(2\pi f_c t) + \frac{1}{2}AM[\cos 2\pi(f_c + f_m)t] + \frac{1}{2}AM[\cos 2\pi(f_c - f_m)t]$$

This equation is far less complicated than it may appear. The equation states that the result of using a sinusoid of frequency f_m to modulate a carrier with frequency f_c is a modulated carrier with three frequency components in its spectrum. The first component is simply the carrier itself at frequency f_c and having an amplitude A. The second and third components are at a frequency f_m above and below the carrier frequency, or $f_c \pm f_m$. These two components both have the same amplitude of $^1/_2 AM$.

The amplitude M of the modulating sinusoid is the index of modulation. The frequency components at $f_c + f_m$ and $f_c - f_m$ are called the upper and the lower *sidebands*.

Spectrum

Signals:

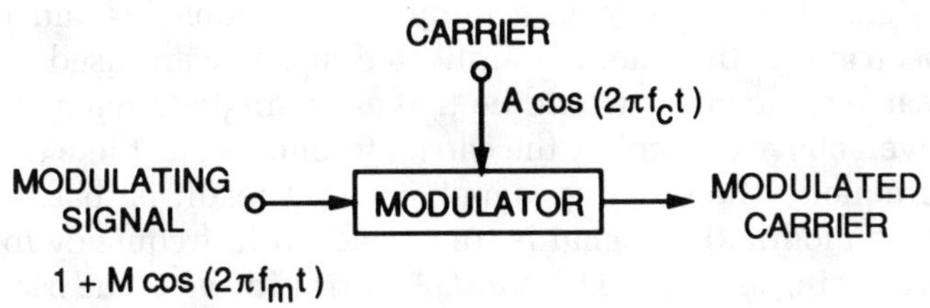

Spectrum:

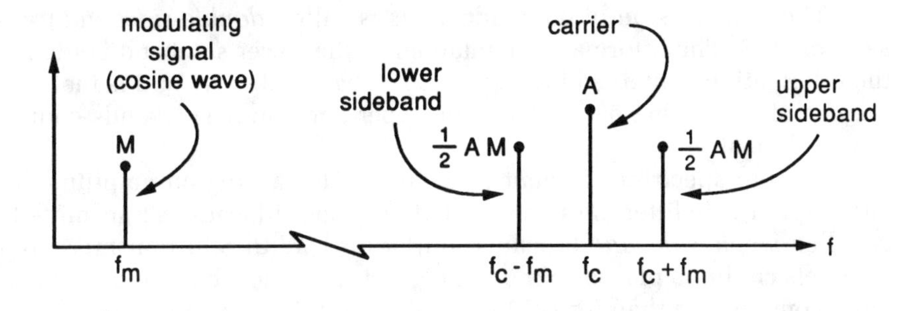

SIDEBANDS

The preceding derivation of the spectrum of an amplitude-modulated carrier was for a sinusoidal modulating signal at a frequency f_m. A more complicated modulating signal is composed of a number of frequencies and has its own baseband spectrum. If this more complicated signal were used to amplitude modulate a carrier, the effect would be that each single-frequency component would be moved above and below the carrier frequency, just like the derivation for a single-frequency modulating signal. The net result is that the baseband spectrum of the modulating signal is translated up in frequency to the carrier frequency, producing an *upper sideband.* A mirror image of the baseband spectrum is also flipped in frequency below the carrier frequency and is called the *lower sideband.*

If the original baseband signal had a bandwidth extending from 0 Hz to W Hz, the amplitude-modulated carrier would occupy a frequency band from $f_c - W$ to $f_c + W$. This is a bandwidth of $2W$ Hz. Hence, the bandwidth of an amplitude-modulated carrier is twice the maximum frequency in the baseband signal.

The transmission of both sidebands is called *double-sideband* transmission, or DSB. The information contained in the lower sideband is identical to the information contained in the upper sideband. Clearly, DSB is inefficient in the use of spectrum space, since the upper and lower sidebands contain the same information.

A more spectrum-efficient scheme of transmitting an amplitude-modulated carrier is to filter the signal so that only one sideband is transmitted. This is called *single-sideband* transmission, or SSB. With SSB transmission, more channels could be placed in the same spectrum space, but the demodulator is more complicated than for DSB transmission. SSB transmission was patented in 1923 by John R. Carson of AT&T.

The transmission of the carrier itself consumes a fair amount of transmission power and clearly does not transmit any information, other than perhaps that the station is still on the air. Another scheme for transmitting an AM signal is to suppress the carrier and transmit either both or only a single sideband. Single-sideband transmission with a *suppressed carrier* (SSB-SC) saves transmission power, but requires a more complicated modulator and demodulator. In the absence of a modulating signal, no carrier is transmitted in suppressed-carrier transmission.

Some of the complexities of modulation and demodulation that are encountered with SSB-SC transmission can be avoided if only a small portion of the lower sideband is transmitted along with the whole upper sideband. Since only a vestige of the lower sideband is transmitted, this method of transmission is called *vestigial* transmission. It is used for transmitting the picture portion of television.

Sidebands

Spectrum:

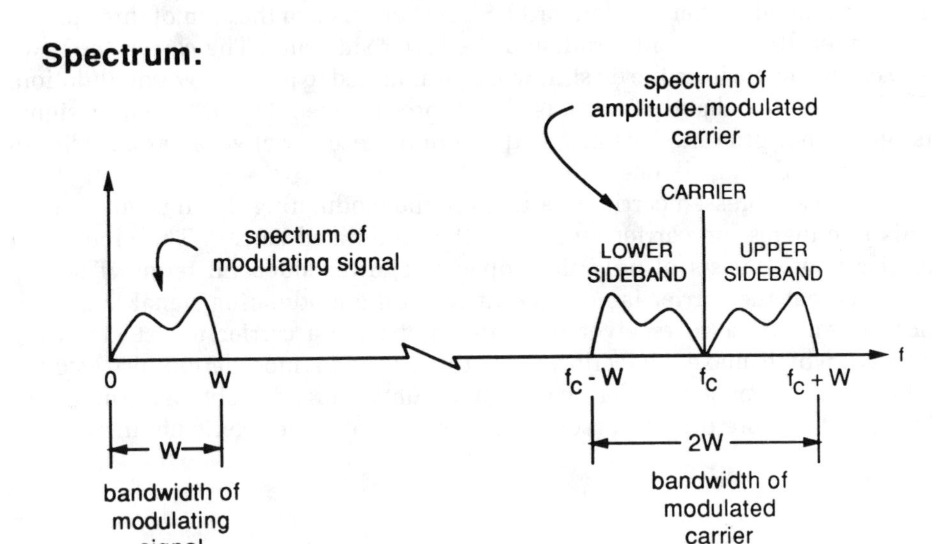

Transmission Schemes:

- Double Sideband (DSB)
  CARRIER + UPPER SIDEBAND + LOWER SIDEBAND
- Double-Sideband Suppressed Carrier (DSB-SC)
  UPPER SIDEBAND + LOWER SIDEBAND
- Single-Sideband Suppressed Carrier (SSB-SC)
  ONE SIDEBAND ONLY
- Vestigial Sideband
  UPPER SIDEBAND + PORTION OF LOWER SIDEBAND
  (+ CARRIER, IF NOT SUPPRESSED)

SUPPRESSED CARRIER

The modulated carrier for standard DSB AM consists of the sum of three terms: the carrier, the upper sideband, and the lower sideband. The carrier term, we saw earlier, came from the dc shift which was added to prevent overmodulation. The end result is that the carrier is always present even if the modulating signal is zero or not present. This means that a radio receiver always has a carrier to detect even during silence.

With suppressed carrier modulation, the modulating signal is not shifted. This eliminates the carrier term from the modulated carrier. The modulated carrier now consists of only the upper and lower sideband terms. The end result is that the carrier is not present when the modulating signal is zero or not present. A radio receiver now does not have a carrier to detect during silence, which makes detection more difficult. Overmodulation also occurs, which means that demodulation of the modulated carrier to obtain the modulating signal is more difficult, requiring more complex electronic circuitry.

Suppressed Carrier

Standard DSB AM:

MODULATED CARRIER = CARRIER + UPPER SIDEBAND + LOWER SIDEBAND

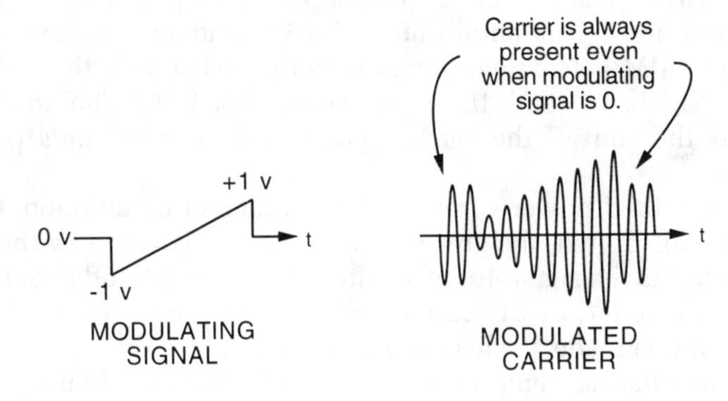

Suppressed Carrier:

MODULATED CARRIER = ~~CARRIER~~ + UPPER SIDEBAND + LOWER SIDEBAND

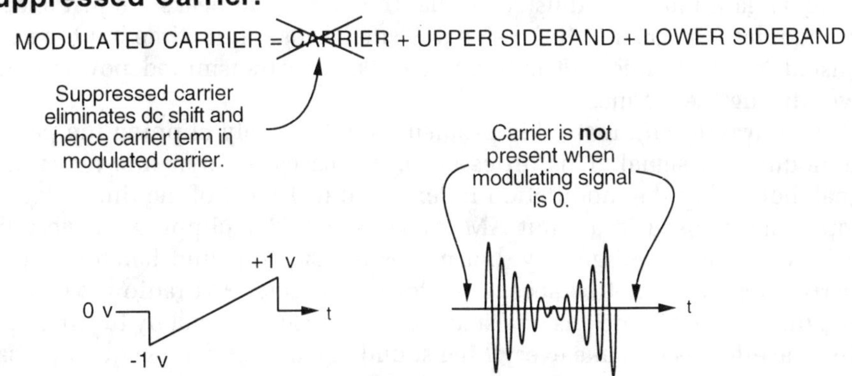

TRANSMITTED POWER

The power transmitted in an amplitude-modulated carrier can be calculated for the case of a sinusoidal modulating signal.

The rms value of a sine wave is the peak amplitude divided by the square root of two. Since power is proportional to the square of the rms value, the power of a sine wave is proportional to one-half the square of the peak amplitude.

Since the carrier has a peak amplitude of A, its power is $\frac{1}{2}A^2$. Each sideband sinusoid has a peak amplitude of $\frac{1}{2}AM$, and so the power in each sideband is $\frac{1}{2}(\frac{1}{4}A^2M^2)$. The total power in both sidebands is the sum of the power in each sideband, and this sum equals $\frac{1}{4}A^2M^2$. The total transmitted power is the sum of the carrier power and the sidebands power or $\frac{1}{2}A^2(1 + \frac{1}{2}M^2)$.

So, the transmitted power varies with the index of modulation. Clearly, the more power transmitted, the greater is the distance covered by the signal, and, also, the better is the signal-to-noise ratio. However, the index of modulation should not exceed 1, or overmodulation would occur. So, the index of modulation should be made as close to 1 as possible.

Since the modulating signal varies in its peaks, particularly if it is speech or music, the index of modulation likewise varies continuously during transmission. In order to have a modulation index as close to 1 as possible, the modulating-signal gain must be adjusted so that the maximum peaks are just about to overmodulate the carrier. If the signal is small most of the time but the gain is adjusted for the infrequent maximum peaks, the transmitted power will be lower during these times.

One way to eliminate this problem is deliberately clipping the peaks of the modulating signal. This, in essence, decreases the dynamic range of the signal, but makes the modulation index equal to 1 most of the time. This type of approach is used in aircraft AM transmission. The clipping has very little effect on speech intelligibility, but makes the signal sound harsh and raspy. Clearly, this would not be acceptable for commercial AM radio broadcasting. Here, the low-level portions of a signal are increased in level by turning up the gain. The effect is to make everything sound equally loud, even quiet passages of music, but maximum power would always be transmitted to ensure reception over the greatest distance.

Power

RMS Review:

$$\text{Power} = \text{RMS}^2$$

$$\text{but: RMS} = \frac{\text{Peak}}{\sqrt{2}}$$

$$\text{therefore: Power} = \frac{\text{Peak}^2}{2}$$

AM Power:

Total Power = Carrier Power + Upper Sideband Power + Lower Sideband Power

$$\text{Total Power} = \frac{A^2}{2} + \frac{1}{2}\left(\frac{AM}{2}\right)^2 + \frac{1}{2}\left(\frac{AM}{2}\right)^2$$

$$\text{Total Power} = \frac{A^2}{2}\left[1 + \frac{M^2}{2}\right]$$

SUPERHETERODYNE RECEIVER

The *superheterodyne* AM radio receiver was invented by U.S. Army Major Edwin H. Armstrong during World War I. The patent was filed on December 30, 1918 and granted on June 8, 1920.

The basic principle of operation is to beat (or heterodyne) the radio-frequency (RF) signal picked up by the antenna with a local oscillator to reduce all received signals to a single signal in a narrow intermediate-frequency (IF) band. A fixed-frequency, highly tuned amplifier can then be used to amplify the received signal before detection.

The AM broadcast band occupies a band of frequencies from 550 kHz to 1,600 kHz, and each station occupies a band 5 kHz above and below its assigned carrier frequency. The total AM band of frequencies is picked up by an antenna. The particular desired station is selected by a tuned RF amplifier operating over a band of frequencies from f_c–5 kHz to f_c+5 kHz. The center frequency f_c is varied, usually by varying the capacitance in a high-Q resonant circuit.

At the same time that the RF amplifier is tuned, the frequency of a local oscillator in the radio is also tuned, but to a frequency f_c–455 kHz. The output of this oscillator and that of the RF amplifier are then multiplied together in a circuit called a *mixer*. The result of this mixing operation is to generate sum-and-difference frequencies of the two input signals in a manner similar to conventional amplitude modulation. The difference frequencies will occupy a range from 455–5 kHz to 455+5 kHz. The result is that the carrier frequency has been eliminated and the mixed signal occupies a fixed band of frequencies that is 10 kHz wide, centered about 455 kHz. This band is amplified by fixed-frequency, highly tuned amplifiers that can be designed to be very efficient.

The output of the IF amplifier is detected by a conventional diode detector, filtered, and amplified by an audio-frequency (AF) amplifier for input to a loudspeaker.

Superheterodyne Receiver

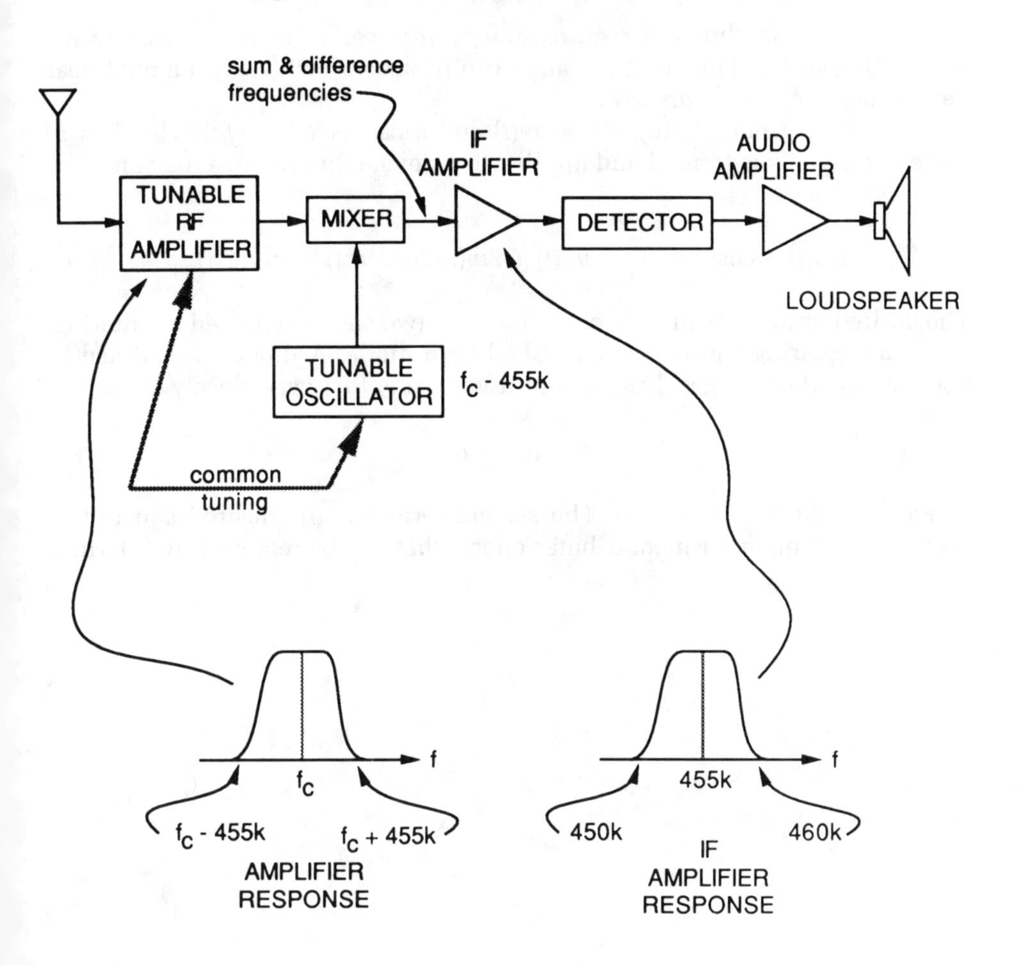

MODULATORS

The process of modulation requires the multiplication of the carrier by the modulating signal. This multiplication can be accomplished with a nonlinear device that obeys a *square law.*

Consider a modulating signal $m(t)$ and a carrier $\cos(2\pi f_c t)$. The sum of these two signals is formed and applied to a device that squares the sum. The result is

$$[m(t) + \cos(2\pi f_c t)]^2 = m^2(t) + 2m(t)\cos(2\pi f_c t) + \cos^2(2\pi f_c t)$$

The desired term is the middle one. The other two can be removed by filtering.

The squaring can be accomplished by a diode. A diode, if confined to operate in certain ranges, has an output voltage e_o that very closely is

$$e_o = a_1 e_i + a_2 e_i^2$$

where e_i is the input voltage. The second term has the desired square-law response. The other term contributes effects that can be removed by filtering.

Modulators

Square-Law Modulator:

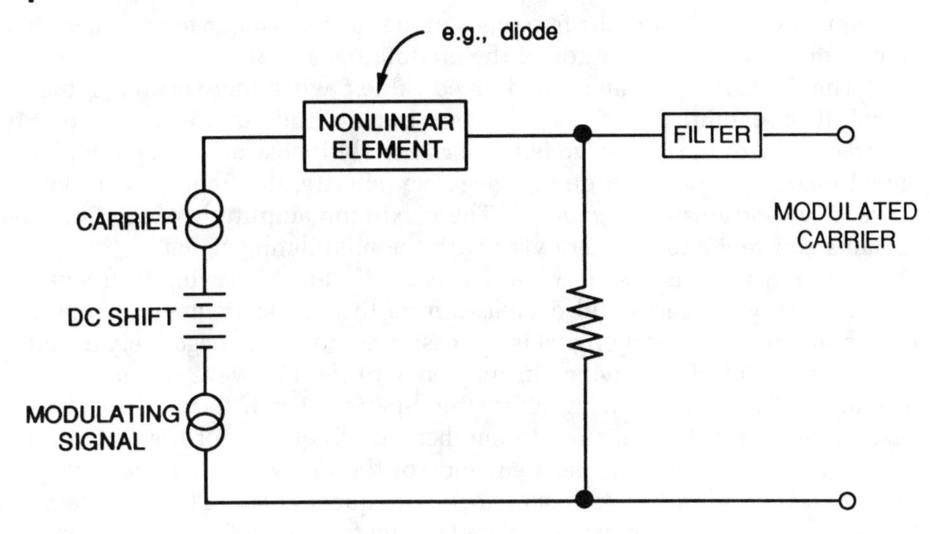

Frequency Modulation

PROCESS

In frequency modulation, the frequency of the carrier is made to vary in proportion to the changing amplitude of the modulating signal.

The FM carrier has an unmodulated value f_c when there is no modulating signal. If the modulating signal increases in amplitude in a positive polarity, the frequency of the FM wave is proportionately increased. If the modulating signal increases in amplitude in a negative polarity, the frequency of the FM wave is proportionately decreased. The maximum amplitude of the FM wave remains constant and does not vary with the modulating signal.

The FM wave is a sine wave that is continuously varying in frequency. The frequency increases and decreases from the carrier frequency, according to whether the modulating signal is increasing positively or negatively in amplitude. The extent of the swings in frequency of the FM wave depends on the extent of the amplitude swings of the modulating signal.

From one instant in time to another, the frequency of the FM wave is constantly varying. The actual frequency of the FM wave at some particular instant in time is called its instantaneous frequency, $f_i(t)$. The instantaneous frequency equals the sum of the constant carrier frequency f_c plus some constant k times the modulating signal $m(t)$, or

$$f_i(t) = f_c + km(t)$$

Since the amplitude A of the modulated carrier is constant, the transmitted power is likewise constant, and, unlike amplitude modulation, does not vary with the modulating signal.

Frequency Modulation

Example:

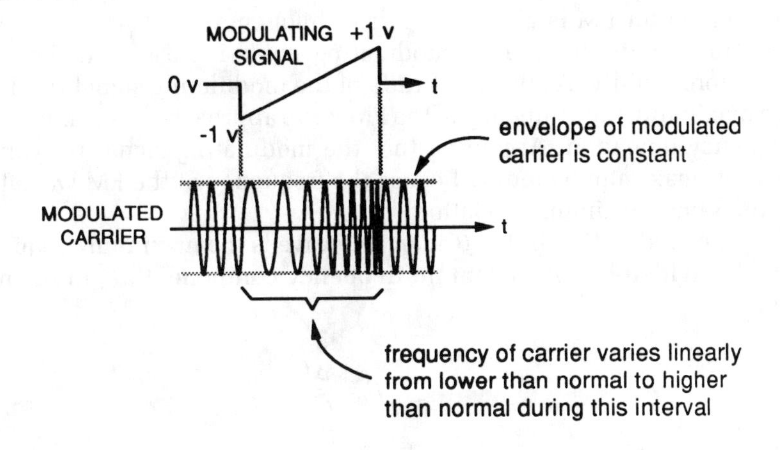

Power:

constant maximum amplitude of carrier

$$\text{TRANSMITTED POWER} = \frac{A^2}{2}$$

MODULATION INDEX

The modulation index for an amplitude-modulated signal was a ratio of amplitudes, since the amplitude of the carrier is varied in AM. There is a similar concept to express the degree of modulation for FM, except that the index of modulation for FM is given in terms of frequency.

As the amplitude of the modulating signal increases, the FM wave oscillates more rapidly. As the amplitude of the modulating signal reaches its positive maximum, the frequency of the FM wave approaches its maximum positive-frequency deviation. Assuming that the modulating signal is symmetric, the negative maximum value will cause the frequency of the FM wave to decrease by the same maximum deviation.

The modulation index β of an FM wave is the maximum frequency deviation Δf divided by the maximum frequency component f_{max} in the modulating signal or

$$\beta = \frac{\Delta f}{f_{max}}$$

The carrier frequency varies from a low of $f_c - \Delta f$ to a high of $f_c + \Delta f$.

The modulation index for FM is a measure of how far the carrier swings in frequency compared to the maximum frequency in the modulating signal.

Modulation Index

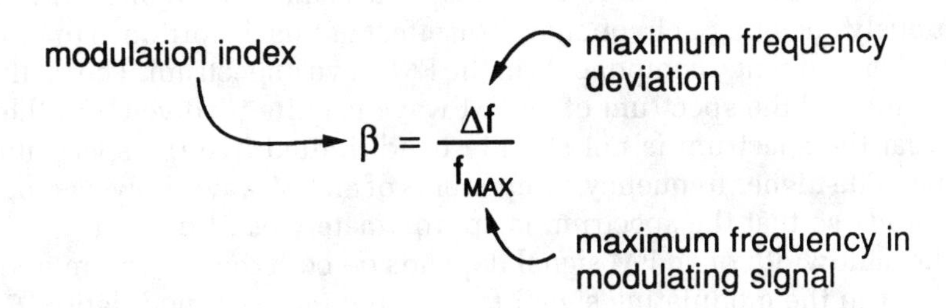

BANDWIDTH

An FM wave has a very complicated shape. It is a distorted sine wave that is continuously changing in frequency. The effect of this distortion is to introduce many higher frequency components in the FM wave's spectrum. The mathematical derivation of the spectrum of an FM wave is quite involved, but the result shows that the spectrum is not sharply band-limited like the spectrum of an AM wave. The higher frequency components of an FM wave, however, decrease in amplitude so that the spectrum is approximately band-limited.

The bandwidth of an FM signal depends on both the maximum frequency component in the modulating signal f_{max} and the index of modulation β. However, if the modulation index is small, typically equal to or less than 0.2, the bandwidth of the FM wave simply is twice f_{max}. For other indices of modulation, the situation is not so simple, and the bandwidth depends on both f_{max} and the frequency deviation Δf.

FM transmission in which the bandwidth is twice f_{max} is called *narrowband FM*. FM transmission for which the bandwidth is greater than twice f_{max} is called *wideband FM*.

A rule of thumb for large indices of modulation ($\beta \geq 1$) is that the bandwidth of an FM wave is twice the maximum frequency deviation plus twice the maximum frequency component in the modulating signal. In mathematical terms, the bandwidth BW of an FM wave is

$$BW \approx 2\Delta f + 2f_{max} = 2f_{max}(1 + \beta)$$

where Δf is the maximum frequency deviation, f_{max} is the maximum frequency component in the modulating signal, and β is the modulation index.

The bandwidth varies with the frequency deviation, and so a limit must be placed on its maximum value so that each FM radio station remains within its allocated frequency space. The FCC fixes the maximum frequency deviation at 75 kHz. Since the maximum frequency component in the baseband signal for FM transmission is 15 kHz, the bandwidth for a commercial FM radio station is about 2(75k + 15k) = 180 kHz. Adding a 10 kHz guard band at each end of the spectrum gives a total frequency space of 200 kHz for an FM radio station.

Bandwidth

Spectra:

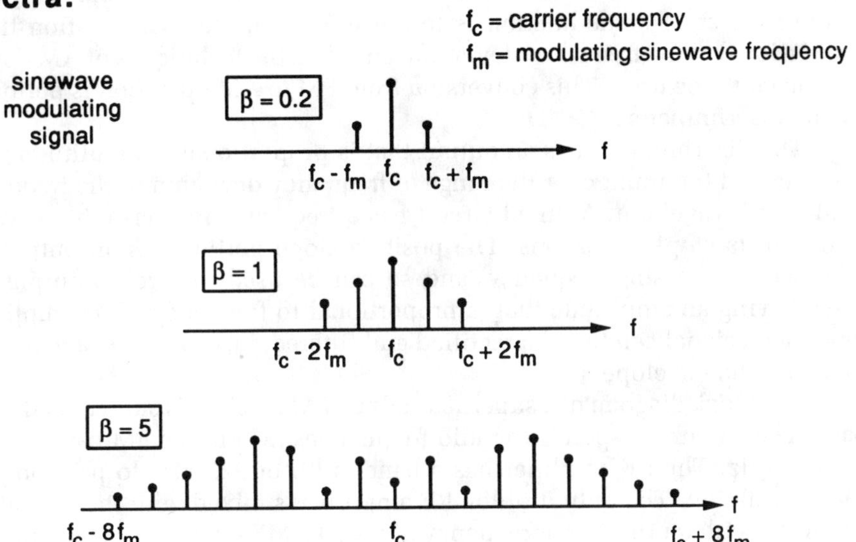

Rule of Thumb:

$$\text{Bandwidth} \approx 2 f_{MAX} (1 + \beta) = 2 \Delta f + 2 f_{MAX}$$

- $2 f_{MAX}(1+\beta)$: for large indices of modulation ($\beta \geq 1$)
- Δf: maximum frequency deviation
- f_{MAX}: maximum frequency component in modulating signal

FM Broadcast Radio:

$\Delta f = 75\text{kHz}$
$f_{MAX} = 15\text{kHz}$

bandwidth $\approx$ 2 (75 + 15) kHz = 180kHz
+ 20kHz guard band
200kHz

4

DEMODULATION

The amplitude fluctuations of the modulating signal are encoded in FM as a change, or deviation, in the frequency of the carrier. One simple approach to the process of demodulation is to create an amplitude fluctuation that is proportional to the frequency deviation, and then the techniques of AM demodulation could be used. This conversion from FM to AM operation is performed by the discriminator.

The discriminator has an output that is proportional in amplitude to the frequency of the input over the range of frequency deviation of the frequency-modulated waveform. A tuned circuit has a frequency response that is nearly linear in its sloping portions. The positive slope portion has an output that rises with increasing frequency, and so can be used to give an output sine wave having an amplitude that is proportional to frequency. This amplitude-modulated signal can then be rectified and filtered, just like an AM waveform, to obtain the envelope.

The block diagram of a superheterodyne FM receiver is shown on the next page. The antenna responds to radio frequencies (RF) in the FM band from 88 to 108 MHz. The RF amplifier has a bandwidth of 225 kHz to pass only the selected station. The output of the RF amplifier is mixed with the output of a variable oscillator that has a frequency of $f_c - 10.7$ MHz, where f_c is the unmodulated carrier of the station to which the RF amplifier is tuned. The difference-frequency output from the mixer is centered about 10.7 MHz with a bandwidth of 225 kHz. This signal is amplified by the intermediate-frequency (IF) amplifier.

Any noise or other amplitude variations in the FM output signal from the IF amplifier are removed by clipping the peaks of the signal, and then filtering it to obtain a very clean, constant-amplitude, frequency-modulated IF signal. This operation is called limiting and is performed by the limiter circuit. This signal is used as input to the discriminator. The final output from the discriminator is an audio-frequency signal that is amplified, and then converted to sound waves by a loudspeaker.

High frequency hiss can sometimes creep into the audio signal in FM transmission. The effect of this noise can be reduced by pre-emphasizing the high frequencies in the baseband signal before modulation of the carrier, and then de-emphasizing the high modulation of the carrier and the high frequencies in the frequency-demodulated audio signal. The final audio signal is the same as the transmitted baseband signal, but the effect of any high frequency noise has been filtered out by this process.

Demodulation

Discriminator:

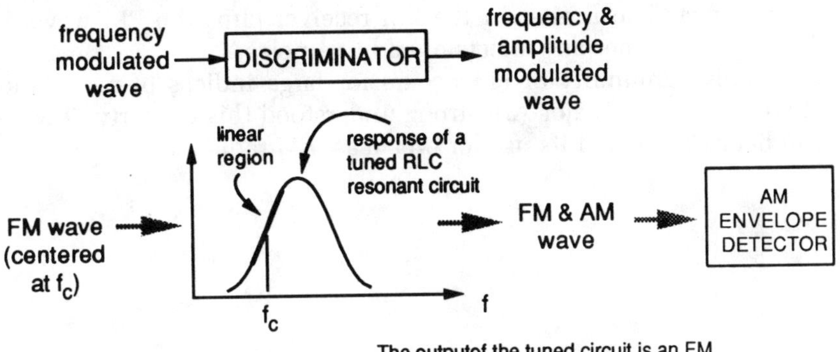

The outputof the tuned circuit is an FM wave that also varies in amplitude. In effect, the FM wave has been converted into an AM wave.

FM Superheterodyne Receiver:

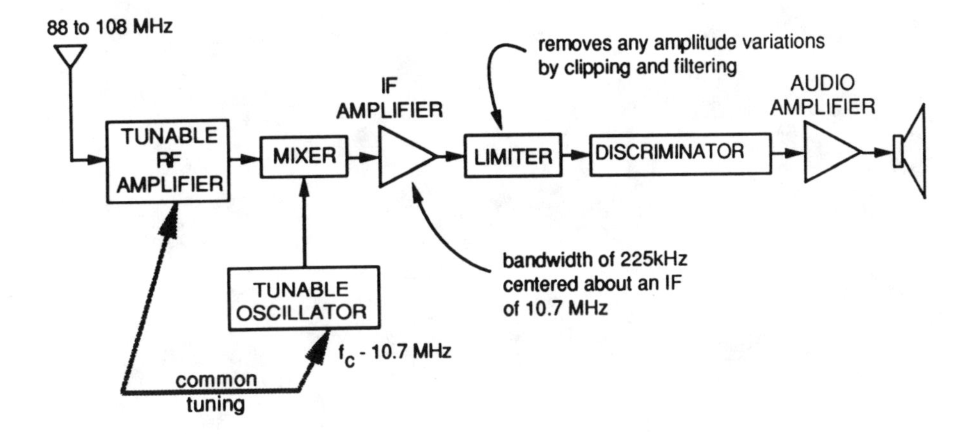

EFFECTS OF NOISE

Frequency modulation is very resistant to the effects of additive noise on the modulated carrier waveform. Since it is frequency deviation and not amplitude variation that carries the information in an FM wave, the additive noise has very little effect. The limiter in the FM receiver clips the FM wave, which eliminates even further any effects of additive noise.

The noise immunity of FM occurs for large indices of modulation, so called wideband FM. Major Armstrong understood this property of wideband FM and hence promoted its use for broadcast FM radio.

Noise

Amplitude Modulation:

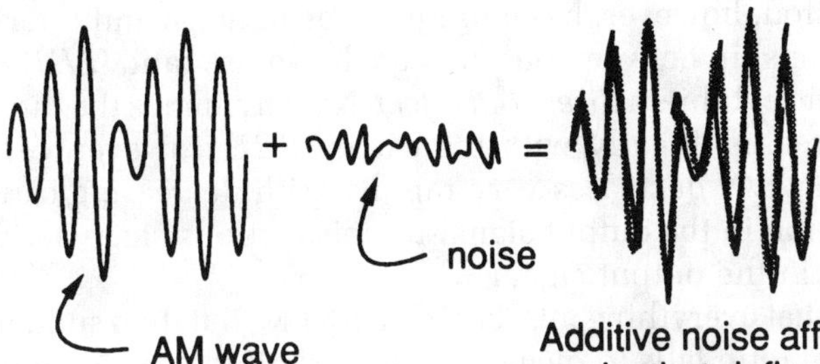

Additive noise affects the peaks that define the envelope. Since the envelope encodes the information, the noise has a large effect.

Frequency Modulation:

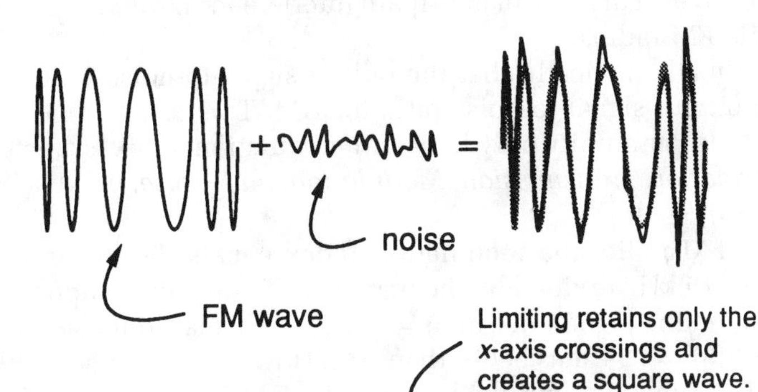

Additive noise affects the peaks but has little effect on the frequency which is determined by where the wave crosses the *x*-axis.

Limiting retains only the *x*-axis crossings and creates a square wave.

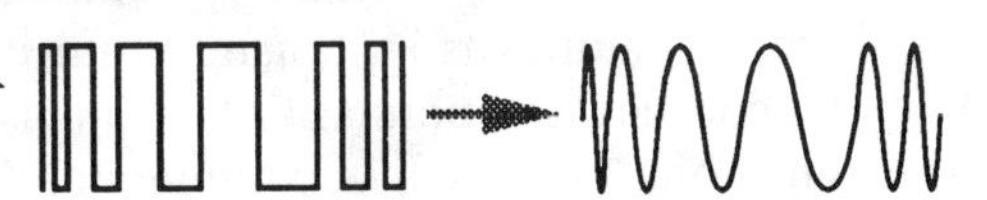

Filtering the square wave makes the original FM wave noise-free.

EFFECTS OF NOISE (cont'd)

AM radio transmission is very susceptible to *impulse noise*, such as that caused by the electromagnetic effects of lightning, but FM radio transmission is not.

FM radio transmission, however, has a unique type of noise and interference problem. The relationship between output signal-to-noise ratio (*S/N*) and carrier signal-to-noise ratio exhibits a *threshold effect*. Namely, above the threshold, the output *S/N* increases linearly with the carrier *S/N*. However, below this threshold, the output *S/N* decreases very rapidly with decreasing carrier *S/N*. This rapid degradation in the output signal-to-noise ratio is characterized by spikes or large clicks in the output signal.

What this means is that everything may be fine with FM, but then suddenly the output signal-to-noise ratio falls to pieces, particularly if the carrier amplitude becomes too close to the noise level. One particular type of interference that can reduce the carrier amplitude relative to the noise is multipath interference.

Multipath interference occurs when the FM radio wave bounces off some object and reaches the receiving antenna by two different paths. If the path lengths differ by the appropriate amount, the reflected signal arrives 180 degrees out of phase with the direct-path signal. The two waves cancel, but the received noise adds noncoherently and does not cancel. The carrier signal-to-noise ratio then suffers abruptly, falls below the threshold, and a sudden burst of noise comes from the FM receiver. This type of multipath interference is often encountered with automobile FM radios.

It can be shown mathematically that the output signal-to-noise ratio for FM is $3\beta^2$ times the output signal-to-noise ratio for AM. This means that FM is superior to AM in noise immunity only if the FM modulation index is greater than $\sqrt{1/3}$. (See *Information Transmission, Modulation, and Noise,* by Mischa Schwartz, McGraw-Hill, 1959.)

For commercial FM radio, the modulation index equals the maximum frequency deviation of 75 kHz divided by the maximum frequency component of 15 kHz in the modulating signal, or 75/15 = 5. So, the signal-to-noise ratio of commercial FM is $3(5)^2 = 75$ times better than AM. However, the bandwidth of commercial FM is 13 times f_{max}, but the bandwidth of AM is only twice f_{max}. In effect, FM achieves its noise immunity at the expense of bandwidth.

Of course, commercial FM broadcasts a baseband signal with a maximum frequency of 15 kHz, while commercial AM radio broadcasts a maximum baseband frequency of only 5 kHz. Commercial FM radio is then much higher in fidelity than AM radio, even if all other factors were equal.

Noise (cont'd)

Threshold Effect:

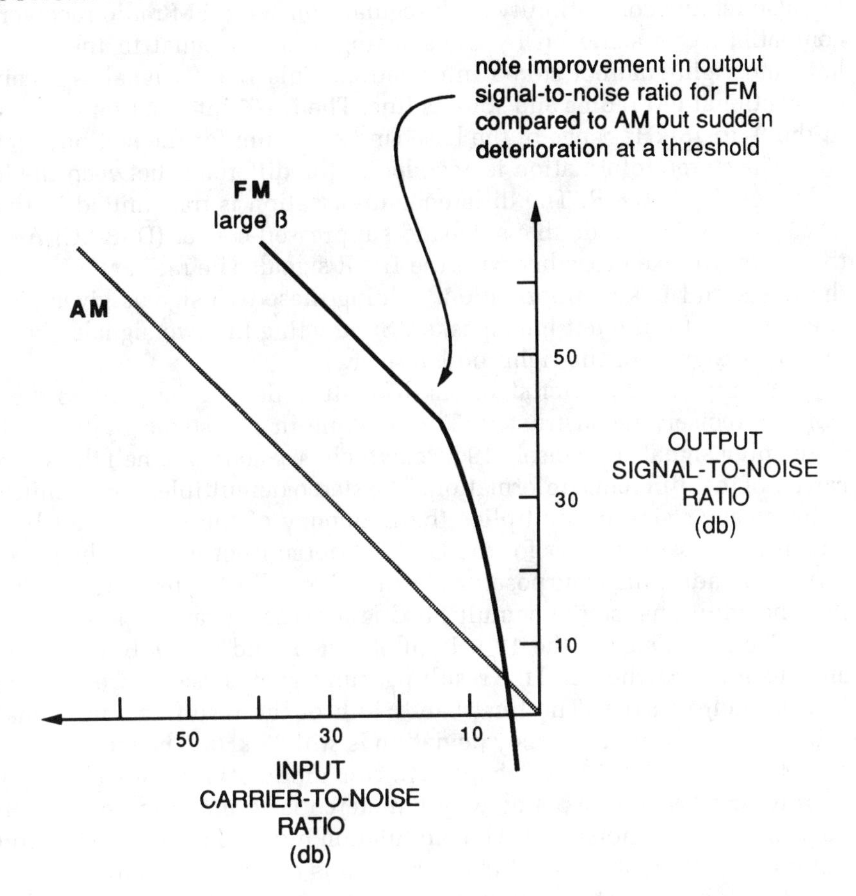

FM STEREO

FM stereo transmission not only transmits a signal that contains the separate left-channel and right-channel information needed for the stereophonic effect, but also retains compatibility with regular nonstereo FM radio receivers. This compatibility is achieved by transmitting a signal equal to the sum of the left- and right-channel stereo information. This L + R signal is received by conventional FM radios and sounds fine. The L + R information is transmitted in the 0- to 15-kHz range of the baseband spectrum for the station.

The stereo information is encoded as the difference between the left and right signals, or L − R. The difference information is transmitted in the range of 23 to 53 kHz as a double-sideband suppressed-carrier (DSB-SC) AM signal that is multiplexed together with the L + R signal. The radio receiver decodes the L − R and L + R information. Adding these two signals gives 2L, or the stereo signal for the left loudspeaker. Subtracting the two signals gives 2R, or the stereo signal for the right loudspeaker.

A suppressed-carrier signal is difficult to demodulate unless the carrier can be precisely reconstructed. This is done in FM stereo by transmitting a stereo pilot signal at precisely 19 kHz, which is exactly one-half the suppressed carrier of the difference information. The stereo demultiplexing circuitry in the radio receiver simply multiplies the frequency of the stereo pilot by two to obtain the missing carrier for the DSB-SC stereo information. The stereo pilot serves the additional purpose of alerting the radio to stereo transmission so that the additional stereo demultiplexing processing can be performed.

The L + R signal, the 19 kHz pilot signal, and the DSB-SC L − R signal are all added together, and the resulting sum signal is used to frequency modulate the radio carrier. The total bandwidth of the resulting sum signal is 53 kHz. The maximum frequency deviation is still 75 kHz. The modulation index of FM stereo is 75k/53k or about 1.4. This modulation index corresponds to narrowband FM and explains why FM stereo is so much more susceptible to noise than conventional FM. The modulation index of conventional, nonstereo FM radio is 75k/15k or 5, which corresponds to wideband FM.

In addition to stereo information, other information can be multiplexed into an FM transmission. The FCC has authorized FM stations to multiplex information into a range of frequencies from 53 to 99 kHz. This subsidiary communications authorization (SCA) can be used to transmit music or digital data. The FM radio stations encode the SCA transmissions, which can then be received by subscribers equipped with decoders to receive the subsidiary channel.

FM Stereo

Waveforms:

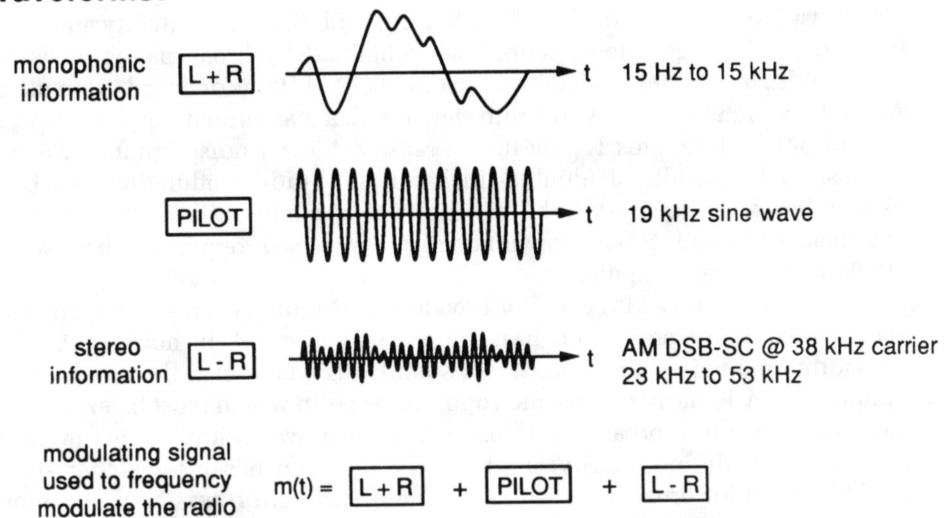

Spectrum:

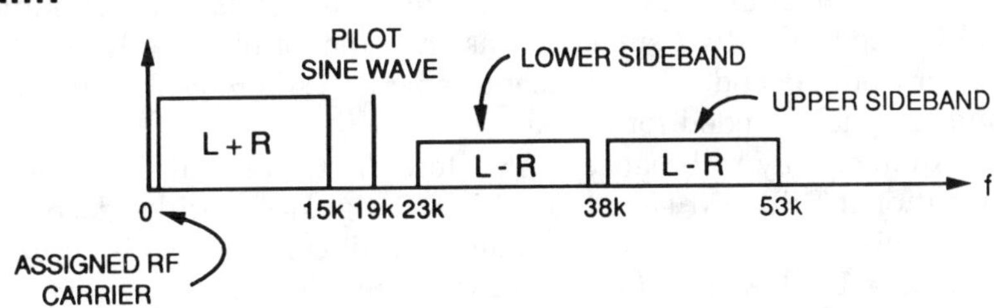

Decoding:

L + R + L - R = 2 L

L + R - L - R = 2 R

Other Modulation Topics

AM STEREO

Broadcast FM radio was designed to offer high-fidelity sound quality and noise immunity. The high-fidelity sound was obtained by broadcasting an audio signal with a maximum frequency component of 15 kHz, as opposed to broadcast AM radio which broadcasts an audio signal with a maximum frequency component of only 5 kHz, clearly low-fi. FM radio achieved noise immunity at the expense of bandwidth. If double-sideband amplitude modulation had been used to broadcast the 15-kHz hi-fi audio, a total bandwidth of 30 kHz would have been sufficient. However, wideband FM was used, requiring a bandwidth of 200 kHz per radio signal.

The extra bandwidth used for broadcast FM radio became an expeditious way to broadcast stereo, though at the expense of a loss in noise immunity. The additional enjoyment from stereo sound was well worth the loss in noise immunity, particularly in large metropolitan areas in which most listeners were close enough to the broadcast antenna that noise was not a serious problem anyway. With the popularity of stereo, the question became whether stereo could be used for broadcast AM radio. The first question was how to achieve stereo transmission while maintaining backward compatibility with all the existing mono radio receivers. A second question was whether consumers would really care given the already low-fi quality of AM radio.

One way to enable stereo transmission is to broadcast the separate left and right signals as separate single-sideband signals. An old radio receiver would decode the two sidebands as the compatible L + R signal. A stereo receiver would decode each sideband separately, thereby obtaining the separate L and R signals needed for stereo.

Another way to broadcast AM stereo is to transmit a L + R signal as a conventional DSB AM signal. This L + R signal would be decoded by older radio receivers and would assure compatibility. The stereo information would be sent as a L – R signal that would frequency modulate the AM carrier using narrowband frequency modulation. The AM carrier would be simultaneously amplitude and frequency modulated with this scheme.

AM Stereo

Sideband Encoding:

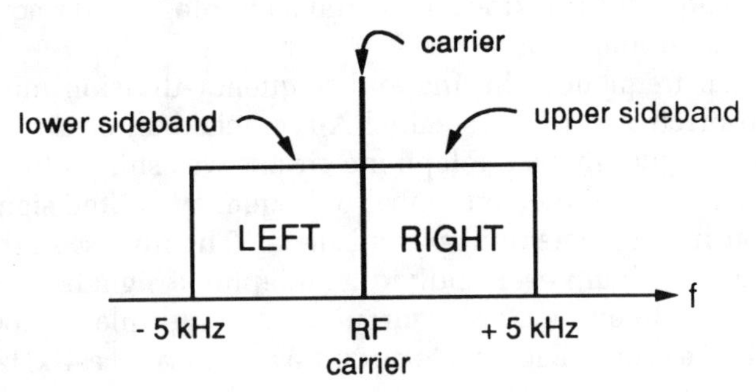

AM-FM Encoding:

L + R → AM-DSB (10 kHz BW)

L - R → FM (10 kHz BW)

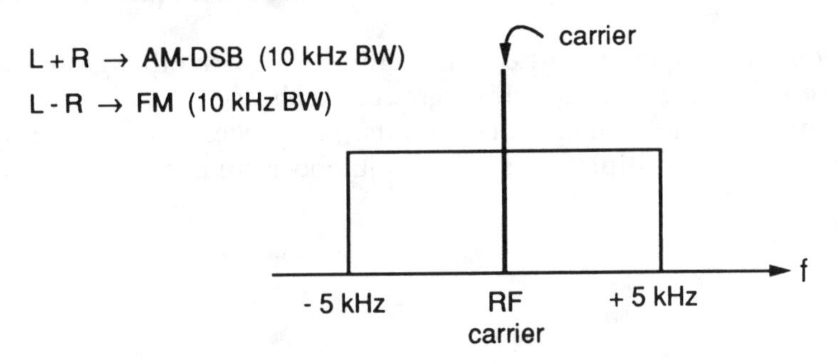

4

TELEPHONY

Frequency-division multiplexing was used extensively in the transmission of telephone signals across the country and between continents. All this changed with time-division multiplexing and digital technology, and frequency-division multiplexing is no longer used.

The actual frequency shifting and frequency-division multiplexing was performed in electronic devices called A-type channel banks. The "A" stood for analog. The signal in each telephone circuit was shifted to its own unique band of frequencies and then a number of frequency-shifted signals were combined together into a single multiplexed signal. The inverse of the process was then performed to obtain each individual telephone signal.

A total of 12 telephone voice circuits were multiplexed together to form what was called a group. Each voice circuit was allocated a 4-kHz band. Single-sideband, suppressed-carrier, amplitude modulation was used to shift each voice circuit to its unique band of frequencies. SSB-SC was used since it is the most bandwidth efficient way to accomplish frequency shifting and multiplexing. The 12 voice circuits in a group occupied a band of frequencies from 60 to 108 kHz.

Groups would be frequency shifted and multiplexed with other groups to form what were called supergroups. Five groups, each of 12 voice circuits, formed a supergroup of a total of 60 voice circuits. The process was frequently continued. The ultimate multiplex was called a jumbogroup combining 3,600 voice circuits.

Telephony

A-type Channel Bank:

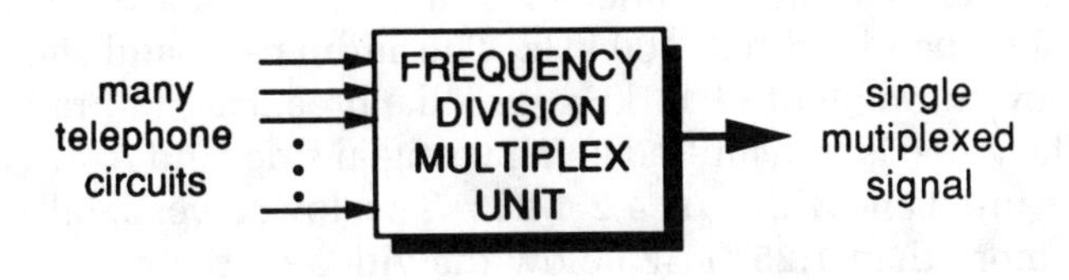

Group:

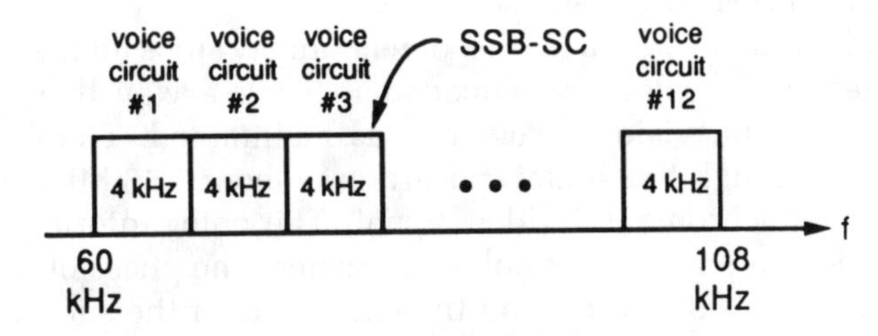

12 circuits multiplexed together is called a GROUP.

TELEVISION

From a technological perspective, television is truly radio with pictures since the video information is sent separately from the audio information. The audio is transmitted as FM at an audio carrier located exactly 4.5 MHz above the video carrier. The audio sidebands are 50 kHz above and below the audio carrier for a bandwidth of 100 kHz. The audio baseband signal has a maximum frequency component of 15 kHz just like broadcast FM radio.

The video is transmitted using vestigial sideband AM. The upper sideband has the full bandwidth of 4.2 MHz. The lower vestigial sideband does not extend more than 1.25 MHz below the video carrier.

The band allocated each television station can be calculated as the sum of the vestigial sideband of 1.25 MHz, the audio carrier of 4.5 MHz, the upper audio sideband of 0.05 MHz, and a guard band of 0.20 MHz. The band allocated each television station thus is 6.0 MHz. The use of vestigial AM saves about 3 MHz if DSB-AM had been used.

Color television is a marvel of technology since all the additional information needed for color was encoded in such a way that compatibility with monochrome television receivers was maintained. The color information is encoded in a high-frequency subcarrier at about 3.56 MHz that is added to the normal monochrome TV video signal. The color information modulates the amplitude and phase of the color subcarrier. The phase of the subcarrier represents the hue of the color, and the amplitude of the subcarrier represents the saturation of the color. If you wish to learn more about monochrome and color television, please read my book *Television Technology: Fundamentals and Future Prospects*, Artech House (1988).

Television

Spectrum:

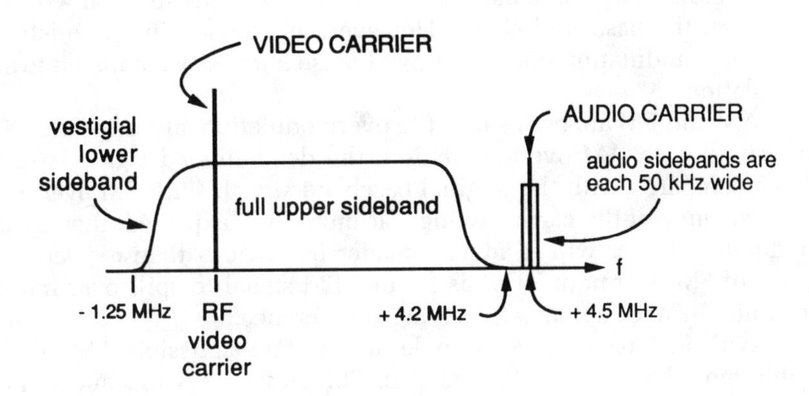

VIDEO: vestigial amplitude modulation
AUDIO: frequency modulation

Bandwidth:

1.25 MHz	vestigial sideband
4.50 MHz	audio carrier
0.05 MHz	audio upper sideband
0.20 MHz	guard band
6.00 MHz	TV station band

AM VERSUS FM

Amplitude modulation and frequency modulation are both ways to shift a signal in frequency from its baseband to another range. Assuming identical baseband signals, either amplitude modulation or frequency modulation will offer equal fidelity of the baseband signal. However, when used in its wideband mode, frequency modulation offers improved noise immunity compared to amplitude modulation.

AM and FM are both subject to overmodulation, although the effect is not the same. With AM overmodulation, the demodulated signal is distorted in shape compared with the original baseband signal. With FM overmodulation, the frequency of the carrier swings far more than expected, thereby creating a modulated carrier with a much broader bandwidth than expected. The end result of FM overmodulation is for one FM signal to spill over into the band occupied by another signal, creating interference.

With such techniques as single sideband transmission, AM can be considerably more bandwidth efficient than FM. However, wideband FM offers the ability to submultiplex together many baseband signals, such as was done with broadcast FM stereo radio.

The transmission of a signal from Earth to a communication satellite located 23,000 miles above Earth requires considerable immunity from noise since the received signal is relatively weak. Hence, FM transmission is used. Cellular telephony also uses frequency modulation for similar noise immunity reasons.

AM vs FM

Frequency Shifting:

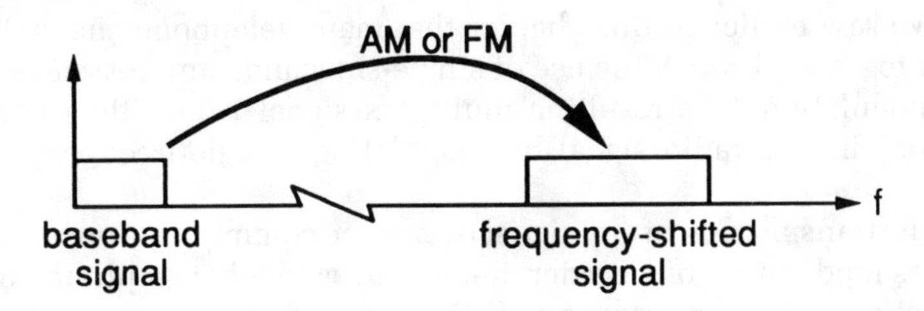

Bandwidth:

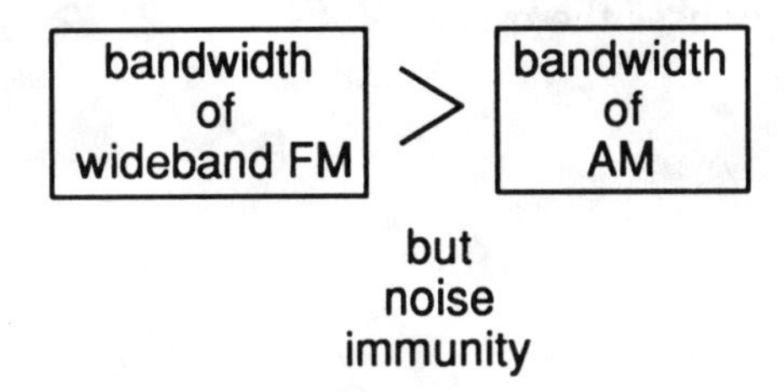

but
noise
immunity

Overmodulation:

AM: distortion of received signal

FM: spillover into adjacent bands

MODULATION AND MULTIPLEXING

Modulation is the means by which multiplexing is accomplished. Both amplitude modulation and frequency modulation shift signals from one frequency range to another, thereby enabling frequency division multiplexing.

We saw earlier in this chapter that many telephone channels are multiplexed together through the use of single-sideband, suppressed-carrier, amplitude modulation. The resulting multiplex signal would then be transmitted over the air as a radio signal by modulating a radio-frequency, carrier sine wave.

The transmission of any signal over some communication medium usually involves modulation of a carrier. Radio waves involve amplitude or frequency modulation of a radio-frequency (RF) carrier. Transmission over optical fiber usually involves amplitude modulation of a light-wave carrier, even when the baseband signal conveys digital information. Thus, even in today's world of digital, analog modulation is an essential technology that should be understood. Again and again, we see that new technologies do not eliminate the old, but instead each technology uses the other.

Modulation and Multiplexing

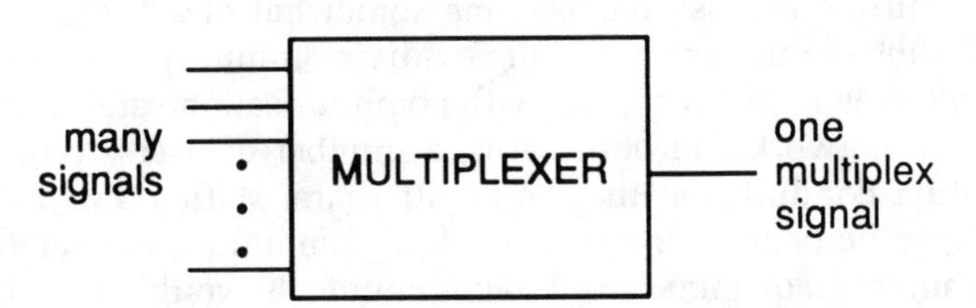

Frequency-division multiplexing is accomplished by amplitude modulation or frequency modulation.

MULTIPLE ACCESS

The term "multiple access" has become somewhat of a synonym for multiplexing. We thus see the use of frequency-division multiple access, or FDMA. The terminology is frequently used with communication satellite systems and also local area networks. In both cases, a number of users want access to a communication conduit that they must all share. With FDMA, each user is assigned a specific band of frequencies. In a communication satellite system, the whole band of frequencies used to communicate would be divided into 12 subbands, called transponder bands, each 36 MHz wide. Each user would be assigned one transponder band which could be used at any time.

Another approach to allowing multiple access is called time-division multiple access, or TDMA. With TDMA, all users use the same frequency band, but at different times.

Conceptually, FDMA is akin to frequency-division multiplexing, and TDMA is akin to time-division multiplexing.

Multiple Access

FDMA:

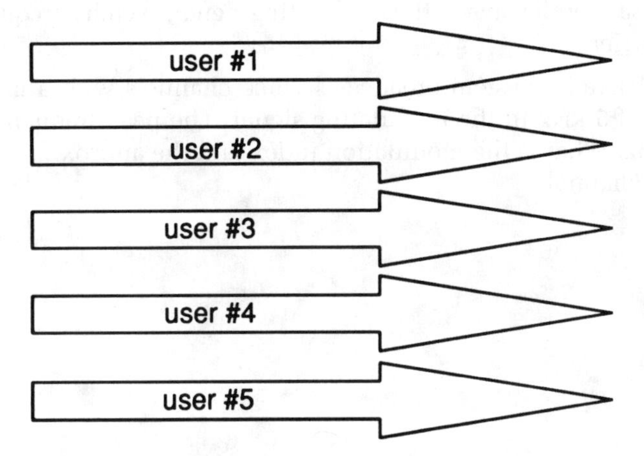

dedicated frequency bands for each user

TDMA:

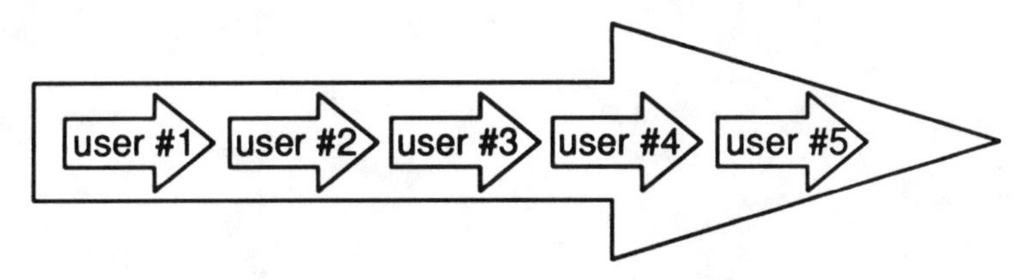

frequency band allocated into dedicated time slots for each user

PROBLEMS

4.1. Using a square wave as the modulating signal, draw the modulated carrier for indices of modulation of 1 and $^1/_2$.

4.2. Using a sawtooth wave as the modulating signal, sketch a frequency-modulated carrier.

4.3. A cellular radio system broadcasts voice channels with a maximum frequency of 5 kHz in the modulating signal. The peak frequency deviation is 12 kHz. What is the modulation index and the approximate bandwidth of each channel?

Chapter 5: Digital Signals

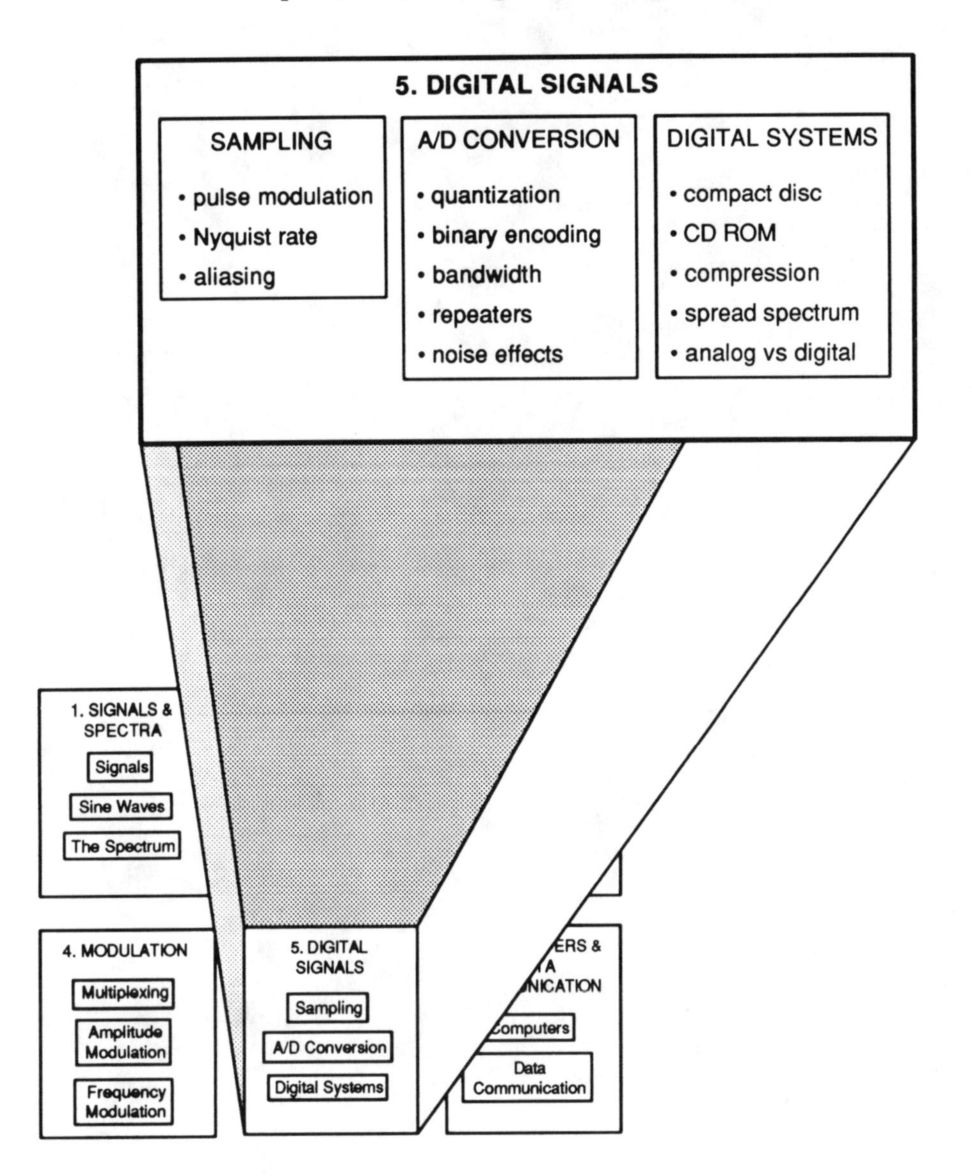

5

Introduction to Digital Signals

The whole world appears to be going digital! The stereo world is being revolutionized by the digital sound of the compact disc. Telecommunication systems are increasingly employing digital switching and such digital transmission media as optical fiber. This module treats digital signals and their practical application to stereo audio and telecommunication.

One concept that is fundamental to digital signals is *sampling* the instantaneous amplitude variation of the original signal at discrete values in time. *Nyquist's sampling theorem* states that if this sampling is performed at the proper rate, no information is lost about the original signal, and it can be perfectly reconstructed later on. Another fundamental concept is *quantizing* the amplitude fluctuation into discrete intervals, which are then digitally encoded by using *binary* numbers. The continuous time and amplitude variation of the original signal is then sampled and quantized into discrete intervals and values.

The process of digitizing a signal is complicated and results in a new binary digital signal that takes on one of two discrete values. This results in great immunity to additive noise. However, the bandwidth of the digital signal is many times that of the original signal. Again, noise immunity has been gained at the expense of bandwidth.

Sampling

TIME-DIVISION MULTIPLEXING

Time-division multiplexing enables a number of communication conversations or paths to occur simultaneously by sharing a medium or channel in time. A small portion of time is allocated to each individual communication signal.

An essential component of time-division multiplexing is the process of sampling a signal in time. In sampling, the instantaneous amplitude of a signal is determined at fixed intervals in time. These instantaneous amplitudes are the sample values, or samples, of the signal waveform. The separate samples from each waveform are then combined together and alternately transmitted. In the example shown here, each sample is transmitted as a short pulse whose height is proportional to the amplitude of each sample value.

At the receiver, the multiplexed samples are separated into the individual samples and the individual waveforms are reconstructed, usually by low-pass filtering.

Time-Division Multiplexing

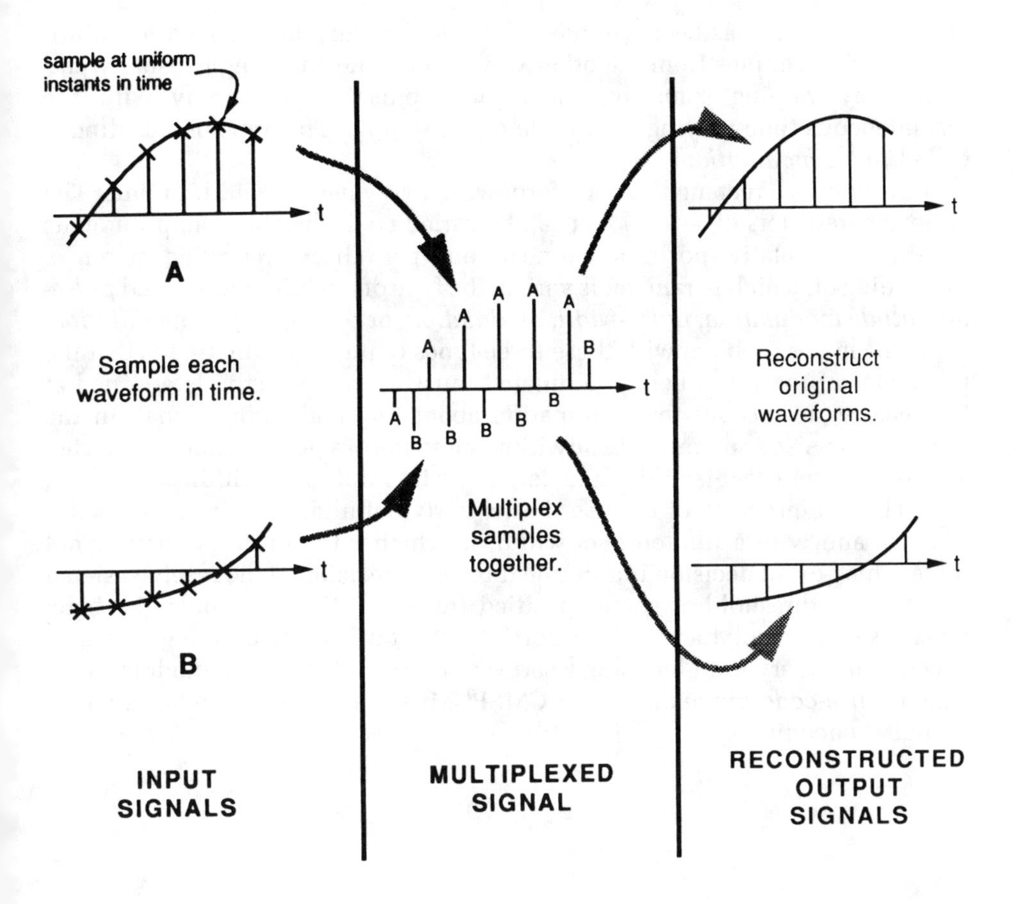

PULSE MODULATION

Once a waveform has been sampled, the samples must be encoded for multiplexing with samples from the other waveforms. The encoding is usually performed by varying some parameter of a pulse in synchrony with the instantaneous time-varying amplitude of the sample. This type of encoding is called *pulse modulation*.

A *pulse* is a rectangular waveform with a very narrow width in time. The various parameters of a pulse that can be varied are its height or amplitude, its width, and its relative position in time to some periodically recurring reference. Depending on which parameter is varied, the type of modulation is called *pulse-amplitude modulation, pulse-width modulation*, or *pulse-position modulation*.

The major problem with these three types of pulse modulation is that the parameter of the pulse being modulated must be very exactly determined at the receiver to recover the information about the modulating signal. In the presence of noise and limited bandwidth, the shape of a pulse becomes distorted and unclear, and so it is difficult to determine its amplitude, width, or position.

The one property of a pulse that is fairly definite, even in the presence of noise and with a limited bandwidth, is whether the pulse is there or not there. This type of decision is a simple threshold decision: if the received signal exceeds the threshold at some specified instant of time, the pulse is there; otherwise, it is not there. The information about the modulating signal is encoded in binary form by using a series of pulses. This type of modulation is called *pulse-code modulation*, or PCM. PCM is the nucleus of today's world of digital encoding.

Pulse Modulation

Pulse:

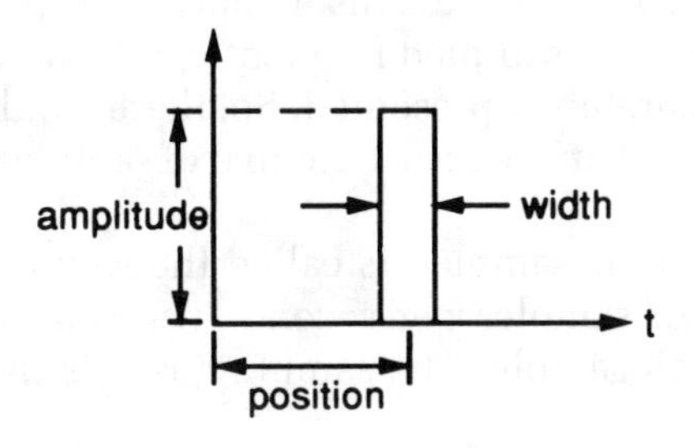

Types:

- Pulse-Amplitude Modulation (PAM)
- Pulse-Width Modulation
- Pulse-Position Modulation
- Pulse-Code Modulation (PCM)

Noise & Distortion:

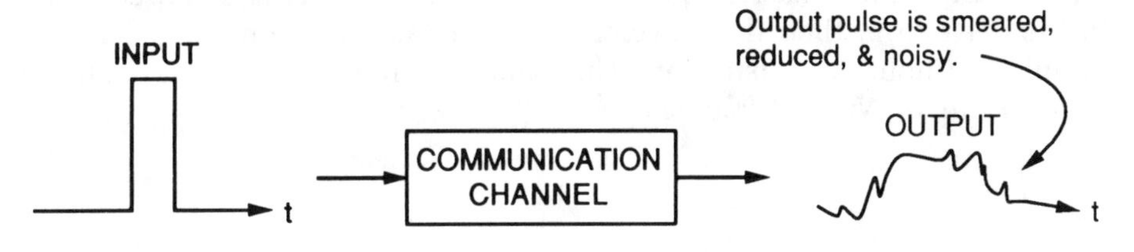

SAMPLING RATE

A key question is, how frequently should a signal be sampled? Consider a sine wave. If the wave is sampled very frequently, then many more samples are obtained than are needed to characterize the instantaneous amplitude variation of the waveform. If the wave is not sampled frequently enough, then its amplitude variation will not be accurately represented. So, there would appear to be an optimum rate in the sense of using the least number of samples for accurately characterizing the waveform.

The rate at which a signal is sampled is called the *sampling rate*, and it is expressed as the number of samples per second. The *sampling interval* is the time interval between each sample. The sampling rate is the reciprocal of the sampling interval.

If a waveform is slowly varying, then fewer samples per second will be required than if the waveform were rapidly varying. So, the optimum sampling rate depends on the maximum frequency component present in the signal. If the maximum frequency component in the signal is F_{max}, then the optimum sampling rate is twice F_{max}, or the sampling rate equals $2F_{max}$. If a signal has a maximum frequency component of 5 kHz, then the sampling rate is 10,000 samples per second. The sampling rate is also sometimes called the *sampling frequency*.

Strictly speaking, the sampling rate should be slightly more than $2F_{max}$, but the sampling rate is often taken to be exactly $2F_{max}$. The optimum sampling rate was first stated by Harry Nyquist, and the Nyquist sampling theorem is named after him.

The Nyquist sampling theorem states that if a waveform is sampled at a rate at least, but not exactly equal to, twice the maximum frequency component in the waveform, then the waveform can be exactly reconstructed from the samples without any distortion. This sampling rate is sometimes called the *Nyquist rate* or *Nyquist frequency*.

Sampling Rate

Sampling:

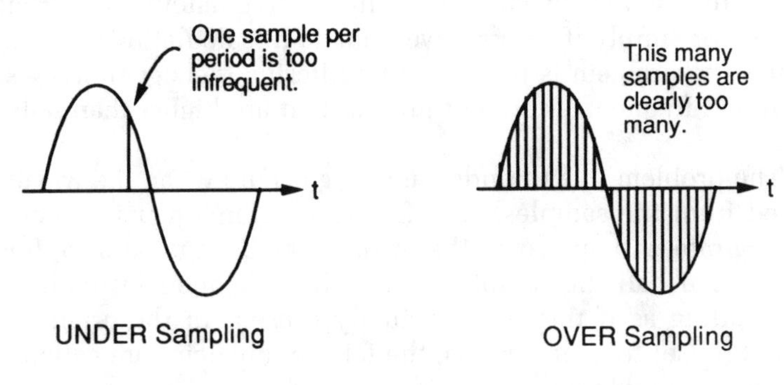

Nyquist Sampling Rate:

F_{MAX} = maximum frequency component in signal

Nyquist sampling rate $\geq 2\,F_{MAX}$

(samples per second)

[sampling rate is usually = $2\,F_{MAX}$]

ALIASING

A problem can occur if a waveform is sampled at a rate less than twice the maximum frequency component in the waveform. This undersampling would occur if the waveform contained higher frequency components than were expected, or simply if an error were made in calculating the sampling rate. In practice, the problem is prevented by filtering the signal being sampled to be certain that frequencies are not present that are higher than half the sampling rate.

The problem due to undersampling occurs when the waveform is reconstructed from the samples. Any frequency components higher than half the sampling rate will appear in the reconstructed signal at false frequencies that are less than half the sampling rate. These frequencies could be said to be masquerading as if they were actually present in the original signal but at different (false) frequencies. So, the false frequencies are called alias frequencies, and the problem is called *aliasing*.

The cause of the aliasing can perhaps be made clear through a specific example. A sampling rate of 20 samples per second is assumed. This means that frequencies over 10 Hz should not be present in the waveform being sampled. Suppose a sine wave at 15 Hz somehow is present. This sine wave will be sampled every 1/20 second or every 50 milliseconds. Since the period of the sine wave is 1/15 second or about 67 milliseconds, there clearly are less than two samples per period, and the sine wave is being undersampled. Consider what happens when these samples are used to reconstruct the original signal.

Upon reconstruction, the samples will appear to have come from a sine wave with a period of 200 milliseconds, or a corresponding frequency of 1/0.2 = 5 Hz. So, a false frequency of 5 Hz has been introduced. This false frequency comes from the 15 Hz sine wave in the sampled signal that now appears to be masquerading as a 5 Hz sine wave in the output of the reconstruction process. The 5 Hz sine wave is an "alias" of the 15 Hz sine wave.

If some other frequency that was higher than half the sampling rate were present in the signal to be sampled, it would appear in the output of the reconstructed signal at an alias frequency. If the input signal had a frequency component F_{high} higher than half the sampling rate F_{samp}, then an alias frequency F_{alias} would appear in the reconstructed signal at a frequency of

$$F_{\text{alias}} = F_{\text{high}} - \frac{1}{2}F_{\text{samp}}$$

If F_{high} were more than twice the sampling rate, then integer multiples of $^1/_2\ F_{\text{samp}}$ would be subtracted from F_{high} until the resulting frequency became less than $^1/_2\ F_{\text{samp}}$.

Aliasing

Aliasing:

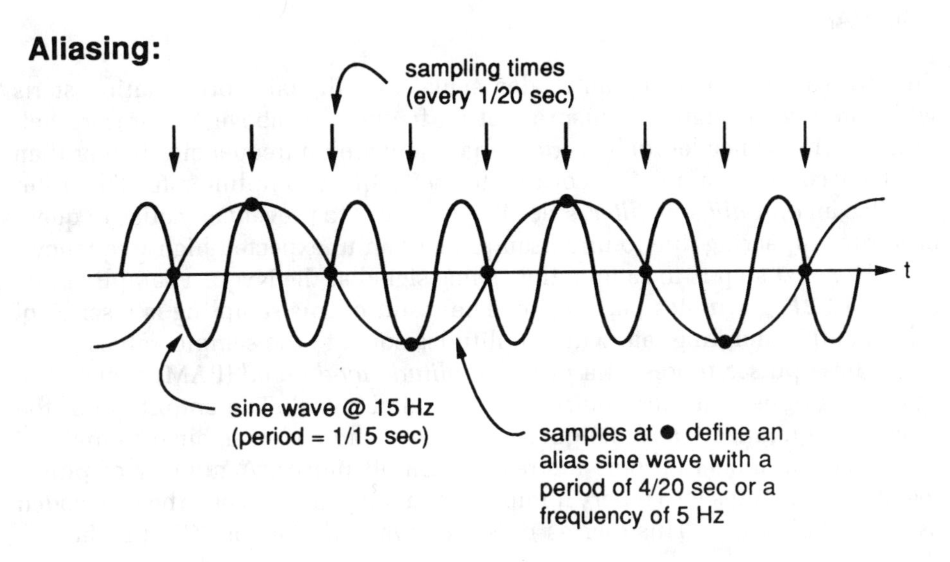

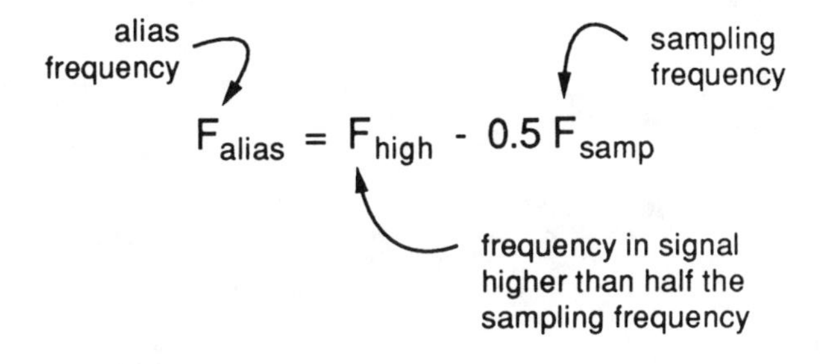

Anti-Aliasing Filter:

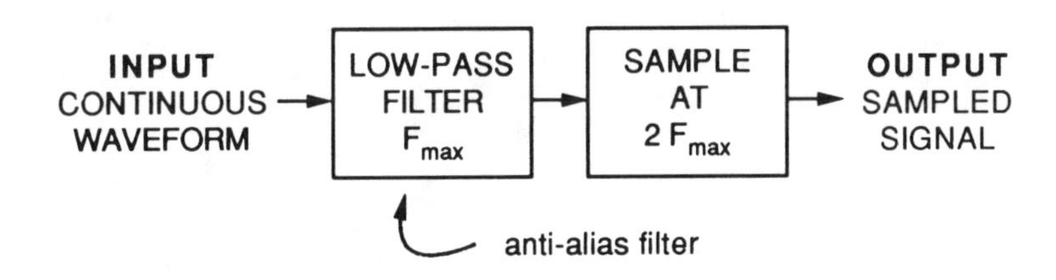

Analog-to-Digital Conversion

SUMMARY

The process of converting an analog signal to a digital representation starts with filtering the signal to ensure that no frequencies above F_{max} are present. This filtering is needed to be certain that there are no frequencies higher than can be accurately sampled according to the Nyquist sampling rate. This filter is called an *anti-aliasing filter*, since its purpose is to prevent any alias frequencies from appearing due to undersampling of an unexpected high frequency.

The next step is to sample the analog signal at the Nyquist sampling rate of at least $2F_{max}$ samples per second. The result of this sampling is a series of pulses at the sampling rate with amplitudes equal to the sample values.

These pulses represent a *pulse-amplitude-modulated* (PAM) signal. The next steps transform these pulses into a digital signal. The amplitude of the pulses are quantized, and the quantized values are coded as binary numbers. The binary numbers become a stream of on-off pulses. A number of pulses together then represent a binary number. The analog samples have been encoded as a series of pulses. This then is *pulse-code modulation*, or PCM for short.

A/D Sampling

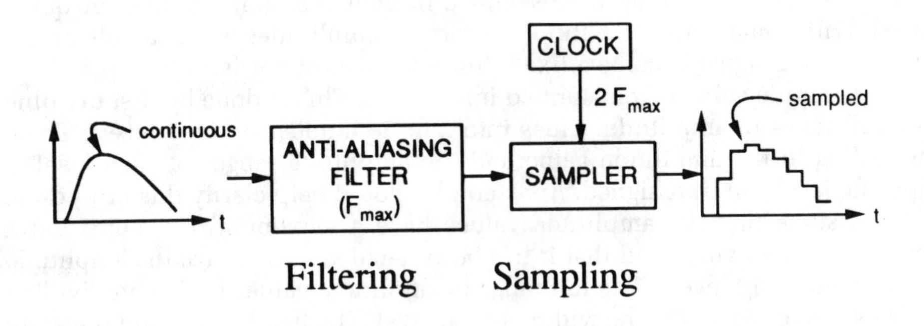
CLOCK
2 F_{max}
sampled
continuous
ANTI-ALIASING
FILTER
(F_{max})
SAMPLER
t
t
Filtering
Sampling

QUANTIZATION

After the analog signal has been sampled in time, the sample values are quantized. With quantization, the infinite variety of amplitudes of the sample values of the analog signal become a fixed, finite number of levels.

A sample value is transformed into a level. This is done by first dividing the full range of amplitude values into a finite number of fixed levels. Then, according to the amplitude value and the amplitude ranges of each level, a specific level can be assigned. An example should help clarify this procedure.

Assume that the amplitude values have a maximum of 8 volts and a minimum of −8 volts, and that it has been decided to quantize the amplitude values into eight levels. The full range of amplitude values is 16, and dividing this by 8 gives 2 volts as the width of each level. The first level would therefore be from −8 to −6 volts, the second level from −6 to −4 volts, and so forth until the eighth level from 6 to 8 volts.

The first level is usually assigned a value of 0, and thus in this example, the eighth level would be assigned a value of 7. If an actual sample value were 4.52 volts, it would be quantized as level 6 when using this procedure.

The step size of the levels in the preceding example were all equal. The step size of the levels could also be made different so that more steps might be available for small amplitude portions of the signal, rather than large amplitudes. If the signal were mostly small amplitudes and only rarely achieved large amplitudes, this type of nonlinear step size would give more amplitude resolution to the portion of the signal that was most often encountered. This type of quantization is called nonlinear quantization.

Nonlinear quantization is frequently performed by first compressing the amplitude of the input signal. With compression, the amplitude of the signal is reduced in size for the more amplitude it has. A linear quantizer can then be used to quantize the compressed signal. Upon reconstruction, the amplitude must be expanded to compensate for the initial compression. The process of compressing the amplitude of a signal and then *expanding* it is called compansion or companding.

The process of quantization destroys some information about the true sample values of the amplitude of the analog signal. The difference between the actual values and the quantized values is an error that contributes a form of noise, called quantization noise, when the analog signal is reconstructed from the digital data. If the step size is kept small enough, this noise will be negligible. If the step size is a, then the rms value of the error is $a/\sqrt{12}$.

Quantization

Example:

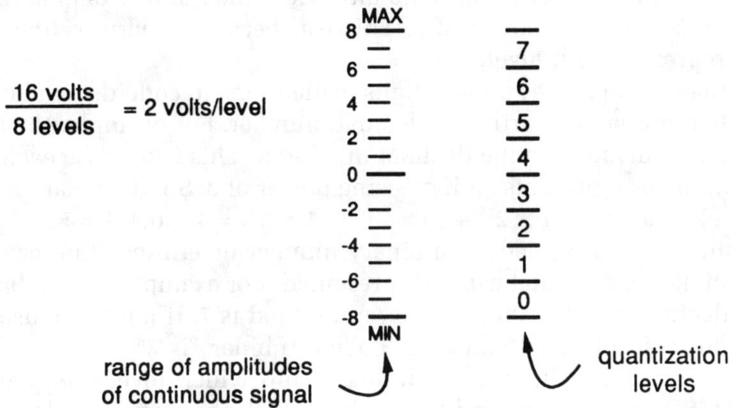

Step Size:

7
6
5
4
3
2
1
0

step sizes are all equal

LINEAR QUANTIZATION

7
6
5
4
3
2
1
0

step sizes are progressively smaller in this middle region

NONLINEAR QUANTIZATION

BINARY CODING

The last step in converting an analog signal into a digital representation is the coding of the quantized sample values to a binary format.

In the preceding description of quantization, the number of quantization levels was chosen to be a power of 2. This was because binary coding would be used to represent each level.

The binary format uses binary digits, called *bits*, to code decimal data. A series of bits represents a particular decimal number. For example, the binary number 101 is equivalent to the decimal number 5. This is because each place in a binary number represents an increasing power of 2. So, the binary number 101 can be expanded as $(1 \times 2^2) + (0 \times 2^1) + (1 \times 2^0) = 4 + 0 + 1 = 5$.

The number of bits used for a binary number determines the maximum size of a decimal number that can be represented. For example, with 3 bits, the maximum decimal number that can be represented is 7. If n bits are used, the maximum decimal number that can be coded in binary is $2^n - 1$.

Binary coding will be used to indicate into which level the quantized sample falls. Since 0 will be used for the first level, the number of levels that can be represented with n bits is one more than the maximum decimal number that can be coded. So, the number of levels that can be coded in binary form using n bits is 2^n.

As we mentioned previously, the choice of the number of levels is determined by how finely the amplitude variation of the analog signal needs to be represented so as to reduce distortion and quantization noise to acceptable amounts. If too many levels are used, then the number of bits used to code the levels will be excessive, leading to a high bit rate and, accordingly, a high bandwidth to pass the binary signal with an acceptable amount of distortion.

A binary number consists of a string of 1s and 0s. These 1s and 0s are represented as a string of pulses. The presence of a pulse indicates a 1; the absence of a pulse indicates a 0. A simple threshold decision is used to decide whether a pulse is present or not at the receiver. The string of pulses is called a digital signal. A digital signal is very impervious to noise and distortion, since, as long as enough of the pulse remains, a simple threshold decision is possible to decide about its presence or absence.

If an analog signal is sampled at a rate of R samples per second and n bits are used to code each sample as a binary number, then the bit rate of the digital signal is nR bits per second. If 10 analog signals were multiplexed together by using digital time-division multiplexing, then the bit rate of the multiplexed digital signal would be $10nR$.

Binary Coding

Binary Numbers:

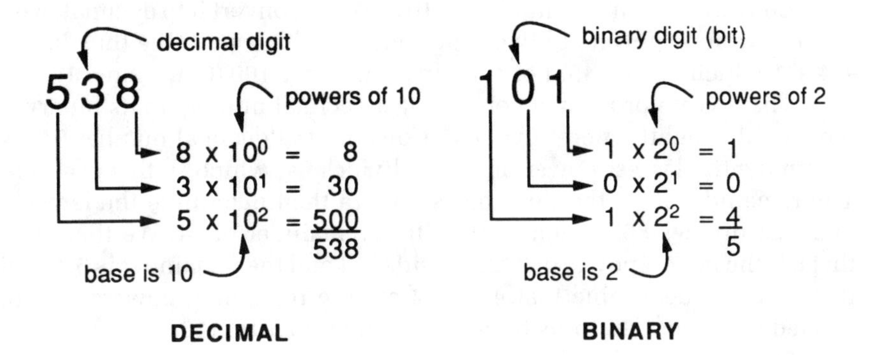

Levels:

n bits

$2^n - 1$ = maximum decimal number that can be encoded with n bits

2^n = number of levels (0 through $2^n - 1$)

Pulse Code:

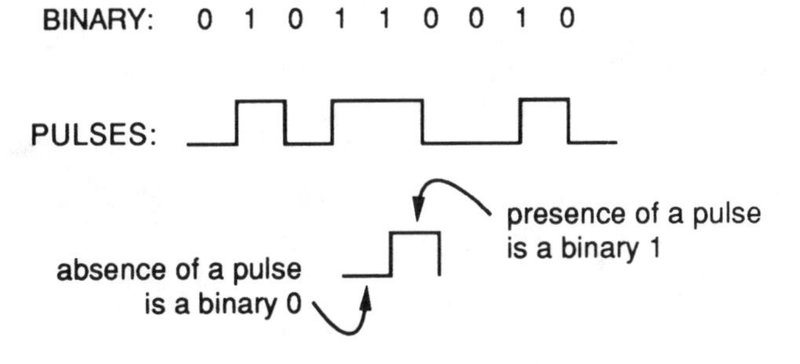

BINARY/DECIMAL CONVERSION

The process of converting numbers from binary to decimal and decimal to binary is made much easier through the use of a table headed by the ascending powers of 2. Starting at the left, we have 1, 2, 4, 8, 16, and so on.

Consider the binary number 00101101. To convert it to decimal, we simply add the various powers of two represented by binary 1s. We thus have 32 + 8 + 4 + 1 which equals 45. For the binary number 10010, we have 16 + 2 = 18.

The reverse process of converting a decimal number to its binary equivalent is only slightly more involved. Consider the decimal number 94. We first determine the largest power of 2 that divides 94, which in this example is 64. The remainder from the division is 30. We then determine the largest power of 2 that divides 30, which is 16 with a remainder of 14. We then determine that 8 is the largest power of 2 that divides 14 and the remainder is 6. Continuing this way, we determine that 4 and 2 are the remaining powers of 2 that are needed to convert 94 to its binary equivalent of 1011110.

Our excursion into binary numbers is complete, and we can now summarize the process of analog to digital conversion.

Binary/Decimal

	2^7 128	2^6 64	2^5 32	2^4 16	2^3 8	2^2 4	2^1 2	2^0 1	DECIMAL
example #1	0	0	1	0	1	1	0	1	45
example #2				1	0	0	1	0	18
example #3	0	1	0	1	1	1	1	0	94

DIGITAL CODING AND DECODING

The process of coding an analog signal into a digital format consists of four stages. The first stage is the low-pass filtering of the analog signal to ensure that no frequencies greater than F_{max} are present. This filtering stage is called an anti-aliasing filter, since its purpose is to prevent alias frequencies upon reconstruction.

The second stage is the sampling of the analog signal at the Nyquist rate of at least $2F_{max}$ samples per second. A clock circuit creates pulses at a rate of $2F_{max}$, which cause the sampler to sample the analog input signal.

In the third stage, the sample values are quantized into 2^n levels. If each level has the same step size, the quantization is linear. If the signal is small most of the time, then more levels could be used for the smaller values. The quantization levels would then not be of equal step sizes. Such quantization is called *nonlinear quantization.*

In the fourth and last stage, the quantized levels are encoded in binary form by using n bits for each quantized sample of the analog signal. The result of this last stage is a digital signal consisting of on-off pulses corresponding to binary 1s and 0s. This digital signal is called a *baseband digital signal.*

The process of converting or decoding a digital signal into its equivalent analog signal consists of three stages. The first stage is to detect the presence of the digital pulses. This detection is performed by examining the digital signal at specific instants in time to determine whether a fixed threshold is exceeded or not. If the digital signal exceeds the threshold at the examining instant, then it is decided that a digital pulse corresponding to a binary 1 is present.

The second stage is to decode the binary data into an analog signal, consisting of a series of pulses with an amplitude equal to the decoded binary value and a time width equal to the sampling interval. In the third and last stage, this rectangular waveform is smoothed by a low-pass filter with width F_{max}.

A device that can both code an analog signal into a digital representation and decode a digital signal into its equivalent analog signal is called a coder-decoder, or *codec* for short.

A/D Coding and Decoding

Coding (A → D):

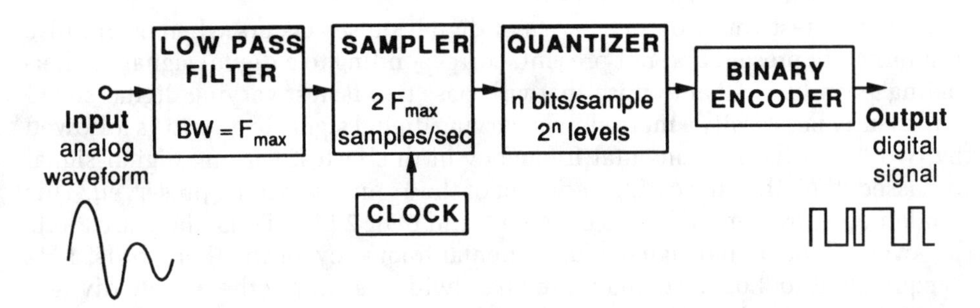

Decoding (D → A):

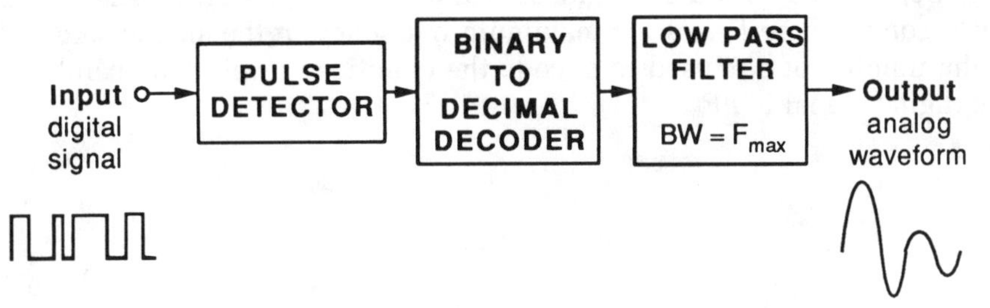

Codec:

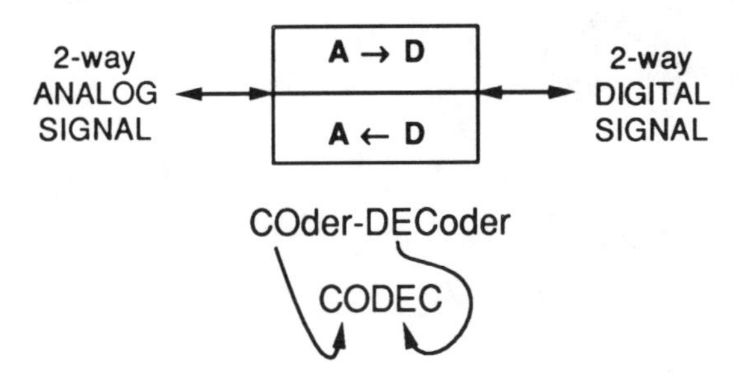

BANDWIDTH

The bandwidth of a digital signal is determined by the bandwidth needed to pass a single pulse or a series of pulses with reasonable fidelity. An intuitive derivation of this bandwidth follows.

The fastest variation of a series of digital pulses occurs when each pulse is alternately present and not present, corresponding to a digital signal of alternating 1s and 0s. A bandwidth that will pass this fastest varying digital signal will clearly pass easily a more slowly varying digital signal (e.g., two 1s followed by two 0s). If the fundamental frequency of this fastest varying digital signal is passed, then the alternating variation of the 1s and 0s will be preserved. This fundamental frequency covers a single 1 and 0, or 2 bits. Thus, there are 2 bits per cycle of the digital signal's fundamental frequency, or in other words, 1 Hz is equivalent to 2 bits. Therefore, the bandwidth is simply the bit rate divided by 2. Or, in formula form, the bandwidth required for a digital signal of R bits per second is $R/2$ Hz.

Some other formulas might be useful. The digital bit rate is $2nF_{max}$ bits per second, where F_{max} is the maximum frequency in the analog signal and n is the number of bits used to encode the quantized levels. The bandwidth of the digital signal is nF_{max}.

Bandwidth

Intuitive Derivation:

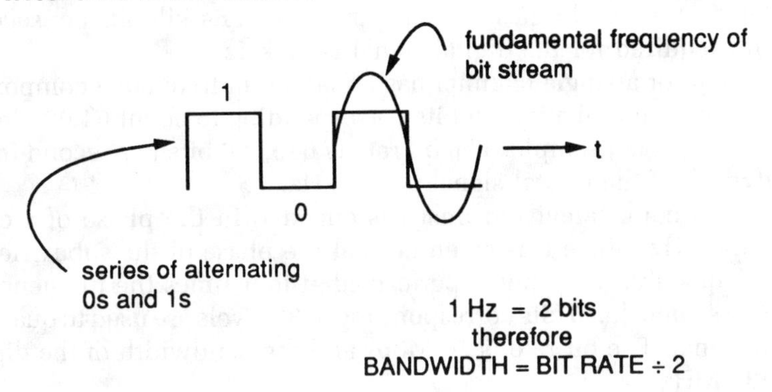

1 Hz = 2 bits
therefore
BANDWIDTH = BIT RATE ÷ 2

Formulas:

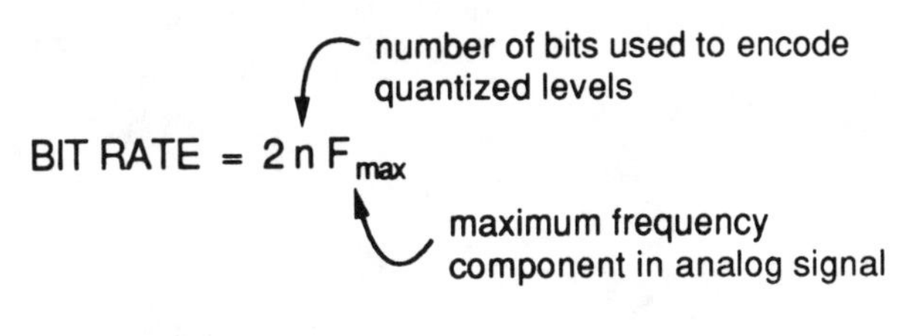

$$\text{BIT RATE} = 2\,n\,F_{max}$$

$$\text{BANDWIDTH} = n\,F_{max}$$

5

BANDWIDTH EXAMPLES

As an example, a telephone speech signal with a maximum frequency component of 4 kHz is digitized by using 8 bits to quantize each sample. The bit rate is 8,000 samples per second times 8 bits per sample, or 64 kilobits per second. The bandwidth required for the digital signal is 32 kHz.

A hi-fi signal for a single channel has a maximum frequency component of 20 kHz. For maximum quality, 16 bits corresponding to about 64,000 levels are used to quantize each sample. The bit rate is 640,000 bits per second (bps), and the bandwidth of the digital signal is 320 kHz.

The hue of a color television image is encoded in the phase of a color subcarrier at 3.5 MHz. Since it is essential that the phase of the subcarrier be represented adequately, sampling is performed at four times the frequency of the subcarrier. Assume that 5 bits corresponding to 32 levels are used to quantize the television signal. The bit rate is 70 Mbps and the bandwidth of the digital bit stream is 35 MHz.

Bandwidth Examples

Telephone Signal:

analog:

F_{max} = 4 kHz

digital:

n = 8 bits (256 levels)
bit rate = 64,000 bits/sec
BW = 32 kHz

Hi-Fi Signal:

analog:

F_{max} = 20 kHz

digital:

n = 16 bits (64,000 levels)
bit rate = 640,000 bits/sec
BW = 320 kHz

Television Signal:

analog:

F_{max} = 4.2 MHz
color subcarrier = 3.5 MHz

digital:

n = 5 bits (32 levels)
sampling = 4 x 3.5 MHz
bit rate = 70 Mbits/sec
BW = 35 MHz

REGENERATIVE REPEATER

A major reason for the digital representation of an analog signal is the noise immunity offered by the digital format.

Consider a digital signal that passes along a transmission line. The input to the line consists of a digital signal with sharp pulses. The effect of the transmission line is to restrict the bandwidth of the pulses, causing a roundness of their corners and a decrease in their sharpness. Resistance losses in the line cause the signal to decrease in amplitude. The pulses then emerge from the transmission line distorted in shape and reduced in amplitude.

Noise also creeps into the line, especially as the amplitude of the pulses decreases, causing a deterioration in the signal-to-noise ratio. The net effect is that noisy, distorted pulses emerge from the transmission line. The absence or presence of a pulse is detected by a *threshold decision*. A fair amount of distortion and noise can be tolerated, since it is not the absolute shape or amplitude of the received pulses that matters, but rather a simple decision about whether a pulse is there or not there.

The decision threshold should be set halfway between the minimum and the maximum digital signal levels. If these levels vary over time, the threshold can be changed in an adaptive fashion.

The detection of the digital pulses must be performed before the distortion and noise become so bad that the threshold decisions cannot be made without appreciable errors. The loss and bandwidth of the transmission line as well as the noise level determine when the detection is to be performed. Once the detection has been performed, new sharp pulses can be generated for transmission over any remaining distance to the final destination. Because the digital signal is regenerated, this operation is called *regenerative repeating*.

Repeater

Noise & Distortion:

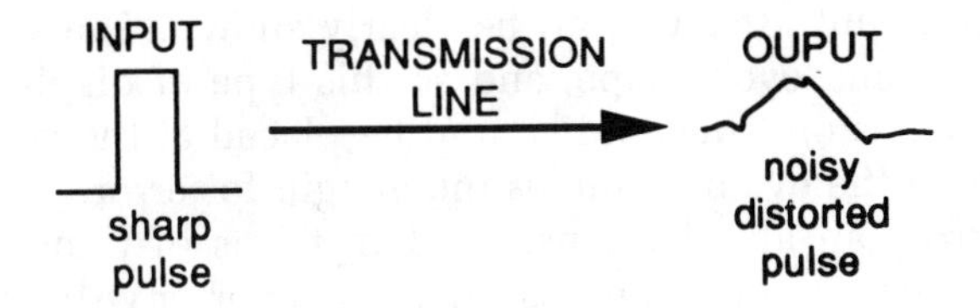

Regenerative Repeater:

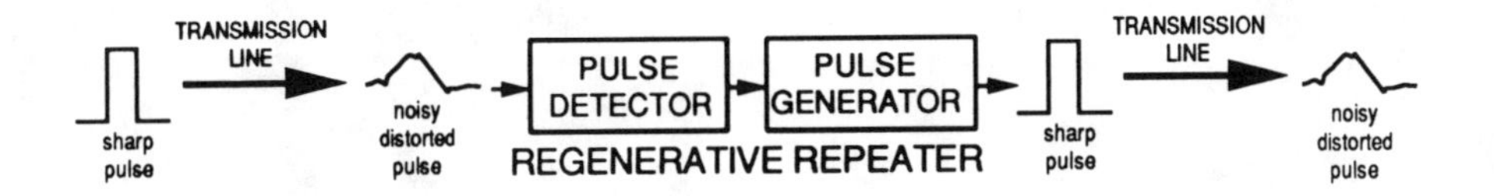

Pulse Detection:

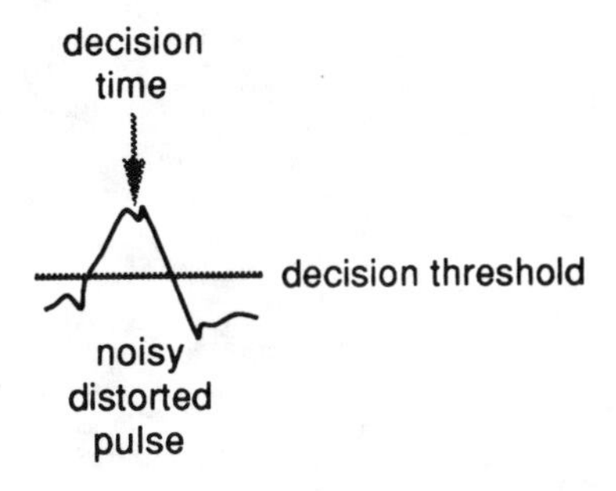

NOISE EFFECTS

The received digital signal can be displayed on an oscilloscope. If the time sweep of the oscilloscope is suitably adjusted, the pulses will be displayed so that they overlap each other and the 1s and 0s can all be seen. The effects of noise, pulse-shape distortion, and jitter will all be clearly shown. The space between the 1s and 0s forms an "eye" shape, and so this type of display is called an eye pattern. The decision threshold should be placed at the center of the eye pattern. The size of the eye determines the margin for error.

The probability of error can also be considered in terms of transition probabilities. If a 0 is transmitted, there is a probability that an error will occur and the 0 will be decoded as a 1. Similarly, if a 1 is transmitted, there will be a probability that an error will occur and the 1 will be decoded as a 0. These errors are shown on a probability transition diagram.

Noise Effects

Eye Pattern:

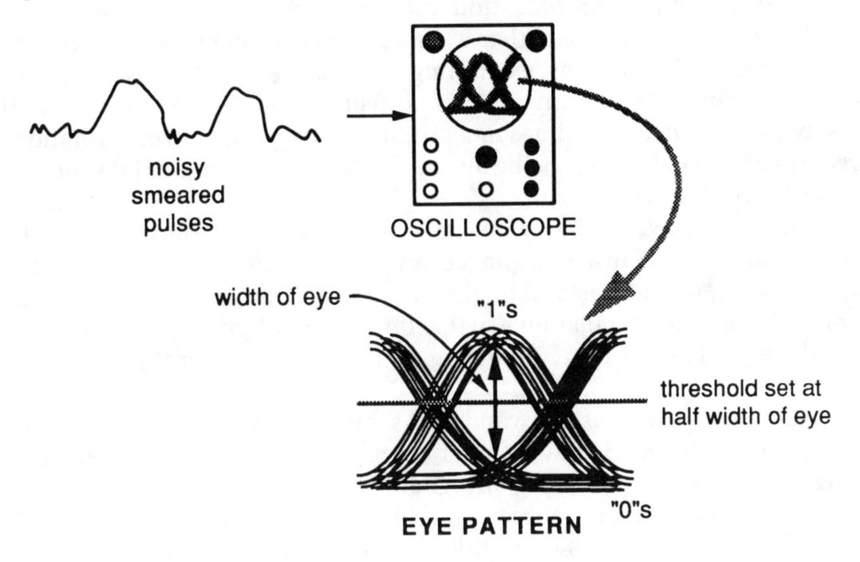

Transition Diagram:

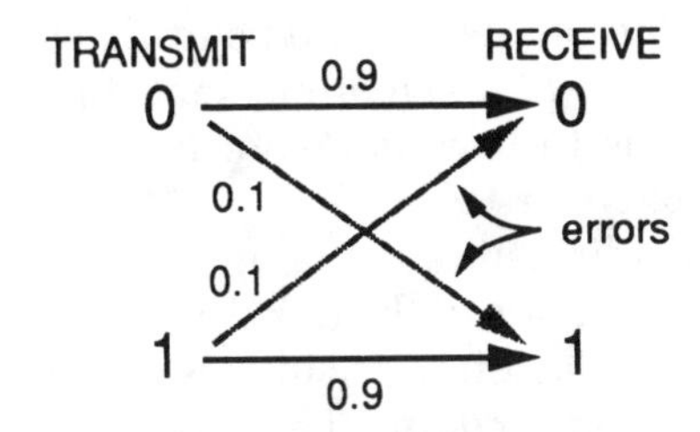

fractions are probabilities, e.g., probability of receiving a "0" if a "1" were transmittied is 0.1

DIGITAL SIGNAL FORMATS

There are a variety of ways that the binary information can be represented in a digital signal. These different digital signal formats reduce the problems associated with bit synchronization and dc wander.

The decision about whether a digital signal element represents a binary 1 or a binary 0 is made by examining the digital signal at exactly the right moment in time. This moment in time usually corresponds to one-half the pulsewidth. The decision times must be precisely synchronized with the actual stream of digital data that is being decoded. This is referred to as bit synchronization or timing recovery.

One way in which bit synchronization could be performed would be to dedicate one channel in a multiplexed system for sending a pattern of alternating bits. However, this is wasteful of channel space. Another way is for the digital signal to have enough transitions so that the timing information can be recovered from the digital signal itself. Some digital formats are better than others in this regard.

Another problem is dc wander. Assume that the decision threshold is determined as the moving average (or dc value) over a number of bits. If a long string of 1s or 0s occurs, the digital signal will have a change in dc value, which could cause problems in correctly performing the threshold decision. Some digital formats reduce this dc-wander problem more than others.

The simplest digital format is to encode a 1 as the presence of a pulse and a 0 as the absence of a pulse. In this simple format, the pulse has a width equal to a digital signal element. However, dc wander and bit synchronization can be problems with this simple format.

The bit synchronization problem could be made easier if the pulse width were one-half the time of a digital signal element. Since, for a binary 1, the digital signal always returns to 0 volts before the next signal element, this format is called *return-to-zero*, or RZ.

With *Manchester encoding*, a transition in voltage level always occurs at the decision time, and the direction of the transition indicates whether a binary 1 or 0 was transmitted. A pulse of half width is always transmitted in each signal element with Manchester encoding. The problems of both timing recovery and dc wander are minimized with Manchester encoding.

Formats

Detection Problems:

- Timing:

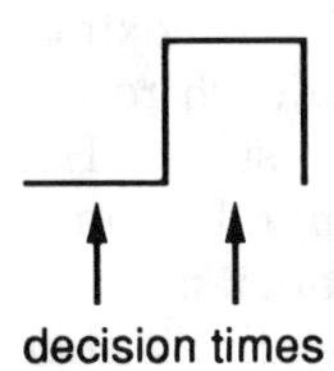

- DC Wander:

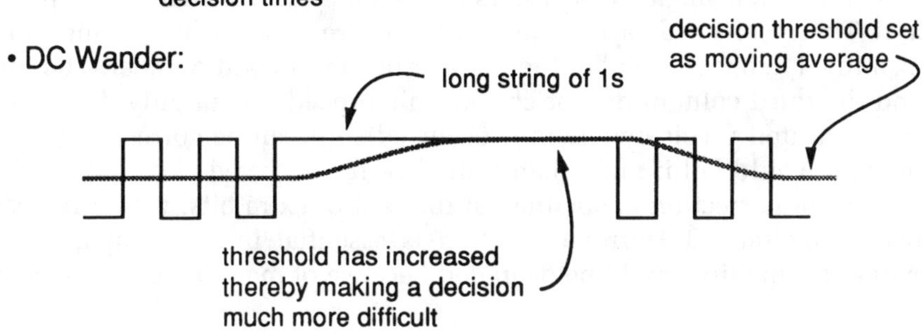

Formats:

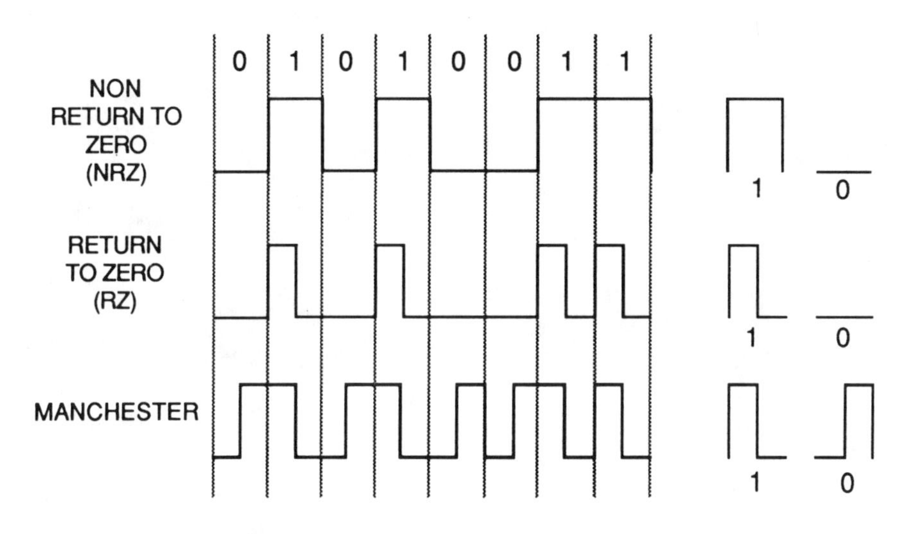

5

ERROR CORRECTION

A powerful aspect of digital is the ability to detect and actually correct errors that occur in transmission. A simple example will clarify the process.

The original bit stream is the following 16 bits: 0110110100101010. These 16 bits are arranged in a table in blocks of 4 bits. An extra bit is added to each row and column to make the number of 1s in each row and column an odd number. For example, there are two 1s in the first row. Hence, a 1 is added to make the total number of 1s three, which is an odd number. The extra 8 bits are appended to the end of the original 16 data bits.

Assume a single error occurs in the eleventh bit changing it from a 0 to a 1. The received data is arrayed like before, and each row and column is examined to check whether the number of 1s is an odd number. The third row and the third column do not check. This procedure not only detects an error but indicated precisely where it occurred so it can be corrected. Clearly this method would fail if more than a single error occurred.

Error correction is obtained at the cost of extra bits, which increases the overall bit rate. But error correction is essential in the compact disc since without it quality would be degraded because of manufacturing defects in the surface of the disc.

Error Correction

Example:

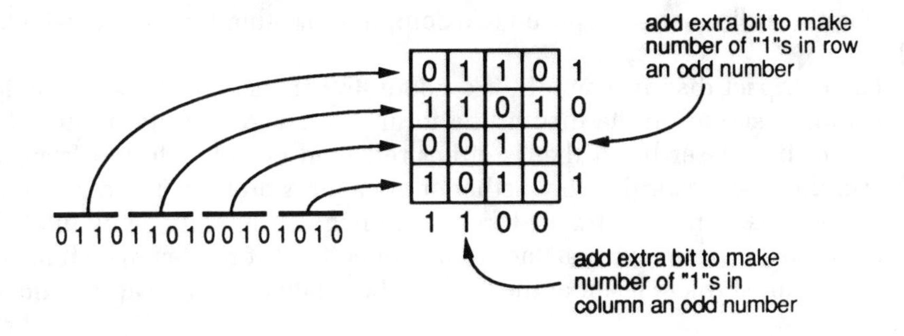

transmitted data stream

0 1 1 0 1 1 0 1 0 0 1 0 1 0 1 0 — original data

1 0 0 1 1 1 0 0 — error correction data

received data stream

0 1 1 0 1 1 0 1 0 0 0 0 1 0 1 0 1 0 0 1 1 1 0 0

error occurred here

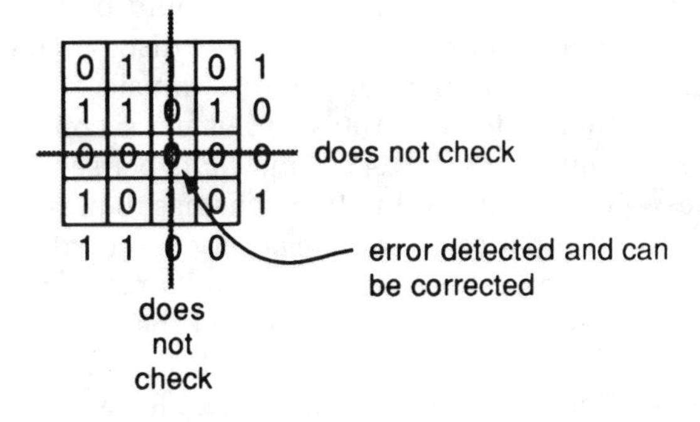

Digital Systems

DIGITAL AUDIO COMPACT DISC

Digital recording has revolutionized stereophonic sound. Nearly all performances are now recorded digitally, ensuring nearly perfect reproduction of the original performance. The original master recordings are made by using PCM on magnetic tape. The magnetic tape is then transferred to a digital audio compact disc (CD) for sale to consumers. (The spelling of "disc" with a *c* refers to the digital audio disc as opposed to a computer or other kind of "disk" with a *k*.)

The compact disc is a small disk, about 4$^3/_4$ inches in diameter. Digital information is stored on the disc in the form of small pits in its surface. The disc is read by a laser beam that is either reflected or not reflected from the surface of the disc, according to whether or not a pit is present. The disc rotates counterclockwise, and it is tracked from the inside to the outside by the laser beam. The disc maintains a constant linear velocity as it rotates, which means that the disc must rotate faster at the inside (about 500 rpm) than at the outside (about 200 rpm).

The frequency response of the signal recorded on the disc is from 20 to 20,000 Hz. The analog signal is sampled at 44,100 samples per second, and 16 bits are used to quantize each sample. Two stereo channels are recorded on the disc. The use of 16 bits to quantize the signal gives a signal-to-noise ratio, dynamic range, and channel separation, all in excess of 90 dB. Harmonic distortion is less than 0.05%.

Sophisticated *error-correcting codes* are used. A burst of noise in the form of dust or dirt on the surface of the disc can cause errors in detecting the bits that have been obliterated. An error burst that is 3,548 bits long can be detected and corrected by using these error-correcting codes. If interpolation is also used, an error burst of about 12,000 bits could be compensated and would not be detectable in listening to the disc. The discs are currently recorded on one side, giving a little more than one hour of music.

The data rate is 44,100 samples per second times 16 bits per sample time times two channels, or 1.4112 megabits per second. The use of error-correcting codes, special data-modulation schemes to prevent track interference and low-frequency bit patterns, and the ability to record other information, increase the final overall bit rate to 4.3218 megabits per second.

The use of a laser beam to read the disc means that there is no wear of the surface of the disc. Thus, the disc will last forever, even after repeated playing. The use of error-correcting codes means that dirt and scratches in the surface of the disc have no effect on playback quality. The use of digital means that each disc is an exact replica of the original digital recording. The concept of "original" then disappears, since each disc is itself an original!

Compact Disc

Performance:

number of channels	2
frequency range	20 Hz to 20,000 Hz
dynamic range	> 90 dB
S/N ratio	> 90 dB
channel separation	> 90 dB
harmonic distortion	< 0.05%

Signal Format:

sampling frequency	44,100 samples/sec
quantization	16 bits/sample (linear)
stereo signal bit rate	1.4112 Mbps
overall bit rate	4.3218 Mbps

Physical:

diameter	120 mm (4.75 inches)
rotation speed	500-200 rpm
playing time	75 min

CD-ROM

The compact disc is a wonderful invention that has revolutionized and reinvigorated the world of recorded sound. The black-vinyl phonograph record is dead, except for a few idealists who revere its background hiss, overemphasized high-frequencies, and album cover art. The phonograph needle traveled a path of nearly 1/2 mile along each side of a phonograph record, gouging destruction along its path and slowly destroying the recorded sound. The laser reads nearly 5 miles of digital data on a compact disc, but takes only the digital information, leaving no degradation of the data stored on the disc.

But beyond its use to store sound, the compact disc can store patterns of bits that represent any information. Unique combinations of 8 bits at a time can encode all the alphanumeric characters and thus text can be encoded and stored digitally. A CD has a capacity to store about 5,500 megabits of data which is the equivalent of about 170,000 full pages of text—or all the text in nearly 900 paperback books. Such a storage capacity is indeed tremendous in a small, thin disc that can be inexpensively stamped from melted plastic. The compact disc thus has started an additional life as a storage medium for vast amounts of text or other data used by computers. Since the compact disc can only be read, it is a read only memory, or ROM for short. We thus have the new acronym: CD-ROM.

The marketing question is what to store on the CD-ROM. One application might be to store computer programs, or software, particularly since many programs now need to be distributed on many floppy discs that are relatively costly to produce. Visual images can be digitized and stored on a CD-ROM. Assuming a resolution of 200 dots per inch, a color 3-inch by 5-inch image requires about 7 megabits when digitized. Hence, nearly 800 color images could be stored on one CD-ROM.

The computer is an interactive tool. When used with a computer, the CD-ROM thus also becomes interactive. Different information can be accessed and displayed from the CD-ROM depending upon the actions and responses of the user. We thus have the interactive CD, or CD-I for short.

CD-ROM

Capacity:

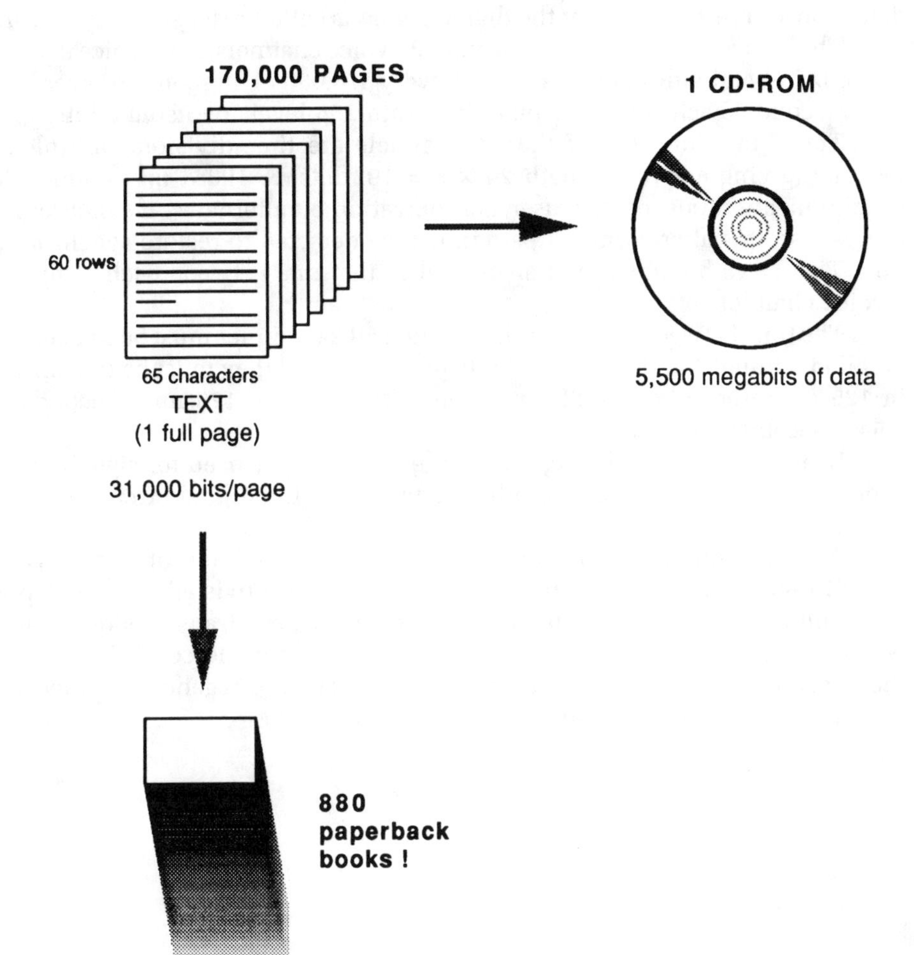

T1 CARRIER

T1 is a digital multiplexing system originally used with twisted pairs of copper wire over short-haul distances of 15 miles in length or so. The digital signal is detected and regenerated every mile along the length of the system. This type of detection and regeneration of the digital signal is called *regenerative repeating.*

The T1 system multiplexes together 24 voice channels. Each voice channel is 4 kHz in bandwidth and is sampled every 1/8,000 of a second, or every 125 microseconds. Each sample is quantized into 256 levels by using 8 bits.

The 8 bits for each of the 24 channels are time-division multiplexed together, giving a run of length $24 \times 8 = 192$ bits. A 193rd bit is added for synchronization purposes. This synchronization bit alternates. If synchronization is lost, this alternating bit pattern can be detected to restore synchronization. These 193 bits must be transmitted in the time between each sample of a voice channel.

The time between each sample in which all 24 samples must be transmitted is called a *frame*. So, there are 193 bits per frame, and they must be transmitted in 125 microseconds. The bit rate then is 193 bits per 125 microseconds, or 1.544 megabits per second.

Every sixth frame, the eighth bit in each sample is used for signaling and supervisory purposes, such as indicating whether the digital circuit is in use or not.

A T1 system uses two channels, one for each direction of transmission. The T1 digital signal can be transmitted over standard twisted pairs of copper wire, and was used by the telephone company when extra transmission capacity was needed between two locations. Rather than incur the cost of tearing up the street to lay new cable, T1 terminal equipment and regenerative repeaters were used to obtain more capacity over existing cables.

T1

T1 System Characteristics:

- digital
- short haul (15 miles)
- regenerative repeater (each mile)
- 24 voice circuits
- two lines required for two-way system

DS1 Bit Rate:

voice channel	= 4 kHz
sampling rate	= 8,000 samples/sec
sampling interval	= 1/8,000 sec = 125 μsec
number of levels	= 256 with 8 bit quantization

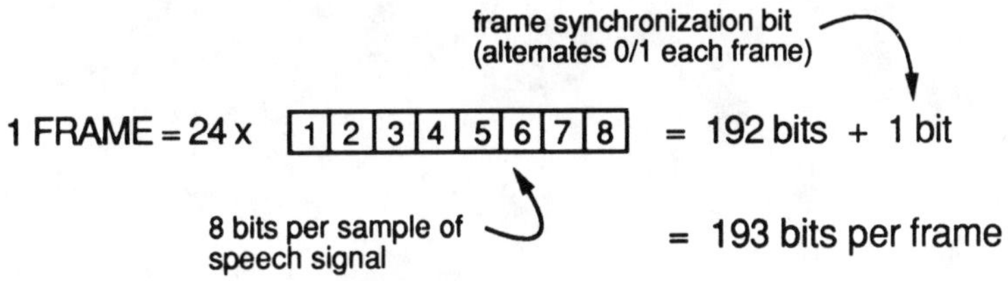

frame synchronization bit (alternates 0/1 each frame)

1 FRAME = 24 x | 1 | 2 | 3 | 4 | 5 | 6 | 7 | 8 | = 192 bits + 1 bit

8 bits per sample of speech signal

= 193 bits per frame

$$\text{bit rate} = \frac{193 \text{ bits}}{125\ \mu\text{sec}} = 1.544 \text{ Mbps} \quad [0.65\ \mu\text{sec/bit}]$$

T1 CARRIER (cont'd)

The T1 carrier system uses bipolar pulses for digital transmission. A 0 is always encoded as the absence of a pulse. A 1 is encoded alternately as a positive or negative pulse. The use of bipolar pulses eliminates any problems caused by momentary direct currents associated with a long string of 1s or 0s.

Time-division multiplexing is accomplished by D-type channel banks, and a digital hierarchy exists depending on the number of channels multiplexed together. A single-channel digital signal at 64 kbps is called a DS-0 signal. Twenty-four of them multiplexed together gives a DS-1 signal at 1.544 Mbps. Four DS-1 signals multiplexed together gives a DS-2 signal at 6.312 Mbps.

Nowadays, digital signals are carried mostly over optical fiber, and light is turned on and off to signify the 1s and 0s of digital. A single strand of optical fiber one-tenth the diameter of a human hair routinely carries data rates of 2 G bps and more. The fiber is so pure that regenerative repeaters are needed only every 100 miles or so.

T1 (cont'd)

Bi-Polar Pulses:

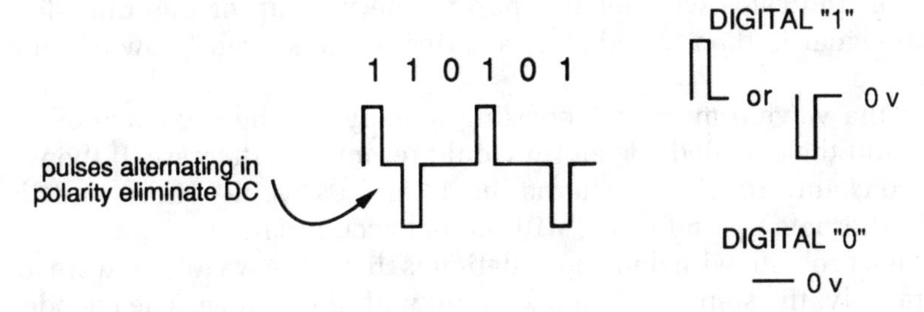

D-type Channel Bank:

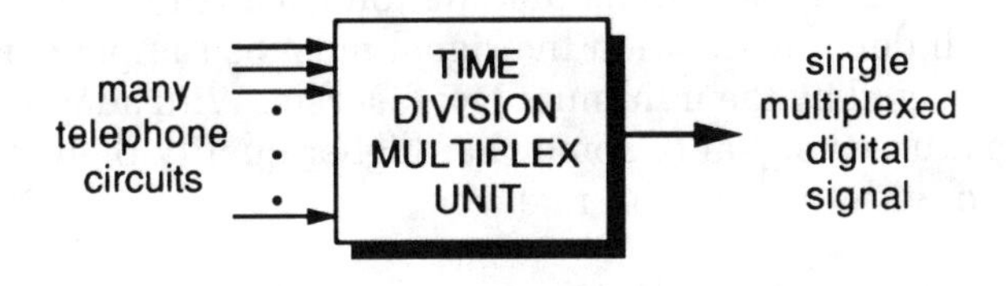

Digital Hierarchy:

DS-0
64 kbps
(single voice channel)

DS-1
1.544 Mbps
(24 channels)

DS-2
6.312 Mbps
(4 DS-1 signals)
or
(96 channels)

DELTA MODULATION

Delta modulation is a technique for digitally encoding a speech signal. In delta modulation, the step size is a single unit which is encoded as a single bit. The single bit indicates whether the step is one unit up or one unit down. The speech signal is then encoded as a series of "ups" and "downs" of a single small step.

If the waveform were increasing quickly, a long sequence of 1s would occur, and the encoded waveform would resemble a staircase. If the waveform were constant, a series of alternating 1s and 0s would occur, and the steps would alternate up and down with no net accumulation.

One problem with delta modulation is that if the waveform were to change very rapidly, the steps could not keep up with the change. The encoded waveform then lags behind the input waveform, creating a form of distortion called slope overload. One solution is to increase the sampling rate. Another solution is to change adaptively the magnitude of the step size. For example, if two positive steps were to occur in a row, then the step size might double for the next step. This type of delta modulation is called *adaptive delta modulation.*

With delta modulation the signal must be sampled at a much higher rate than indicated by the minimum Nyquist rate. With adaptive delta modulation, a reconstructed signal of somewhat higher quality than standard PCM can be achieved with the same bit rate.

Delta Modulation

Process:

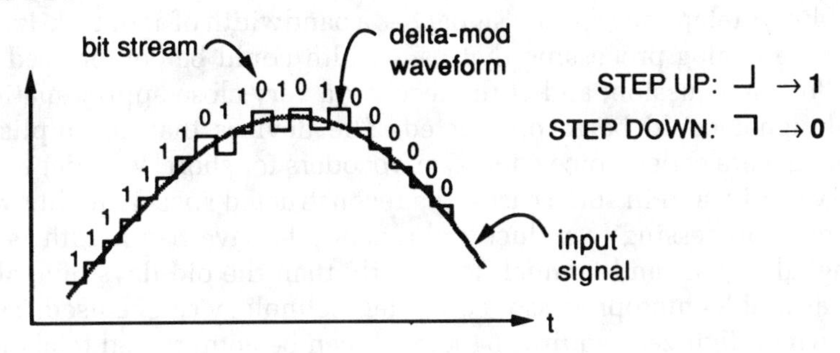

Slope Overload:

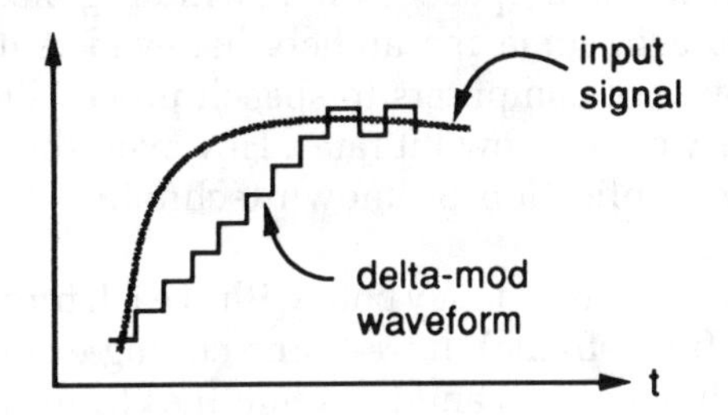

Adaptive Delta Modulation:

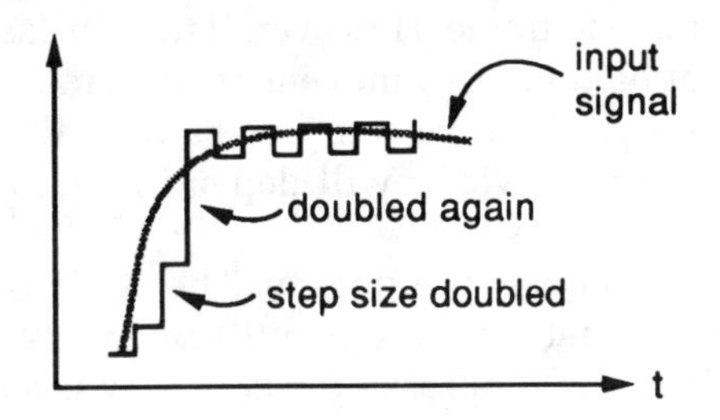

COMPRESSION

It has long been known that there is much redundancy in most signals and that with appropriate techniques the redundancy can be removed, thereby saving bandwidth. A telephone speech signal has a bandwidth of about 4 kHz. Through appropriate analog processing that bandwidth could be compressed to about 400 Hz for transmission, and at the receiver a very close approximation to the original signal could be reconstructed. The devices that accomplished this processing were called voice coders, or vocoders for short. Vocoder technology was very costly, and in some cases, the reconstructed speech quality was poor.

Signal processing to reduce redundancy to save bandwidth is possible with digital signals and is much less costly than the old days of analog since readily available microprocessor computer technology can be used. Telephone speech when digitized requires 64 kbps. It can be compressed to about 8 kbps with reasonable speech quality, although some buzziness and roughness is frequently present. With further processing, speech can be reduced to a bit rate of about 1.2 kbps, but the reconstructed speech, though intelligible, sounds very machine like. A few years ago, some researchers believed that simply applying the ever increasing power of computers to speech processing would solve the problems of poor quality at very low bit rates. However, they discovered that the solution is not the application of known technology but rather the need for new basic knowledge.

Video images have considerable redundancy both within each frame (intra-frame) and from frame to frame (inter-frame). Intra-frame coding examines bit patterns along a scan line (intra-line coding) and between lines (inter-line coding). A repertoire of specific patterns can be constructed and used to encode their frequent occurrence (pattern coding). Unless there is very fast motion, the information from frame to frame changes little. Inter-frame coding examines and sends only the changes from frame to frame. However, if there is fast action, the reconstructed images might blur or have jerky movement. Full motion video requires about 70 Mbps. The video signal can be compressed to as little as 800 kbps, but the quality of the reconstructed video will depend much upon the nature of the image and action.

Compression is always a compromise with quality. The study of human factors affecting quality can help illuminate the acceptability of quality degradation for different compression schemes. Compression technology is advancing rapidly and changing frequently, and this makes the choice of standards quite difficult.

Compression

Video:

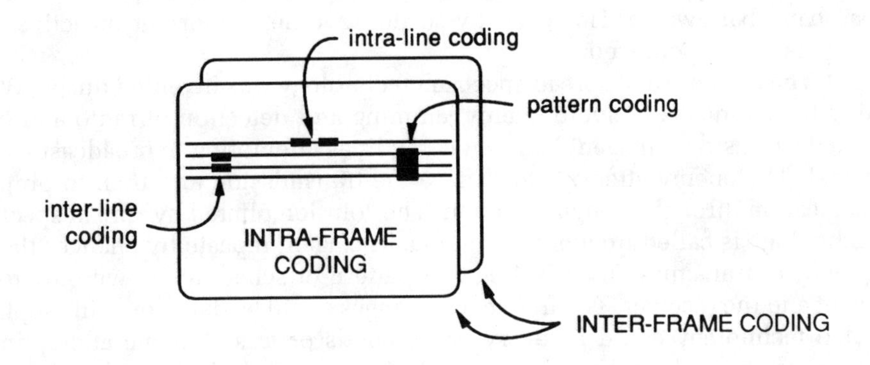

Table:

	full bandwidth	medium compression	highly compressed
• telephone speech	64 kbps	8 kbps	1.2 kbps
• video	70 Mbps	6 Mbps	1.5 Mbps 800 kbps

SPREAD SPECTRUM

In telecommunication, one usually attempts to conserve or save bandwidth. The technology of spread spectrum is exactly the opposite philosophy. The bandwidth required for the transmitted signal is greatly increased over its baseband bandwidth. However, by so doing, other important objectives and advantages are achieved.

The basic idea of spread spectrum technology was invented during World War II as a means to avoid enemy jamming and detection of radio and other signals. Consider an agent behind enemy lines attempting to broadcast a radio signal. The enemy attempts to receive the transmission and then to pinpoint its location through triangularization. The solution offered by spread spectrum technology is called frequency hopping. The agent repeatedly changes the frequency of transmission according to a pattern or schedule known only to the agent and the receiver. The frequency changes could be daily, or using sophisticated technology could be every few seconds or less. Anyone attempting to eavesdrop would not be able to search the band and pinpoint the actual frequency being used for a short instant of time.

Code multiplication is another form of spread spectrum technology that is used with digital signals to enable them to share a transmission medium or channel. It therefore is a form of code division multiplexing or code division multiple access (CDMA). Consider a baseband digital signal at 1 kbps with a bit width of 1 ms. The bandwidth of this baseband digital signal is about 500 Hz. This baseband digital signal is multiplied by a random digital signal (called the code signal) at 1,000 times the rate of the baseband digital signal, or a code rate of 1 Mbps. The code signal has a bandwidth of about 500 kHz. The effect of the multiplication is a new signal that has a bandwidth of 500 kHz, and this new signal would then be used to modulate an appropriate carrier for transmission. Thus the information in the baseband digital signal has been spread over a bandwidth 1,000 times greater.

At the receiver the inverse process is performed to obtain the original baseband digital signal. After demodulating the carrier, the demodulated signal is multiplied by the same identical random code in exact synchrony with its original application. Since +1 times +1 and −1 times −1 are both +1, the multiplication at the receiver by the identical code recreates the original baseband digital signal. If any other signals are present, their codes will be different and they will not pass through the inverse multiplication process. In this way, a number of baseband digital signals all with their own unique patterns of random codes can share the same communication channel.

Code multiplication is still somewhat novel and is currently being considered for use in digital cellular telephony. The technical challenge is the requirement of absolute synchrony of the transmitter code with the receiver code.

Spread Spectrum

Frequency Hopping:

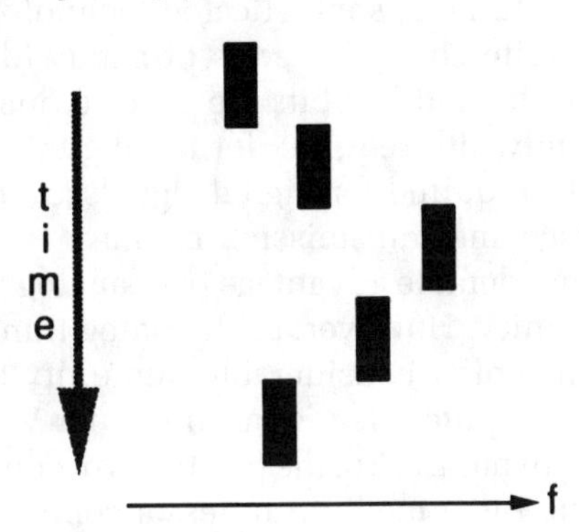

Code Multiplication:

- **TRANSMITTER PROCESS**

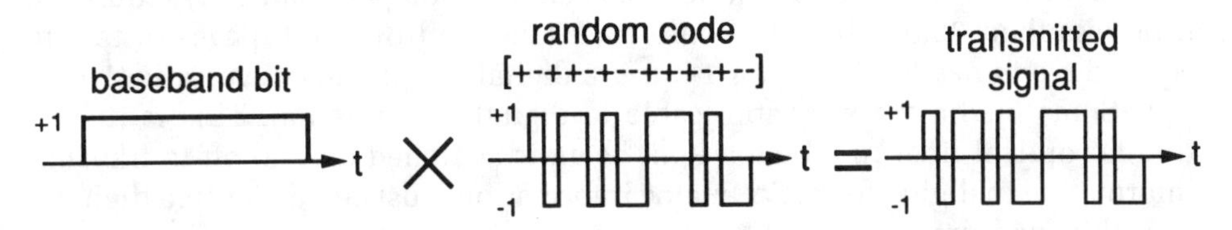

- **RECEIVER PROCESS**

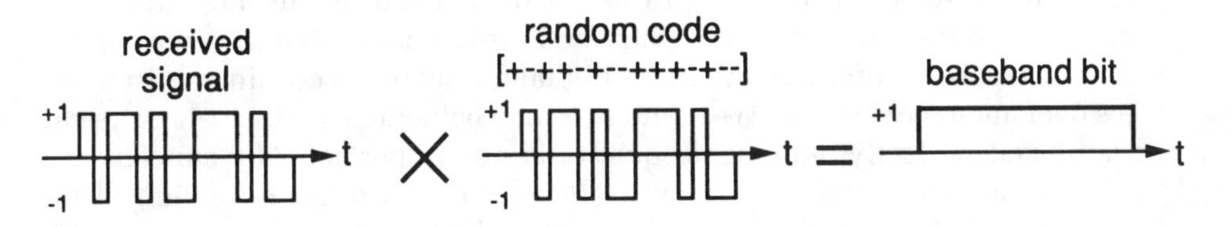

ANALOG VERSUS DIGITAL

The whole world seems to be going digital! But the decision to utilize a digital format for an analog signal should be an engineering-economics decision, and not one motivated by a fascination with new, sophisticated technology.

The use of digital used to be quite costly in terms of bandwidth. If the bandwidth of the analog signal is W Hz, and if n bits are used to quantize the amplitude of the signal, then the bandwidth required for the digital version is at least nW Hz. The large bandwidths required for digital signals are no longer a problem with today's optical storage and transmission media.

Digital transmission offers a considerable advantage over analog baseband transmission in terms of noise immunity. However, such analog transmission schemes as frequency modulation also offer considerable immunity to noise.

Data communication between computers is digital in nature. Voice communication between people is analog in nature. So, the two types of communication are difficult to combine over the same medium unless a common format is used for both. The conclusion is that this common format should be digital. Indeed, modern long-distance telephone networks are all digital. However, the telephone signal carried over the local loop from the home to the serving telephone company is still analog.

From a traffic consideration, voice traffic exceeds such text traffic as e-mail. When converted to digital, 1 second of speech requires 64,000 bits. A page of text (60 rows of 65 characters, each at 8 bits per character) requires a little less than 32,000 bits. It would then take well over 100 pages of text to equal the bits needed for just a 1-minute digital telephone conversation!

Some media are very applicable to digital transmission. This is indeed true for optical fiber in which a light beam is switched on and off to transmit information in digital form. So, analog information must be transmitted digitally over this medium.

Another factor favorable to digital is the availability of inexpensive integrated circuits to perform digital processing of data and signals. Linear, analog, integrated circuits are still quite costly or simply unavailable in some cases.

The recording of information in a digital format offers considerable advantages over analog recording. Wear and tear may be nonexistent, most errors can be corrected perfectly, and signal quality is nearly perfect. The concept of a "copy" disappears, and each copy itself becomes an original version of the "original." So, the decision of analog versus digital depends on the specific application.

Analog vs Digital

Communications:

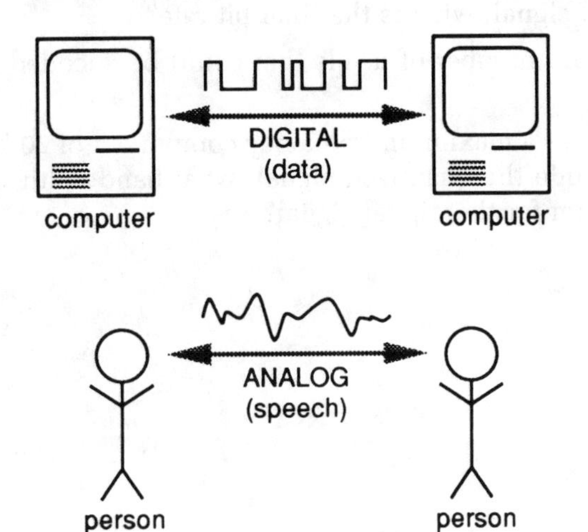

Text-Speech Equivalent:

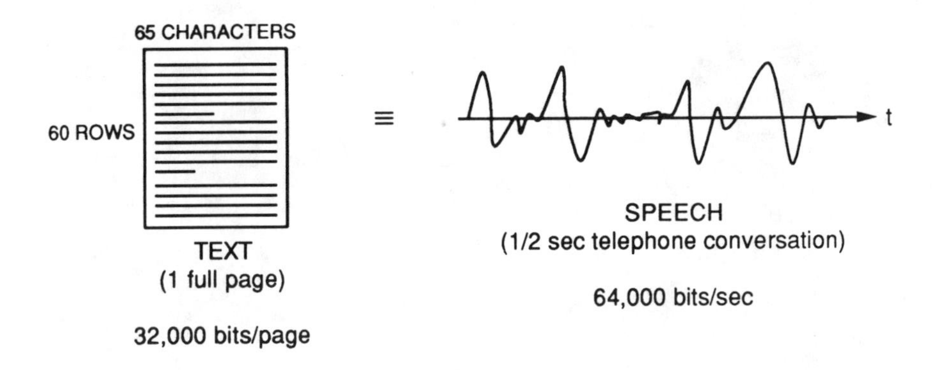

PROBLEMS

5.1. A monochrome television signal has a maximum frequency of 4.2 MHz. What is the minimum sampling rate for this signal? If 8 bits are used to encode the sampled signal, what is the final bit rate?

5.2. What is the maximum number of levels that could be encoded by using 8 bits?

5.3. High-fidelity audio has a maximum frequency component of 20 kHz. If 16 bits are used to encode the quantized signal, what bandwidth would be needed as a minimum for the digital signal?

Chapter 6: Computers and Data Communication

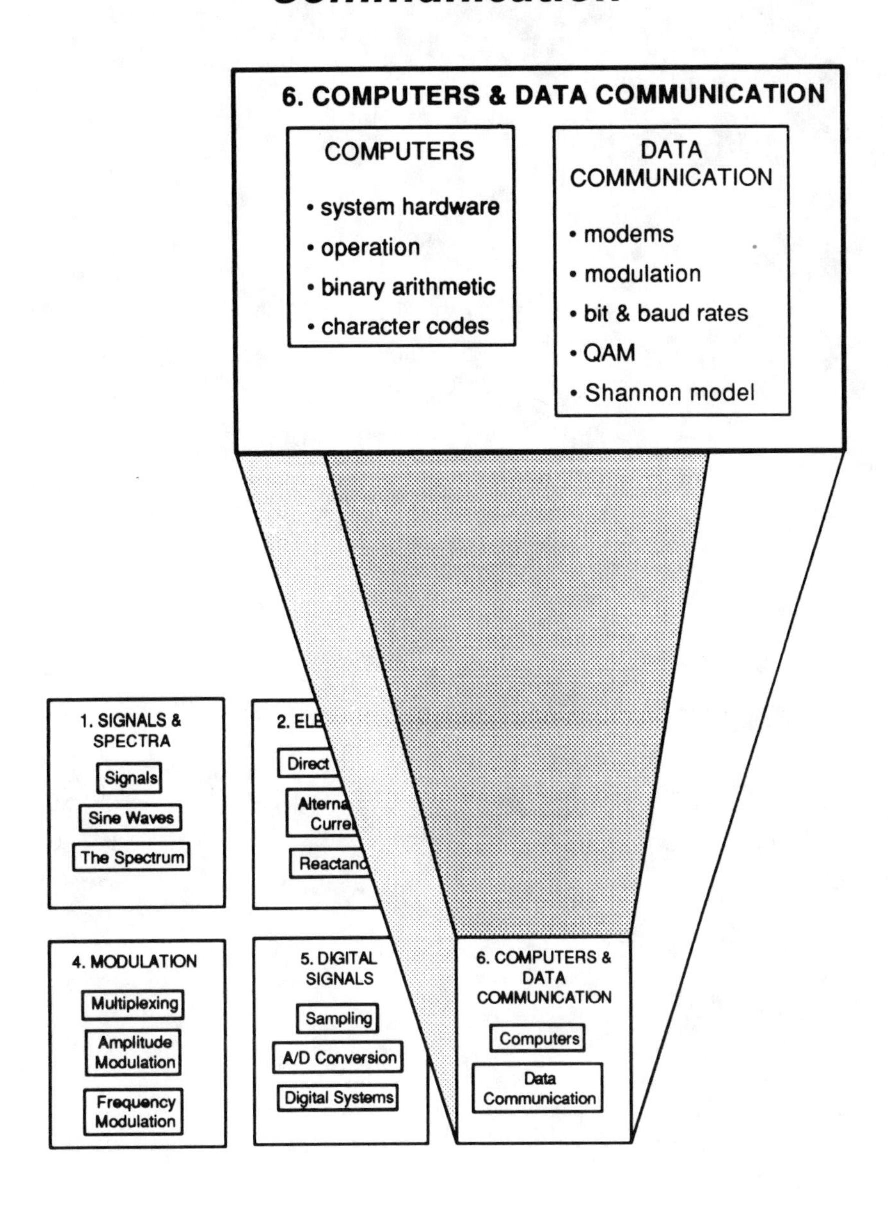

6

Introduction to Computers and Data Communication

The computer with its ability to process information in digital form has had a tremendous effect on the modern world of communication and information processing. The information stored digitally in computers can be accessed and retrieved remotely via the telecommunication of digital data. This module treats computers and data communication.

A digital computer consists of electronic circuits that manipulate digital signals which change between two values to signify the binary digits, or bits. A unique aspect of the computer is that it is a general-purpose machine that can be instructed to perform different functions in response to a set of instructions called a *program*. The aspect of computers that deals with their electrical circuits and physical devices is called *hardware*, while the term *software* refers to the programming aspects.

This module begins with an explanation of the hardware aspects of a digital computer. The hardware aspects are then expanded to include an example of a simple computer system and a simple program written in machine code. This example is then expanded to show how higher-level programming languages have evolved. The computer portion of the module ends with some descriptions of the fundamental workings of the actual digital circuit elements that comprise a computer.

Data communication is treated in the last portion of this module. Data communication involves the transmission of binary and digital data over telecommunication lines. This is accomplished by modulating a sine wave with the digital signal. The various forms of digital modulation are described along with the characteristics of the modulator-demodulators, called *modems*, that perform the operations.

Computers

COMPUTERS AND SOCIETY

It is virtually impossible to avoid computers in our modern society. At work, they are used to manipulate our writing in word processors, to calculate our earnings, and to control manufacturing and inventory processes. On the way to work, they monitor and control the functioning of our automobiles and mass transit systems. At home, they control our microwave ovens and washing machines, maintain the proper functioning of our heating and cooling systems, and interact with us when playing games on our TV sets.

Computers can be incorporated into and hidden within another product, like an automobile or a clothes washer. Or, a computer can be clearly recognized by its keyboard and visual display, and may be as close as your colleague's office next door.

Personal computers for general-purpose use are appearing in the homes of consumers, where they are used for such applications as games, word processing, and spread-sheet calculations. The personal computer in the home is no longer the toy of the computer hobbyist, but instead is being purchased and used by regular folks like you and me.

Computers are very powerful tools when used for their computing capabilities and the wealth of information that can be stored locally. However, computers can also be used as a communication tool to access and retrieve vast amounts of information stored distantly at other computerized data banks. This form of remote data access is made possible by a device called a *modem*, which allows the computer to use the switched, voice, telephone network for digital data communications.

With remote access, it is now possible to gain access to databases carrying detailed information on virtually any topic from airline schedules to legal opinions. Another exciting data communications service utilizing computers is the ability to send and retrieve textual messages, a form of electronic mail. It is also possible to retrieve software—the programmed sets of instructions that control the operation of computers—from distant locations.

Computers & Society

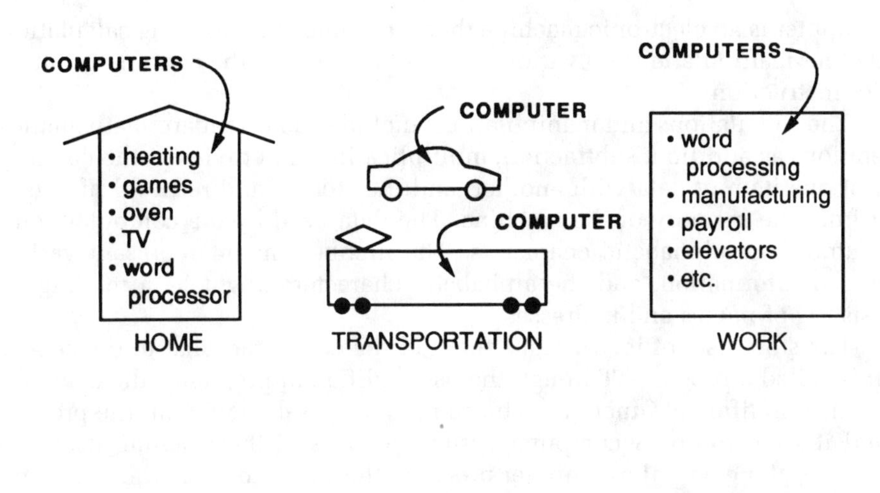

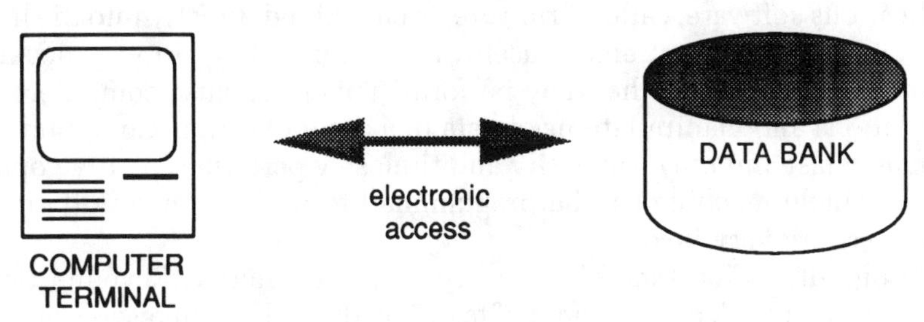

THE COMPUTER—DEFINITION

A computer is an electronic machine that is capable of performing calculations and other manipulations on various types of data under the control of a stored set of instructions.

The calculations and manipulations include such standard mathematical operations as addition, subtraction, multiplication, and division; the comparison of one item of data with another; and the storage and retrieval of data to and from the memory of the machine. The data used by the computer could be numbers or alphabetic characters. The numbers might represent various financial information, and the alphabetic characters might be a mailing list consisting of names and addresses.

The stored set of instructions that tell the computer what to do step-by-step is called a *program*. Through the use of different programs, the computer can perform different functions. The computer is useless without the program to tell it what to do. A computer system consists of the machine itself, the *hardware*, along with the computer program, the *software*. Sometimes, specific applications software, called *firmware*, is embedded directly into the hardware.

Computers are extremely useful tools, since they are very accurate and exact in the operations that they perform. This is because computers perform calculations and manipulations of data in a digital fashion. Computers are also extremely fast and they can easily and tirelessly perform repetitive operations. Since a single machine can be programmed to perform many different tasks, computers are very flexible.

Computers, for example, are used by the Internal Revenue Service to calculate and check our payment of taxes, by the airlines to reserve and confirm our seats on airplanes, by department stores to calculate our bills, by our secretaries and word processors to type our letters and memoranda, and by our manufacturing plants to control operations on assembly lines. Computers have become essential to the workings of our society and industry. Rather than having dehumanized society, computers have enabled a multitude of services to meet our individual needs that otherwise would be impossible.

The Computer

Definition:

A computer is:

AN ELECTRONIC MACHINE

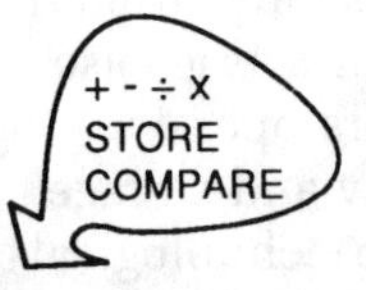

THAT PERFORMS CALCULATIONS AND MANIPULATIONS

ON DATA

UNDER THE CONTROL OF A STORED SET OF INSTRUCTIONS.

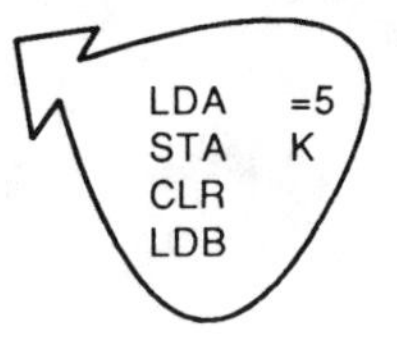

COMPUTERS—SIZES AND TRENDS

Computers come in many sizes and shapes. Some are large machines (called *mainframes*) costing hundreds of thousands of dollars. Others are medium-sized *minicomputers* costing tens of thousands of dollars. Still others are small *microcomputers* costing less than a few thousand dollars. These smaller machines range from desktop personal computers to portable laptop machines that fit in a briefcase and calculators that fit in a shirt pocket. All of these computers operate under the same basic principles, but differ in computing capability and storage capacity.

The technological trend has been for storage capacity and computing speed to increase, while cost and size have been dramatically decreasing. Computers of the past were used mostly by industry. However, the costs of computers have now decreased to the point where they are being purchased by consumers for use in their homes. Home computers are used for such applications as games, word processing, and spread-sheet calculations. Some futurists predicted that by 1990 three-quarters of American homes would have their own personal computer. This heavy penetration will most likely occur early in the next century.

Computers

Sizes:

- **Large** -- MAINFRAME
- **Medium** -- MINI & WORK STATIONS
- **Small** -- MICRO
  desktop
  portable, laptop, notebook
  calculator, calendar, organizer

Trends:

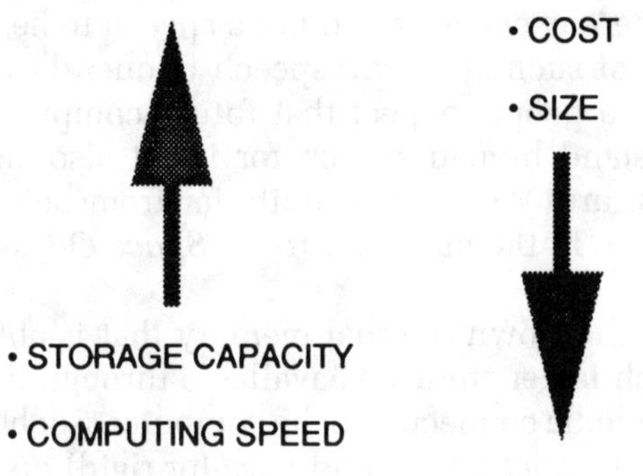

COMPUTER SYSTEM HARDWARE

A computer by itself is useless without the means to communicate with the outside world through various input and output devices.

Input to the computer could be with an alphanumeric keyboard similar to a conventional typewriter keyboard, with punched cards, or with knobs that are turned by the user. These and other forms of input are converted into discrete digital data used by the computer for various calculations and other manipulations.

The results of the calculations and manipulations performed by the computer are transmitted as output to a variety of devices. Output from the computer is frequently visual in nature and may be displayed on a TV-like *cathode-ray-tube* (CRT) screen in the form of rows of text, or perhaps a computer-generated drawing or graph. The computer output could be to a printer that generates a *hard copy* of textual or graphical information on paper. The output on a CRT display is called *soft copy.*

Computers can also generate the electrical signals needed to control a speech synthesizer, thereby enabling synthetic speech to be used as an output medium. The quality of such synthetic speech frequently leaves much room for improvement. Some people expect that future computers will be able to recognize and understand human speech for input also, although that time might be far in the future. We are still quite far from being able to create a computer that, like HAL in the movie *2001—A Space Odyssey*, is able to read lips!

Computers have their own internal memory that is able to store limited amounts of data. Much larger storage is available through the use of external, or *peripheral*, storage media connected to the computer. Such peripheral storage includes *floppy disks, magnetic tape*, and *hard* (or rigid) *disks*. Computers are increasingly using storage media based on optical technology, including lasers that read and write digital data.

To summarize, a computer system consists of four major hardware elements:

- The computer itself;
- Various input services;
- A variety of output devices;
- Peripheral storage.

There is one other aspect of a computer system that is as important as the computer hardware itself, namely, the computer program, or software, that makes the computer functional. The operation of a computer, including the interplay between the hardware and the software, is described in the following section.

Basic Computer System

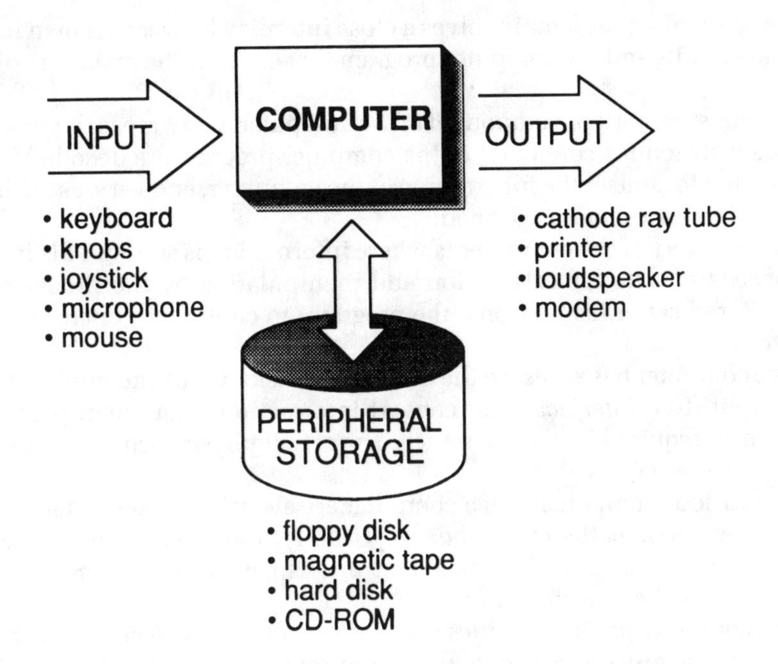

COMPUTER OPERATION

The operation of a computer involves a close interplay between its own internal electrical circuits and the computer program that controls the operation of these circuits.

The most important component of the computer hardware is the *processor.* Here, the instructions contained in the computer program are decoded, signals are generated to control the functioning of the computer, and various arithmetic and logic operations are performed.

The *memory* of the computer is where information is stored. This information consists of data for calculation and manipulation by the processor and also the stored set of instructions (the program) to control the operation of the computer itself.

The computer has access to the outside world for input and output through input-output (I/O) *interfaces* that convert information from the outside world into the form required by the computer. The actual physical connections to the outside world are through I/O *ports.*

The various components of a computer are electrically connected together by sets of wires or paths called *buses.* Usually, these paths simultaneously carry a number of electrical signals, and this is called *parallel transmission.* More will be said about this later.

The computer program resides in a portion of memory. A single instruction is fetched by the processor, decoded, and appropriate control signals are issued to control the operation of the computer. For example, control signals may be issued to cause the computer to fetch numerical data from some location in memory and add the data to other data in the adder portion of the processor. The next instruction may cause control signals to be issued to store the contents of the accumulator in some other location in memory in preparation for final output to the printer.

Each instruction in the program is fetched and processed in sequence, one instruction at a time because a computer is a sequential processing machine. Researchers are investigating some newer computers that can simultaneously process a number of instructions, called parallel processing, offering considerably greater speed for certain highly repetitive calculations.

Internal Operations

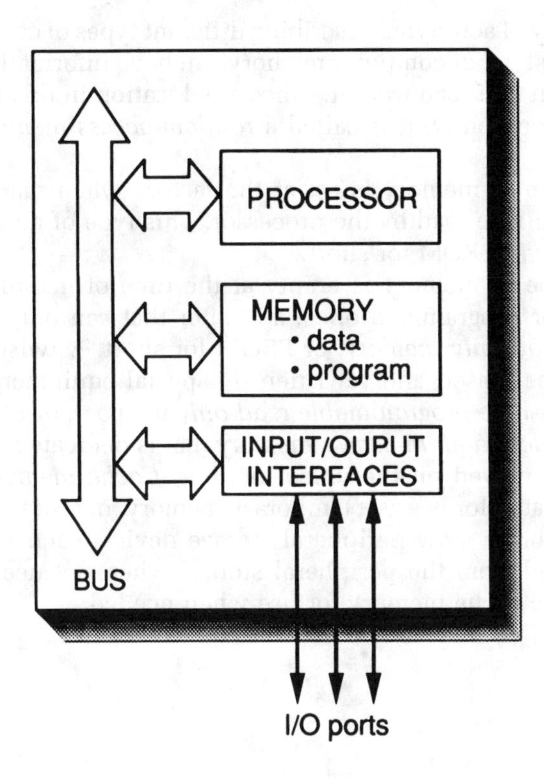

COMPUTER MEMORY TYPES

There are a variety of acronyms describing different types of computer memory. The most flexible type of computer memory can have information both written (or entered) into it and read from it. Since any location in this type of memory can be accessed randomly, it is called a *random-access memory*, or RAM for short.

Another type of memory is set at the factory when manufactured, and afterwards can only be read by the processor. This type of memory is called a *read-only memory*, or ROM for short.

Another type of memory is empty at the time of manufacture and can be written into or programmed once, and after that can only be read. It is a *programmable read-only memory*, or PROM for short. A twist on this PROM is one that can be erased and rewritten by special equipment. This type of memory is an *erasable programmable read-only memory*, or EPROM for short.

VRAM is *video random-access memory* that is dedicated to the storage of data which is displayed on the video monitor. *Cache memory* is RAM that typically is dedicated for use as a temporary memory, or buffer, to speed access to information from a slow peripheral storage device, such as a floppy disk drive. Data is read from the peripheral storage when not needed and stored temporarily in the cache memory for use when needed.

Memories

RAM *Random-Access Memory*

ROM *Read-Only Memory*

PROM *Programmable Read-Only Memory*

EPROM *Erasable Programmable Read-Only Memory*

VRAM *Video Random-Access Memory*

BINARY ARITHMETIC

The numerical decimal information stored and processed within a computer is represented in binary form. This type of representation of information is very precise and affords a high degree of accuracy to the calculations and operations performed by the computer. The computer is a binary machine operating on discrete digital data.

We saw earlier that, as contrasted with the decimal numbering scheme which is based on powers of 10, the binary numbering scheme is based on powers of 2. For example, the binary number 101 is equivalent to 1 times 4, plus 0 times 2, plus 1 times 1. So, 101 in binary is the same as 5 in decimal. The encoding of decimal information in this fashion is called *binary coded decimal* (BCD) encoding. A binary digit is called a *bit* for short. So, in the preceding example, the binary number 101 has a length of 3 bits. A bit can be a 0 or a 1. A binary number is a series, or *string*, of 0s and 1s.

Binary addition is very simple. Zero plus 0 equals 0; 0 plus 1 equals 1; 1 plus 1 equals 0 and a carry of 1. Like decimal addition, we start at the far right with binary addition and apply the preceding simple rules.

The representation of bits with electrical circuits is very easy, since a simple threshold decision is all that is needed to decide whether a 0 or a 1 is indicated. If the signal exceeds the threshold, it is a 1; if it is less than the threshold, it is a 0. Binary circuits employ *Boolean algebra*, which is based on the simple presence or absence of a signal.

An electrical switch is a form of digital memory, since it is in one of two positions or states, either "on" or "off." A transistor circuit that is either conducting or not conducting electricity can then store 1 bit of binary information. Many such circuits can be combined together on a small chip of silicon, forming memories capable of storing hundreds of thousands of bits.

A digital computer can be as small as a single integrated circuit that might be used to calculate and display the time on a wristwatch. Or, a digital computer can be a massive installation with many disk drives and many large processors used to update and maintain an airline reservation system. Whatever their size, all digital computers are based on the principles of binary arithmetic, and they operate according to the previously described principles.

Binary

Binary Representation:

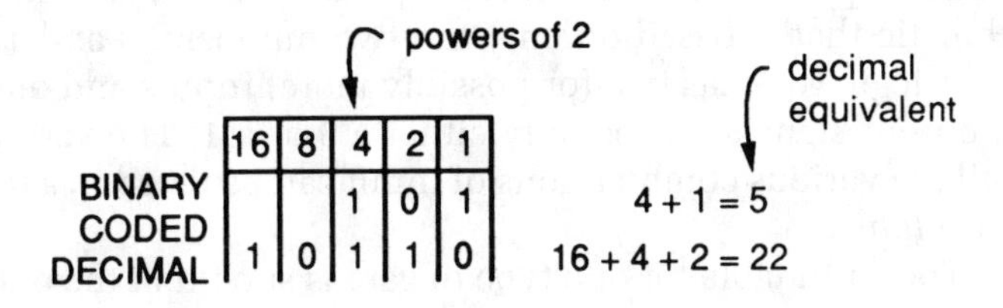

Binary Arithmetic:

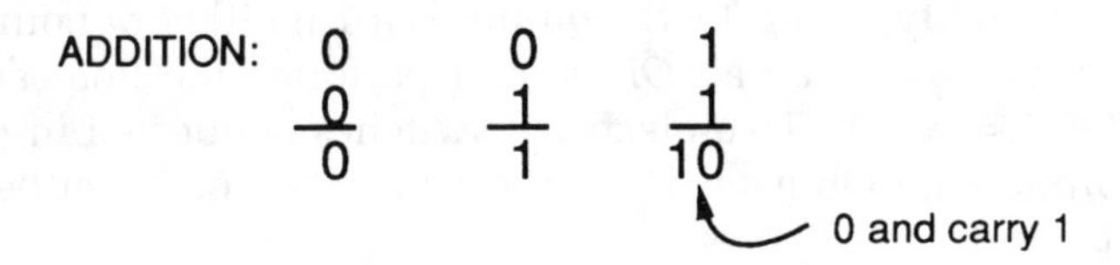

EXAMPLE:

carry 1

1 0 1	→ 5
1 0 1 1 0	→ 22
1 1 0 1 1	→ 27

LOGIC GATES

The actual electrical circuits of a digital computer are composed of elements called *logic gates*. These logic gates perform the elementary addition and multiplication operations of Boolean algebra. Boolean algebra deals with binary arithmetic that is based on only the two numbers, 0 and 1.

A logic gate has two (or possibly more) inputs and one output. The input and output signals can be only either a 0 or a 1. The various possible outputs for all the various combinations of input can be listed in a table, which is called a *truth table*.

The truth table for one type of gate is such that the output can be a 1 only if both inputs are 1s. In other words, the output is a 1 only if input A *and* input B are each a 1. This type of gate is called an *AND gate*. Two electrical switches connected in series would be an example of an AND gate. The lamp in the circuit would be "on" only if both switches were closed or in the *on* position. The Boolean operation of multiplication is performed by an AND gate. In Boolean terms, OUT = AB.

For a second type of gate, the output is a 1 if either *or* both inputs are 1s. This type of gate is called an *OR gate*. It performs the Boolean operation of addition, OUT = A + B. Two electrical switches connected in parallel would be an example of an OR gate. The lamp would be "on" if either switch were in the *on* position.

Another Boolean operation is the negation of a Boolean signal. The negation of a 0 is equivalent to stating "not a 0," which obviously must be a 1. In Boolean algebra, the operation of negation is indicated by drawing a line above the symbol to be negated. The operation of negation simply flips a signal to its opposite value. If the output of an AND gate is negated, the result is called an *NAND gate* (for "NOT AND"). Similarly, the negation of an OR gate gives a *NOR gate* (for "NOT OR").

Various symbols are used to represent AND and OR gates in computer and other digital circuits. The gates themselves are made from transistors biased in such a way that they are either conducting or not conducting, depending on the input signal. A transistor biased in this way performs as a controllable switch.

There is a second type of OR gate that is called an *exclusive-OR gate*, or XOR. The output of an exclusive-OR gate is 1 when either input is 1, but not when both inputs are 1s.

Logic Gates

AND Gate:

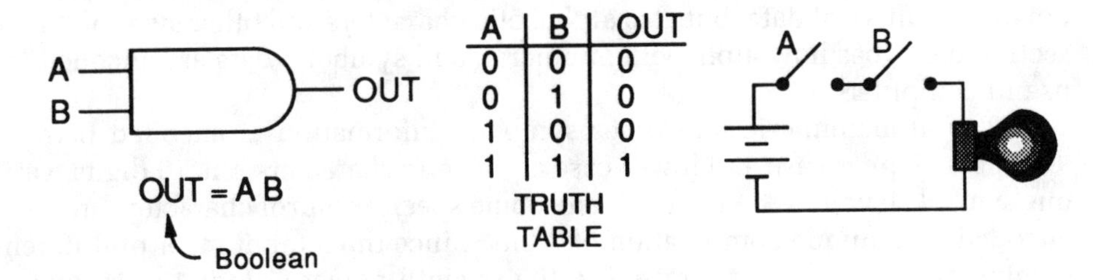

A	B	OUT
0	0	0
0	1	0
1	0	0
1	1	1

TRUTH TABLE

OR Gate:

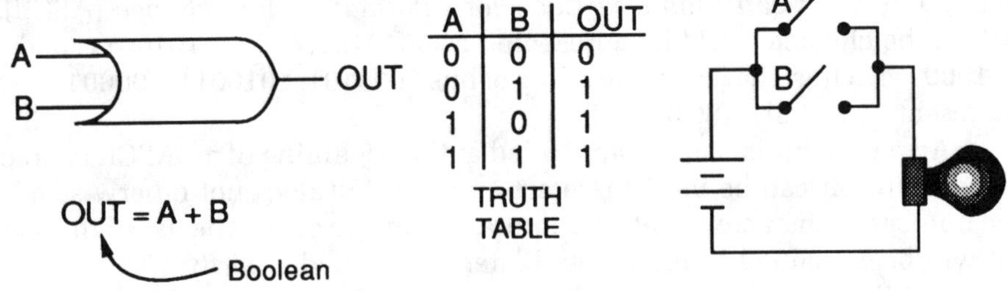

A	B	OUT
0	0	0
0	1	1
1	0	1
1	1	1

TRUTH TABLE

Inverter:

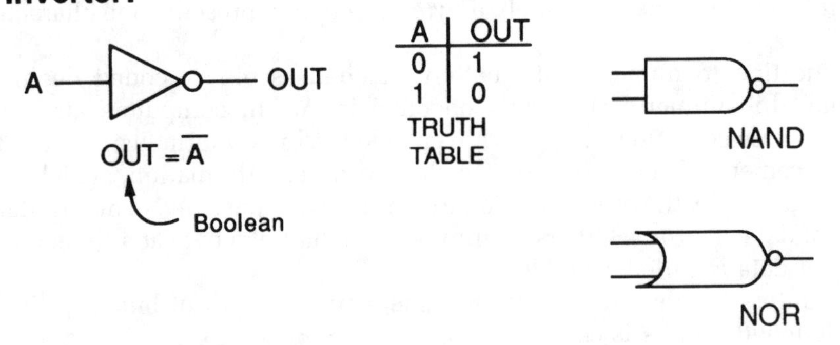

A	OUT
0	1
1	0

TRUTH TABLE

CHARACTER CODES

As we just described, numerical decimal information is encoded in a digital computer in binary form. For example, the decimal number 13 would be encoded as 1101 in binary form (8 + 4 + 1). However, computers can process not only numerical data, but also alphabetic characters and other symbols. This section describes how alphabetic, numeric, and symbolic data are encoded for use in computers.

The alphanumeric and other symbolic information is encoded using a 7-bit code. Upper-case and lower-case alphabetic characters, the 10 digits, various symbols (such as $, &, +, etc.), and some special control characters are each encoded as a unique combination of 7 bits. Since the total number of different combinations of 7 bits is 2 raised to the seventh power (2^7), or 128, a total of 128 different characters can be represented by using a 7-bit code.

The standard 7-bit code for representing alphanumeric and other characters is the *American Standard Code for Information Interchange* (ASCII). In ASCII, the character "U" is represented as 1010101, "S" as 1010011, and "C" as 1000011. Thus, the following string of bits 1010101, 1010011, 1000011 would represent "USC" in ASCII.

An eighth bit is usually appended at the beginning of an ASCII character. This eighth bit can be used to detect an error, but does not otherwise add to the number of characters that are represented by the code. The use of this eighth bit will be explained in more detail later. It is called a *parity bit.*

ASCII is usually envisioned as an 8-bit code, even though only 7 bits are actually used to encode information. A group of 8 bits is so common in computers that it has its own name, namely a *byte.* One byte represents one character in ASCII.

In addition to numeric information encoded as binary coded decimal (BCD) and alphanumeric characters encoded in ASCII, computers also use programs. These computer programs are encoded in binary form, using formats for the placement of the instruction code and other information, which are decoded and used by the processor. Some computers are designed to manipulate one byte of data at a time. Others manipulate one-half of a byte at a time. One-half byte of data is called a *nibble.*

Computers are designed to manipulate a fixed length of bits at a time. This basic length of bits is called a *word,* and the longer the word length, the more bits that are manipulated per single instruction. Computers typically come in word sizes of 8, 16, or 32 bits.

Character Codes

AMERICAN STANDARD CODE FOR INFORMATION INTERCHANGE

ASCII

7-BIT CODE:

- upper & lower case alphabetic characters
- ten digits
- symbols (+=-&*%#$!,etc.)
- control characters

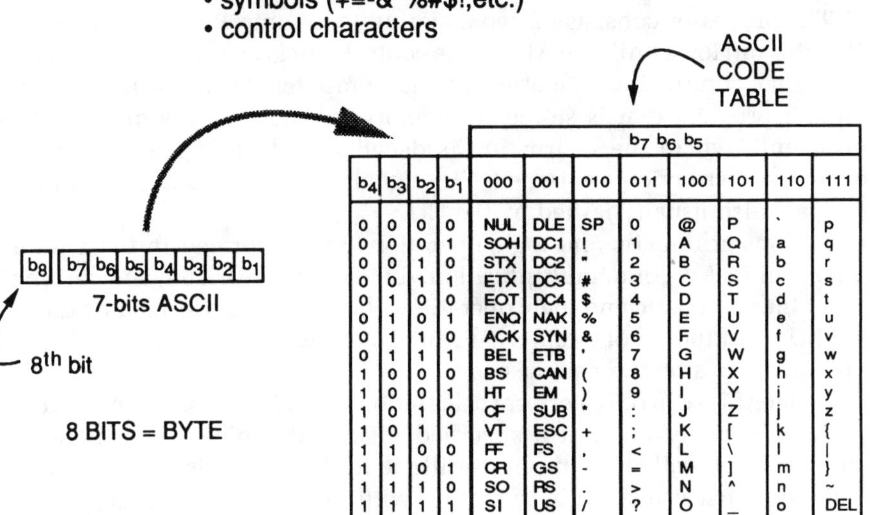

b_4	b_3	b_2	b_1	$b_7 b_6 b_5$ = 000	001	010	011	100	101	110	111
0	0	0	0	NUL	DLE	SP	0	@	P	`	p
0	0	0	1	SOH	DC1	!	1	A	Q	a	q
0	0	1	0	STX	DC2	"	2	B	R	b	r
0	0	1	1	ETX	DC3	#	3	C	S	c	s
0	1	0	0	EOT	DC4	$	4	D	T	d	t
0	1	0	1	ENQ	NAK	%	5	E	U	e	u
0	1	1	0	ACK	SYN	&	6	F	V	f	v
0	1	1	1	BEL	ETB	'	7	G	W	g	w
1	0	0	0	BS	CAN	(	8	H	X	h	x
1	0	0	1	HT	EM	)	9	I	Y	i	y
1	0	1	0	CF	SUB	*	:	J	Z	j	z
1	0	1	1	VT	ESC	+	;	K	[	k	{
1	1	0	0	FF	FS	,	<	L	\	l	|
1	1	0	1	CR	GS	-	=	M	]	m	}
1	1	1	0	SO	RS	.	>	N	^	n	~
1	1	1	1	SI	US	/	?	O	_	o	DEL

CONTROL CHARACTERS

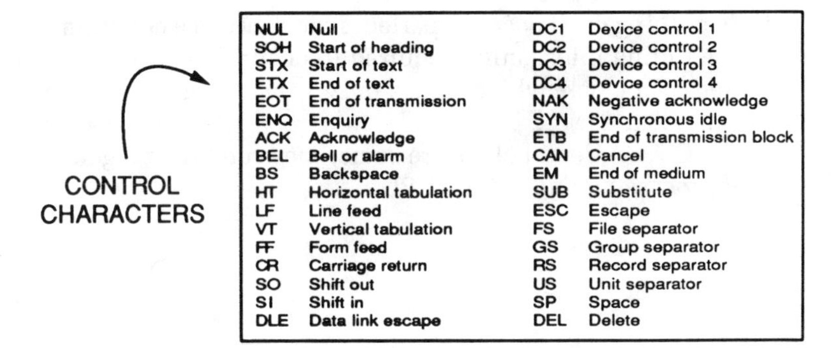

NUL	Null	DC1	Device control 1
SOH	Start of heading	DC2	Device control 2
STX	Start of text	DC3	Device control 3
ETX	End of text	DC4	Device control 4
EOT	End of transmission	NAK	Negative acknowledge
ENQ	Enquiry	SYN	Synchronous idle
ACK	Acknowledge	ETB	End of transmission block
BEL	Bell or alarm	CAN	Cancel
BS	Backspace	EM	End of medium
HT	Horizontal tabulation	SUB	Substitute
LF	Line feed	ESC	Escape
VT	Vertical tabulation	FS	File separator
FF	Form feed	GS	Group separator
CR	Carriage return	RS	Record separator
SO	Shift out	US	Unit separator
SI	Shift in	SP	Space
DLE	Data link escape	DEL	Delete

INTERNAL HARDWARE CONFIGURATION

The electronic equipment and circuitry that forms a digital computer is called the computer hardware. The instructions that comprise the program that controls the operation of the computer is called the computer software. A computer system consists of both hardware and software, and one without the other is useless.

The internal organization of the computer circuitry is more involved than we have previously described. To understand the interplay between the hardware and the software, the internal configuration of this circuitry must be described in more detail, particularly regarding the processor portion.

The processor consists of two major units: a control unit (the CU) and an arithmetic and logic unit (the ALU). The control unit issues the various electrical signals that control the operation of the computer. Each instruction of the computer program that is stored in memory is entered individually into the control unit, where the instruction is decoded and the signals necessary to execute the instruction are issued. The signals necessary to initiate input and output are also usually issued by the CU.

Various arithmetic and logic operations are performed in the arithmetic and logic unit. Addition, multiplication, division, comparison, and shifting are some of the arithmetic and logic operations that may be performed by the ALU. The ALU is composed of circuits called registers and accumulators, where data are temporarily stored for processing.

Some ALUs can only perform calculations with integers. Since the decimal point is effectively fixed at a specified place with integer arithmetic, such calculations are called fixed-point arithmetic. Other ALUs can also perform calculations that use scientific notation with powers of 10. Because here the decimal point is not fixed, but in effect can float anyplace, these calculations are called floating point.

Everything must occur in perfect synchrony in a computer and at a precise rate. The timing information required to ensure this synchrony and precision is created by a clock, which issues pulses that determine when various operations can occur.

The CU, ALU, and clock are sometimes combined together and called the *central processing unit*, or the CPU.

Internal Configuration

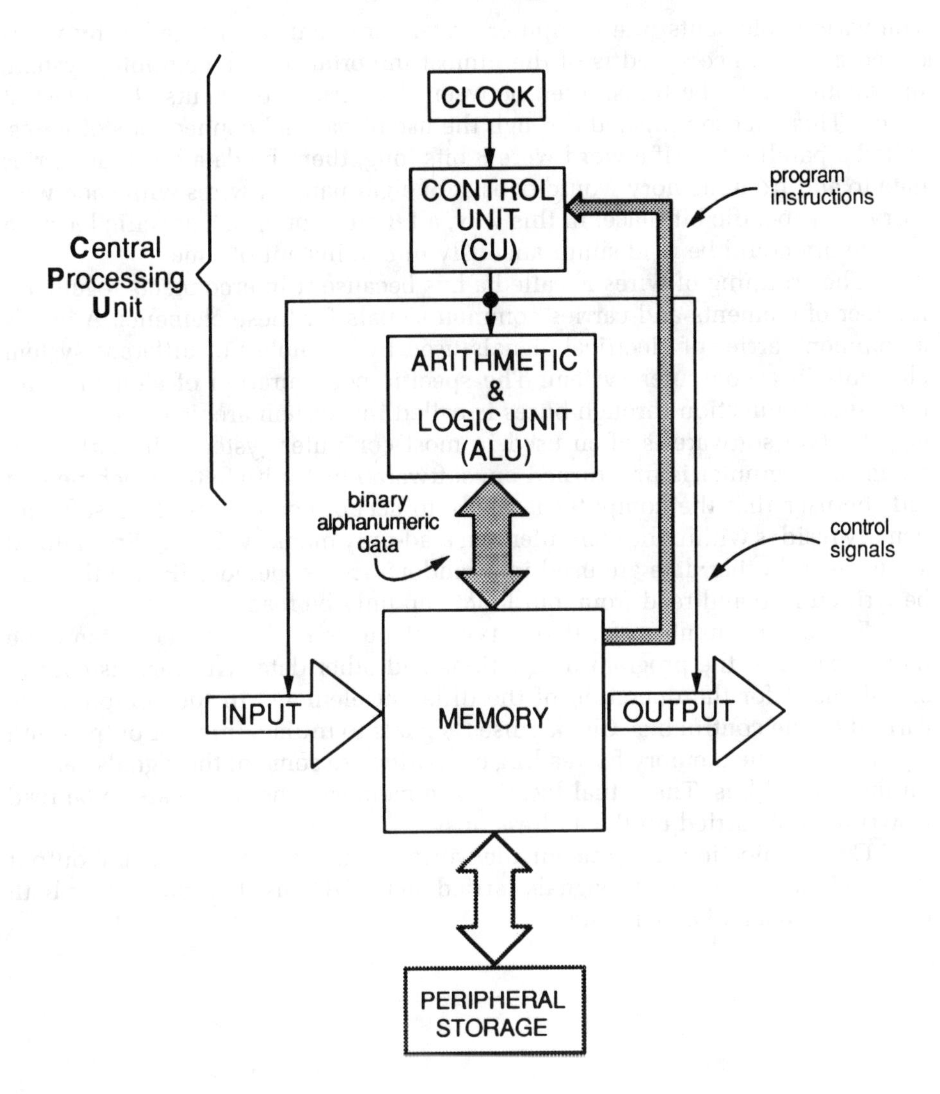

SYSTEM ARCHITECTURE

The various elements of a computer system are connected together by wires called a bus. Since speed is of the utmost importance in a computer system, information must be transmitted between the various elements at the fastest speed. This is accomplished through the use of parallel connections of wires, called a parallel bus. If a word were 8 bits long, then the data bus that carries data to and from memory would consist of eight parallel wires with each wire carrying a specific bit place. In this way, all 8 bits stored in a specific location in memory could be read simultaneously in one instant of time.

The grouping of wires is called a bus because it interconnects together a number of elements and carries common signals for these elements. A bus is a common carrier of electrical signals used by a number of different system elements in a computer system. The specific configuration of elements and their interconnection through buses is called the system architecture.

Certain software is often used in most computer systems. In particular, when the computer is first turned on, software must initiate the machine and tell the user that the computer is ready to accept commands. This software usually resides within the computer in a read-only memory (ROM). Specialized software and other data are used in a random-access memory (RAM) that can be written into and read from, but ROM can only be read.

There are a number of different types of buses in a computer system. The data bus carries the program instructions and other data. The various control signals used for the operation of the different elements in the computer are carried by the control bus. Clock pulses, signals to initiate input or output, and signals to set the memory for reading or writing are some of the signals carried on the control bus. The actual locations in memory where data are to be read or written are carried on the address bus.

Communication with the outside world through various input and output devices is accomplished by signals carried on the I/O bus. This bus carries both control signals and actual data.

System Architecture

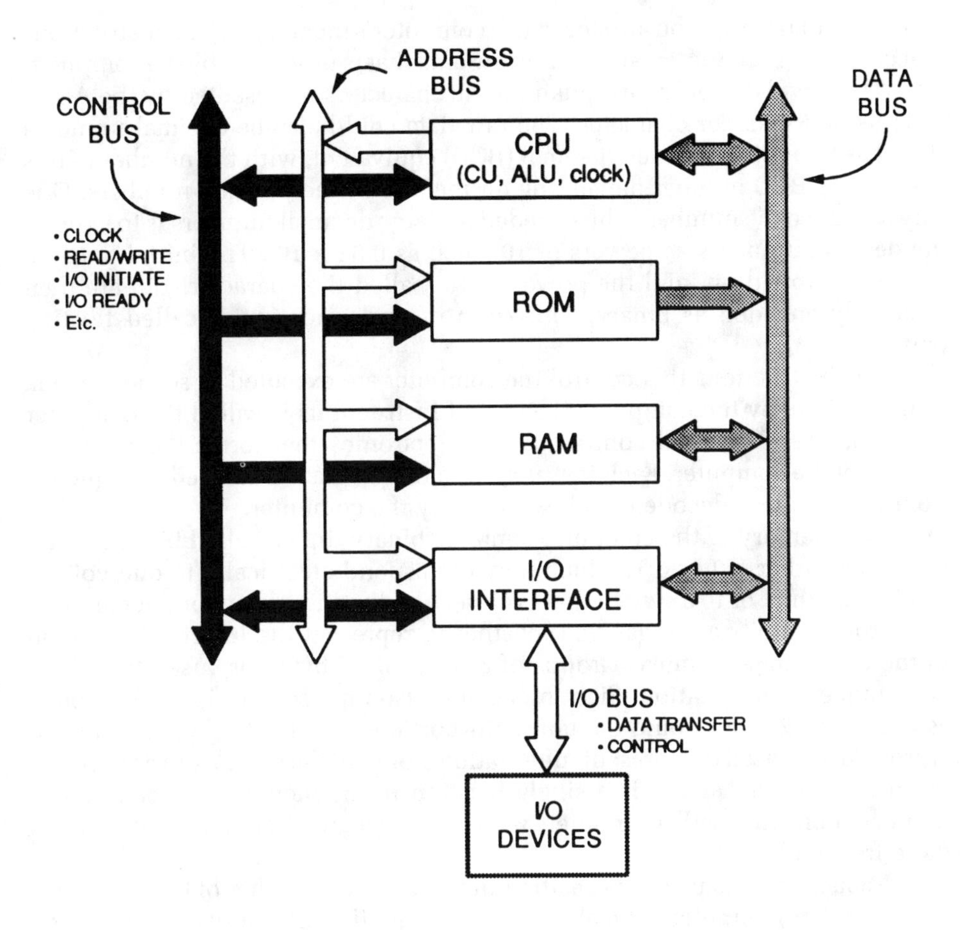

MEMORY

Two types of information are stored in a computer's memory: (1) the instructions that control the computer, and (2) the raw data that is processed by the computer.

The raw data could be alphanumeric characters represented by the ASCII code using 8 bits per character. The raw data could also be decimal numbers stored as their binary coded decimal (BCD) equivalent, with the number of bits needed per BCD number depending on the size of the decimal numbers. One way to reduce the number of bits needed for large decimal numbers is to express the decimal numbers as powers of 10, such as 0.52×10^3. The initial portion, suitably rounded off, and the power of 10 (called the characteristic) are then separately encoded as binary. This form of representation is called floating point.

The instructions that control the computer are executed in sequence, one step at a time, by the computer. The set of instructions is called the computer program, or software, in contrast to the electronics that forms the hardware portion of the computer. Each instruction in the program is fetched in sequence from the memory, decoded, and executed by the computer.

The memory of the computer contains binary digits, called bits. A single bit can be either a 0 or a 1, which may correspond electrically to one voltage level or another, or to a switch that is either conducting current or not conducting. Groups of bits are organized together to represent the information stored in the computer's memory. Groups of 8 bits, called bytes, are used to encode alphanumeric information. Eight bits could also represent an integer encoded as binary, or 2 bytes could contain both portions of a floating point number. A group of bits used to represent information stored in the memory of a computer is called a word. A word is a single location in the memory and can be any number of bits in length. Computers with a word length of 8 bits are encountered quite frequently.

Memory size is usually specified in terms of the number of bytes that can be stored. For computers, the abbreviation K signifies a quantity of 1,024, or 2 raised to the tenth power. So, a computer memory of 1K would be able to store 1,024 bytes. When specifying the storage capacity of a chip, the capacity is given in bits.

When information is read from memory or any other portion of the computer, the information stored there is not destroyed during the process of reading. However, if new information is entered into memory or some register, the prior information there is replaced by the new information.

Memory

Contents:

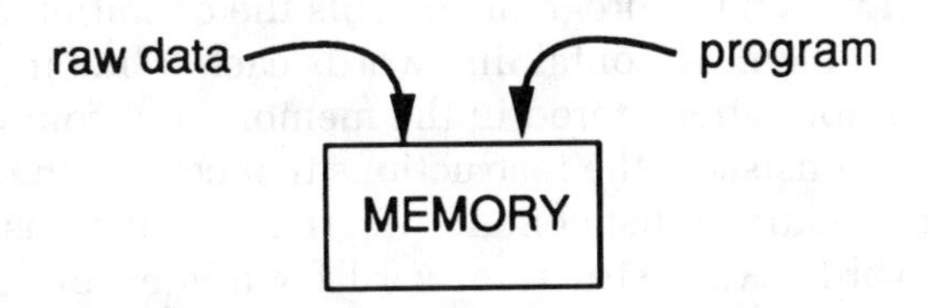

Data:

ASCII (American Standard Code for Information Interchange)
BCD (Binary Coded Decimal)
Floating Point

characteristic

0.52×10^3
FLOATING
POINT

Program:

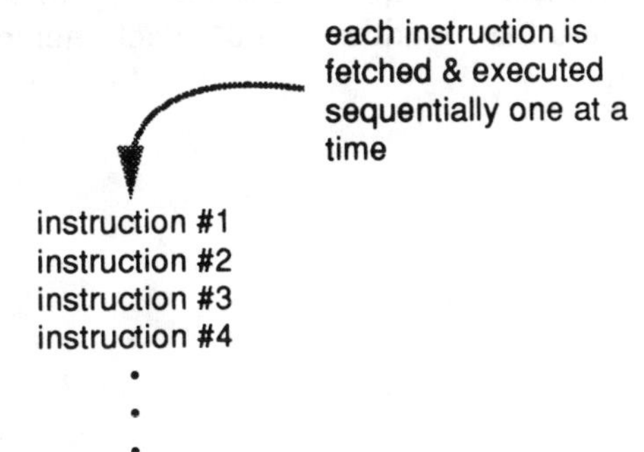

COMPUTER EXAMPLE

There is an interplay between the program and the hardware configuration of the computer. The following example, though simplified, shows this interplay and the manner in which the program controls the operation of the computer. Our computer has a memory containing words each 8 bits in length.

One type of information stored in the memory of a computer is the computer program that consists of the instructions that control the operation of the machine. Memory locations that contain program instructions are said to contain instruction words. An instruction word for this example will consist of two portions: an instruction *code* and an *address*. The instruction code will consist of the first 3 bits and represents the operation to be performed by the computer. The address portion will consist of the last 5 bits and defines the location in the memory where certain operations can retrieve or store data. The instruction code specifies the operation, and the address specifies the operand.

The other type of information stored in the memory of a computer is raw data. Our computer stores integer numbers encoded as binary in words 8 bits in length. The first bit of each data word represents the sign and the remaining 7 bits represent the magnitude of the integer. The largest integer that can be represented with 7 bits is 2 raised to the seventh power minus 1, or 128 – 1 = 127. So, with 8 bits used in this way, it is possible to encode integers from –127 to +127. Any integers outside this constrained range cannot be handled by our computer.

With 5 bits, a total of 32 memory locations can be specified in the address portion of the instruction word. The first memory location has an address of 0 and the last location has an address of 31. Each memory location is 8 bits or 1 byte long.

Computer Example

Words:

• INSTRUCTION WORD

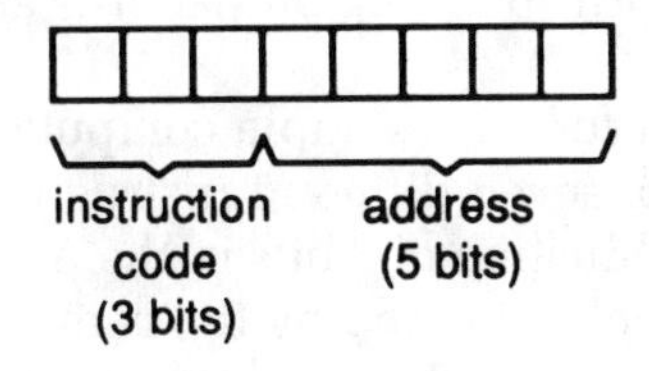

• DATA WORD

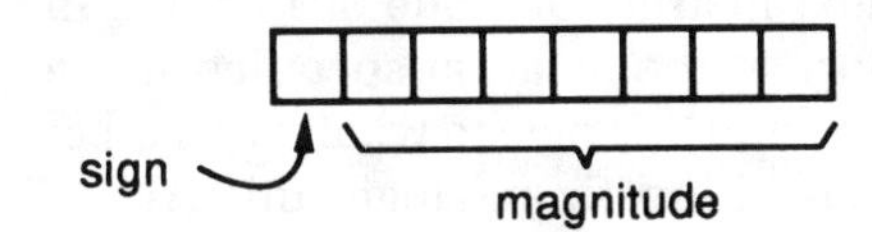

Memory:

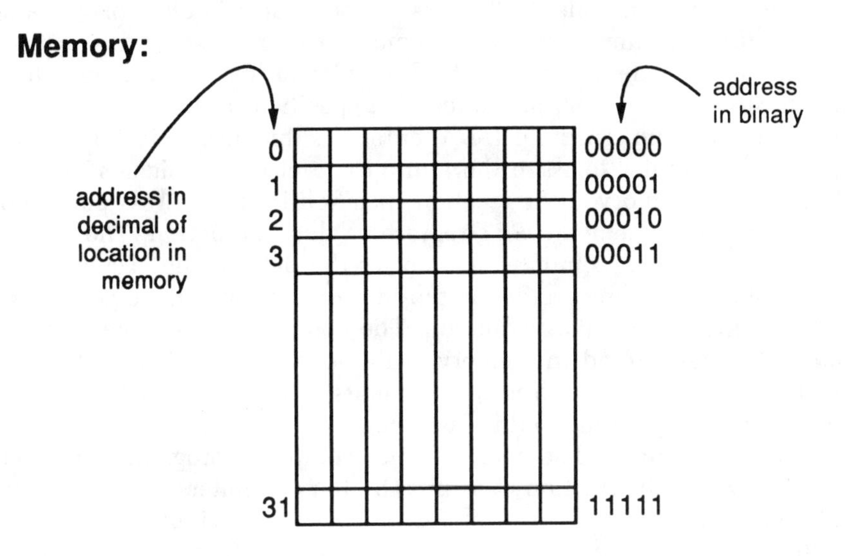

INSTRUCTION SET

The computer must know which operations it is to perform. The information to specify the operation is encoded as a particular bit pattern. The number of bits allocated for specifying the instructions determines the total number of instructions or operations that the computer can perform. The actual instructions along with their corresponding bit patterns are called the instruction set for the computer.

The instruction set for our example computer will consist of only seven instructions. The total of seven different instructions can be represented with 3 bits. Each unique combination of 3 bits in the instruction code will represent a specific operation to be performed by the computer. The simplest code 000 causes the computer to stop and to perform no further operations. The mnemonic *NOP* is used as an aid to help people remember the meaning of this instruction code. The next instruction code 001 causes the computer to fetch the data stored in memory at the location specified in the address portion of the instruction, and then to add this data to whatever has been previously loaded into the accumulator in the arithmetic and logic unit (ALU). The mnemonic *ADD* is used for this instruction.

The instruction code 010 causes the computer to clear or zero-out the contents of the accumulator. The mnemonic *CLR* represents this instruction. The instruction code 011 causes the computer to load the accumulator with the data stored in memory at the location specified in the address portion of the instruction. The mnemonic *LDA* represents this instruction. The code 100 causes the computer to issue the appropriate electrical signals to print the data stored in memory at the location specified in the address portion of the instruction. The mnemonic *PRT* represents this instruction. The instruction code 101 causes the computer to transfer the data in the accumulator into memory at the location specified in the address portion of the instruction. *STA* is the mnemonic for this instruction. The code 110 causes the computer to subtract the data stored in memory at the location specified in the address portion of the instruction from the contents of the accumulator. *SUB* is the mnemonic corresponding to this instruction.

The instructions that comprise the computer program are processed sequentially, one instruction at a time, by the computer. Each instruction is read from memory into the control unit (CU), where the instruction is decoded and the appropriate signal is issued to cause the operation to occur. The control unit has a special register, called the program counter, which contains the address of the location in memory containing the instruction being processed. After the instruction has been completed, the program counter is increased by increment 1, and the CU then fetches the next instruction from that address.

In some circumstances, it is desirable to skip around in a program and to jump to a specific location to process the instruction contained there. This operation is accomplished by the instruction code 111, represented by the mnemonic *JMP*. This instruction causes the computer to jump to the specified address and to continue the program from there. This instruction is particularly useful in specifying the beginning location of a program in memory.

Instruction Set

Binary Code	Operation	Mnemonic
000	stop processing (no operation)	NOP
001	add contents at address to accumulator	ADD
010	clear accumulator	CLR
011	load accumulator with contents at address	LDA
100	print contents at address	PRT
101	store accumulator at address	STA
110	subtract contents at address from accumulator	SUB
111	jump to address & begin processing	JMP

EXAMPLE PROGRAM

As an example of the interplay between the computer hardware and the software, a simple program will be entered into the memory of the previously described computer. The program starts at location 0. The program could have been loaded from a disk, the keyboard, or some other input medium.

Although the program is now in the memory of the computer, the computer does not know that it is to transfer control there and to start the program. Somehow, the computer must be instructed to set the program counter to 0 and to start processing the program. This can be accomplished at the control panel of the computer, where a register can be set by hand to contain the information 11100000, which means JMP to location 0 and start processing. The next step would be to hit the button labeled "execute," which causes the computer to execute the information contained in the panel register. Assume that this has been done, the program counter is now set at 0, and the computer starts processing.

The contents of location 0 are fetched from memory by the CU and loaded into the instruction register in the CU. The instruction code, the first 3 bits, is examined. The code 011 indicates that the accumulator is to be loaded with the contents of location 10000. The CU then issues the appropriate signals to cause the contents of location 10000 to be loaded into the accumulator. The accumulator now contains the contents of location 10000, which is the binary data 00001101. The program counter is increased by increment 1, and the contents of location 1 are loaded into the CU.

The instruction contained in location 1 causes the computer to add the contents of memory location 10001 to the data already in the accumulator. So, the binary data 00001001 is added to 00001101, giving the result 00010110. The next instruction causes the computer to store the contents of the accumulator in memory location 10010. Location 10010 now contains 00010110. The next instruction causes the computer to send the data stored in location 10010 to the printer, where it is converted to the decimal number 22 and printed. The next instruction has an instruction code of 000, which causes the computer to stop processing.

This simple program has caused the computer to add together two numbers and to print the result. Although this is not a particularly involved program, nor is our example computer particularly sophisticated, the principles of interplay between the program and the operation of the computer are identical to more complex programs and computers.

Example Program

Contents of Memory

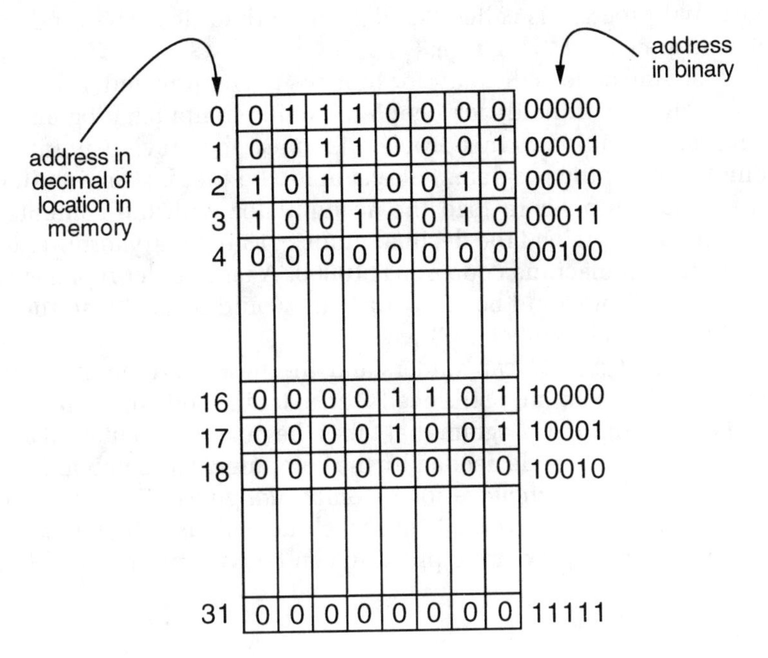

Control Panel:

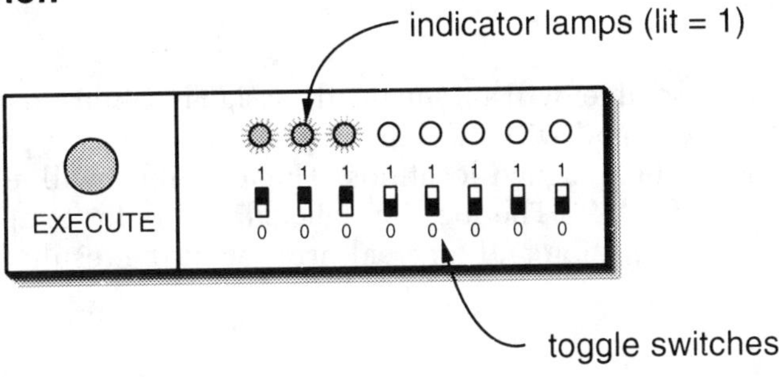

ASSEMBLY LANGUAGE

The instructions that comprised the computer program in the preceding example were in the form of binary information, which was decoded by the computer. This type of computer program is called *machine code* or *machine language*. The actual program is called "code," while the rules governing the writing of the code are called "language."

Machine code is difficult for humans to comprehend. It is also very time-consuming and difficult for people to write computer programs in machine language. A solution to this problem is to use the previously introduced mnemonics to write the program. For example, the first instruction in the program caused the computer to load the accumulator with the contents of memory location that contained the decimal number 13 in binary form. This instruction was written in machine code as 01110000. A far simpler representation of this same instruction would be LDA=13. This would mean, "load the accumulator with the decimal number 13."

Although people can understand the new representation, the dilemma now is that the computer still needs the machine code to operate. The solution is a special computer program that takes the new representation and assembles it into the machine code needed by the computer. The symbolic representation is called *symbolic language*, or *assembly language*. The program that takes symbolic code and converts it into machine code is called an assembler.

The preceding example program can be written in assembly language as follows:

```
LDA =13
ADD =9
STA Z
PRT Z
NOP
```

The assembler will automatically assign locations in memory for the numbers 13 and 9, and will store the appropriate binary representations of these numbers in the assigned locations. The assembler will also assign a location in memory for "Z." The location labeled "Z" is for temporary storage of the result of the addition and for final printing of the result.

Assembly Language

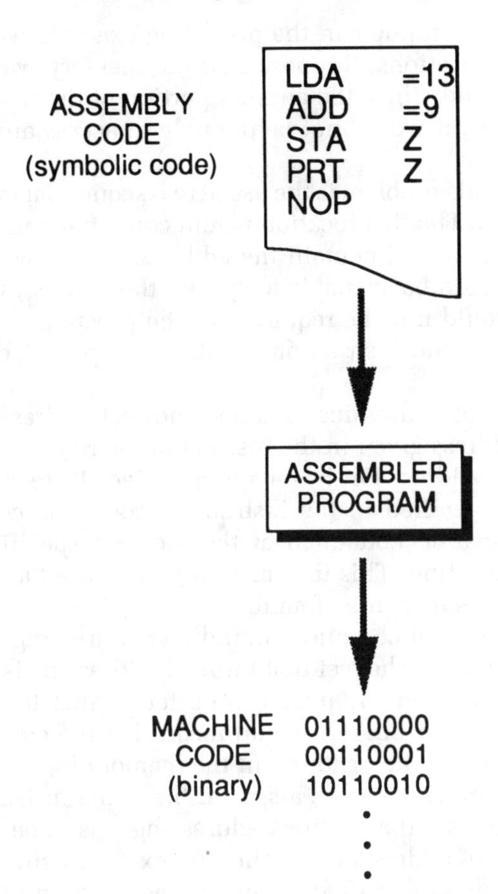

ADDRESSING

The memory of the computer in the preceding example was very small, consisting of only 32 locations. Suppose that the memory was expanded to 256 locations, for example. Since the address portion of the instruction word consisted of only 5 bits, it would not be possible to access any memory location beyond 32 locations!

A solution to this problem is the use of two sequential locations in memory for each instruction. The first location would contain the instruction code, and the second location would contain the address. In this way, all 8 bits in the second location would be available to specify the address. Of course, twice as many locations would now be required for the program. Since the address is specified directly in the instruction word, this type of addressing is called direct addressing.

Another type of addressing is called indirect addressing. With indirect addressing, the address given in the instruction word contains the location in memory where the address for the data is specified. To retrieve the actual data for the operation specified in the instruction code, the computer must first retrieve the information contained at the location specified in the address portion of the instruction. This information is then used as the actual address in which the actual data can be found.

Although somewhat complicated, indirect addressing is useful in certain programming situations. The fact that indirect addressing is to be used is indicated in the assembly code by placing an asterisk after the operation. So, for example, the instruction LDA*X would mean that the computer should load the data from the accumulator found in the memory location with its address stored in the location X. Usually, a special bit in the instruction word would be set to one to indicate that indirect addressing was to be used.

Another type of addressing is called index addressing. With this type of addressing, the address specified in the address portion of the instruction is modified by the contents of a register called an index register. For example, suppose that the instruction had an address of 95 and that indirect addressing had been activated. If the index register contained 5, the final address used for the instruction would be 95 + 5 or 100. Special circuitry is used in many computers so that index registers can be easily increased by increments and used to modify the address portion of the instruction being processed. Most computers contain many index registers. If indirect addressing using index register 3 were specified, the instruction would be written in assembly language as LDA, 3.

Index registers are particularly useful for repetitive calculations that are performed in a loop within the program. The computer can be designed to branch within the program, depending on the contents of an index register, and then to terminate the loop at the conclusion of the calculations.

Addressing

Direct:

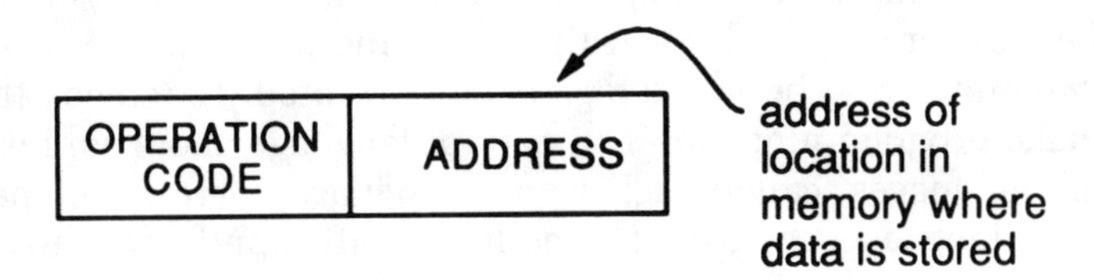

Indirect:

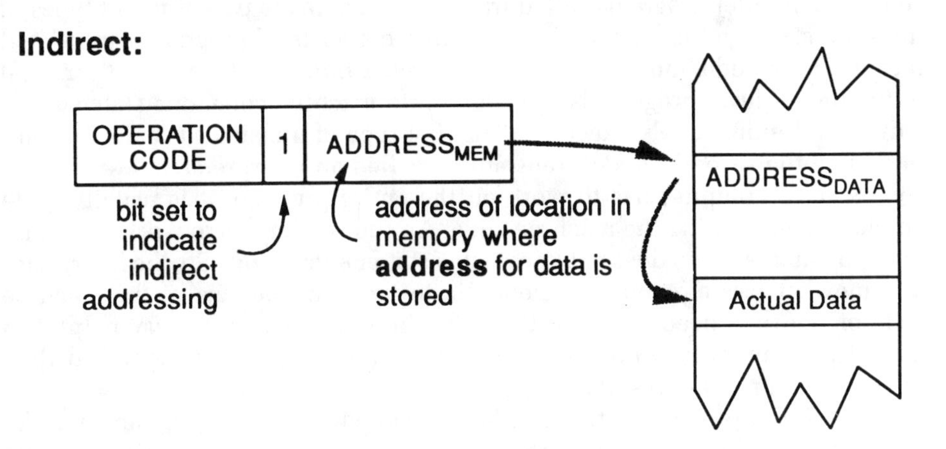

Index:

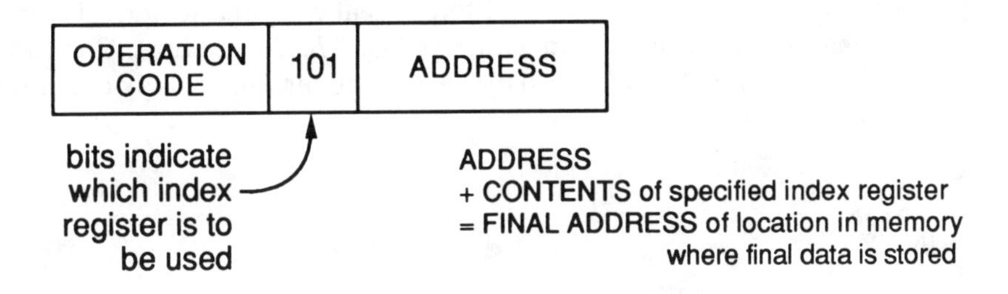

HIGHER-LEVEL LANGUAGES

Although assembly language is a big improvement over machine language, it is still far from the natural language used by humans. The nonprogrammer would be hard-pressed to look at the preceding program and to know that it causes two numbers to be added together and printed. To remedy this situation and to make programming closer to natural language, more sophisticated programming languages, called higher-level language, were developed. Specific higher-level languages were developed for specific application areas and disciplines.

Programs written in higher-level languages are converted into machine code by computer programs called translators. There are two different types of translators, depending on whether the entire higher-level program is translated into machine code and then executed, or each individual statement (or line) in the higher-level program is translated into machine code, one statement at a time, and immediately executed. The first type of translator is called a compiler, and the second type of translator is called an interpreter.

Since a compiler has to work on the entire program, it is usually quite efficient in generating the machine code. An interpreter has to work with only a single statement and does not know what statements follow. So, the interpreter is somewhat less efficient. Efficiency is in terms of the size of the machine code or the time needed to run the code. Interpreters can be very helpful to novice programmers, who can learn much by writing a statement and then seeing the immediate results.

The input program to the translator is called the source program, and the output machine code is called the object program. Some translators first generate an assembly program as an intermediate step, which must then be used with an assembler to generate the machine code.

A popular translator has been written to deal with mathematical equations and formulas. It is a formula translator, or *FORTRAN* for short. Our example program if written in FORTRAN and made more general to add any two numbers would be as follows:

```
READ,X,Y
Z=X+Y
PRINT,Z
```

This program is instantly readable by even the nonprogrammer!

The FORTRAN translator is a compiler, which means that the entire program must be translated. In some instances, it is desirable to write one statement in a program and to have it translated and executed. An interpreter does exactly this, and a subset of FORTRAN has been written as an interpreter, and also supposedly simplified for the nonexpert. This interpreter is called *BASIC,* for beginners all-purpose symbolic instruction code.

A programming language written for business application is called *COBOL,* for common business oriented language.

Higher-Level Languages

Translators:

• COMPILER

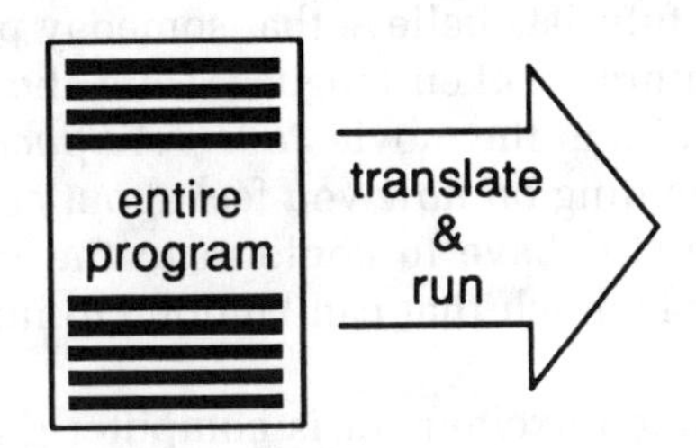

• INTERPRETER

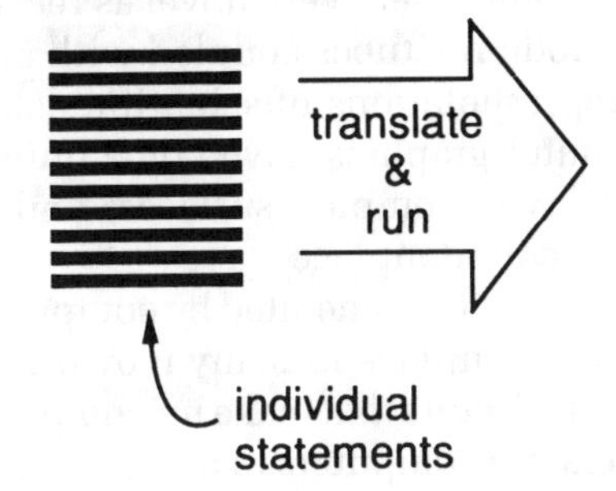

FORmula TRANslator → FORTRAN

PERSON-MACHINE COMMUNICATION

The computer is a tool to help humans. Computers are used by people, and so these tools should be easy or friendly to use. The ease of use is determined at the communications *interface* between the user and the computer. Hence, it is essential that this interface be designed to be as natural as possible for the human user. Some futurists believe that someday people will be able to speak to computers in natural spoken language, and the computer will understand and respond, like HAL in the movie *2001—A Space Odyssey.* Unfortunately—or fortunately, depending on how you feel about computers—this day is far in the future, and we still have to conform to the limitations of the machine. Nevertheless, there is much that can be done to improve the person-machine interface.

One area of much excitement is computer graphics, namely, the use of the computer to generate visual displays of data and other information. Although the use of a computer to plot data is very much taken for granted at present, computer graphics were very novel as recently as the 1960s and 1970s. Pen plotters and cathode-ray tubes coupled with automatic cameras were used to produce visual representations of scientific data, management reports, and even artwork. Computer graphics have come quite far from their early years. Color, shading, shadows, complex surfaces, and the elimination of hidden elements are all now commonplace.

TV commercials are now generated by computers, and computer-generated special effects are commonplace in many movies. The complex diagrams used to create the integrated circuits that make up computers are drawn by computers. Automobile designers use computers to depict new designs. Artists are working with computers. Indeed, computer graphics have become one of the most prevalent and friendliest ways of communicating with computers.

Human Interface

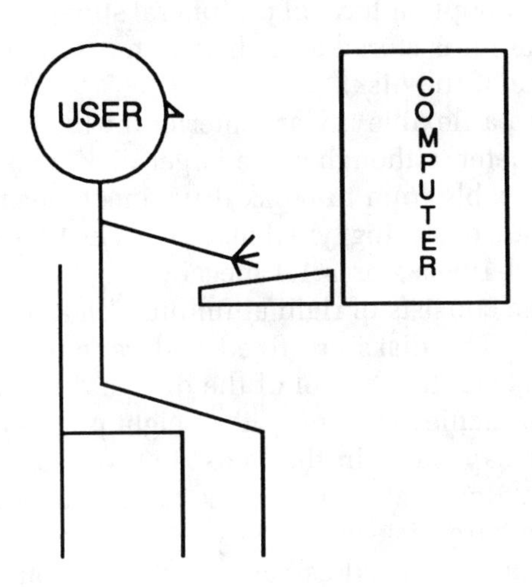

PERIPHERAL STORAGE

Undoubtedly, the most popular form of peripheral storage is the use of magnetic disks. Digital information is stored as a change in magnetism of a metal-oxide coating on the surface of the disk.

Floppy disks use a flexible mylar material for the backing. Floppy disks are 3.5 inches in diameter, although some larger disks existed in the past. The floppy disks are removable from the disk drive mechanism, and so are useful for distributing software or storing typed material. The two-sided, high-density, floppy disk of the mid-1990s stores 1.4 megabytes.

Hard disk storage consists of rigid aluminum disks that are sealed within a head-disk assembly. The disks are fixed and cannot be removed from the drive mechanism. The careful control of the disk and heads allows very large storage capacities to be achieved. From one to eight platters are in a single hard disk drive with drive capacities in the mid-1990s of from 30 megabytes to as much as 8 gigabytes. Removable hard disk drives store a few hundreds of megabytes on a single hard disk as small as 3.75 inches in diameter. Magnetic tape cassettes are sometimes used as back-up storage, but tape is not as easy to access randomly as disk storage. Magnetic storage can be easily erased and rewritten quickly but can also be damaged by stray magnetic fields.

Laser light can be used to retrieve information stored as reflective pits in the surface of a disk, similar to the technique used for the compact audio digital disc. The same technique when used for computer data offers storage capacities in the realm of hundreds of megabytes to many gigabytes on a disk less than 5 inches in diameter.

Optical storage used in this manner cannot be erased or rewritten and hence is read-only memory (ROM). However, the CD-ROM can be easily mass-produced and thus is an inexpensive form of mass storage for distributing large programs and other software.

Optical technology has been developed that uses a laser beam to burn data onto an optical disk, but once written, the data cannot be erased and rewritten. Such storage is write-once read-many, or WORM. Other optical technology is based on magneto-optical principles in which a laser beam heats selected portions of the disk causing a reversal of magnetic field. The disk is read by a low-powered laser beam. Magneto-optical is slow compared to magnetic media but the disks are not subject to damage by stray magnetic fields.

Peripheral storage is an area of fast moving technological advances. Capacities and access speeds continue to increase while costs decrease. The fast advances in optical storage are one example of this technological innovation. Whatever might be stated as the state of the art today will undoubtedly be passé in a few years.

Peripheral Storage

Types:

- MAGNETIC
  - floppy disks
  - hard disks
  - removable hard disks
  - tape
- OPTICAL
  - CD-ROM
  - erasable optical
  - recordable (WORM)

PERFORMANCE SPECIFICATION

A variety of parameters determine the performance specifications of a personal computer.

The size of memory for ROM and RAM is specified in terms of the number of bytes that can be stored. RAMs with capacities of 8 megabytes are standard for most personal computers. The larger the memory, the more data can be stored and processed. Also, larger memories are required for more elaborate and sophisticated programs. Since storage chips are continuing to grow larger in capacity, we can expect to see larger and larger RAM capacities for personal computers. One problem with this hardware trend is that software also continues to grow in its demand for more RAM rather than becoming more storage efficient. The speed with which the memory can be addressed and data can be fetched is also an important consideration.

The basic speed of a computer is specified by the number of instructions that can be processed per second. However, fixed-point hardware is much simpler than floating-point, and so the type of arithmetic hardware must be known to interpret the meaningfulness of a specified number of instructions per second. The question is, what type of instructions? Since some instructions require more time than others, the basic cycle time of the computer is sometimes given. Some instructions then require more cycles than others, although a simple instruction will usually be performed in one cycle.

Some microprocessors work with only 8 bits at a time. Other, faster processors work with 16, or even more, bits at a time, and so are superior to their smaller cousins. A longer word length for the microprocessor usually means that the instruction set can be more elaborate and rich, offering greater flexibility. Also, the address length can be longer, reducing the need for indirect addressing.

The speed of access to input and output devices is another performance specification. The speed with which information can be transmitted to and from input and output devices is yet another specification.

Clearly, comparing different computer systems can become very involved. A solution is to compare the times required to run standard simple programs. Unfortunately, agreement is hard to reach on what constitutes "standard."

Performance

Parameters:

- MEMORY SIZE
- PROCESSING SPEED
- INPUT/OUTPUT SPEED
- HARD DISK SIZE

OCTAL AND HEXADECIMAL REPRESENTATIONS

In the early days of computers, word lengths of 36 and 24 bits were quite frequently encountered. Since it was difficult to remember and to represent information in binary form, an easier scheme was used to represent the binary information. This scheme represented the binary information in groups of 3-bit clusters. Each 3-bit cluster could then be encoded as its decimal equivalent. The maximum decimal digit that can be represented with 3 bits is seven. So, including the digit 0, a total of eight numbers could be represented with 3 bits. This 3-bit clustering was called *octal representation*.

An example may help to clarify octal representation. Assume that the binary number 101111001 is to be represented in its octal equivalent. The binary number is first divided into 3-bit clusters as follows: 101/111/001. Each cluster is then represented as its octal equivalent: 5/7/1. So, the binary number 101111001 can be represented as 571. A subscript is sometimes used to indicate that a number is given in octal, for example, 571_8.

Present-day computers are byte oriented. Unfortunately, the 8 bits that comprise a byte do not correspond with the 3-bit clusters that comprise octal representation. The conceptual solution is to break the 8 bits into two clusters of 4 bits each. Incidentally, a cluster of 4 bits is called a "nibble." So, 2 nibbles make 1 byte! The maximum decimal number that can be represented with 4 bits is 15. This is unfortunate, since it would be desirable to use a single character to represent each 4-bit cluster. The solution is to use alphabetic characters after the digit 9 is reached. Since a total of 16 4-bit clusters can be represented, the 4-bit clustering is called *hexadecimal representation*. The 16 hexadecimal digits are, in order, 0,1,2,3,4,5,6,7,8,9,A,B,C,D,E,F.

An example may help to clarify hexadecimal representation. Assume that the binary number 11011001 is to be represented in hexadecimal notation. The 2 nibbles that comprise the byte are 1101 and 1001, and the hexadecimal equivalents to each nibble are D and 9. So, the hexadecimal equivalent to 11011001 is D9.

Octal and Hexadecimal

Octal:

binary	octal
000	0
001	1
010	2
011	3
100	4
101	5
110	6
111	7

Hexadecimal:

binary	hexadecimal
0000	0
0001	1
0010	2
0011	3
0100	4
0101	5
0110	6
0111	7
1000	8
1001	9
1010	A
1011	B
1100	C
1101	D
1110	E
1111	F

Data Communication

DATA TRANSMISSION

Computer systems have data stored either directly in ROM or RAM, or in peripheral memory, such as a floppy disk. These data are often generated directly by the user of the computer.

However, some information changes very frequently, such as stock prices and the weather, or is very large in volume, like a complete schedule for airline flights. Local storage of these types of information within the user's computer system would be impractical. Instead, it would be more practical to generate, update, and store this information at some central place, which could then be accessed remotely by the user's computer. This is called remote access to a centralized data base of information.

The most ubiquitous remote access to a centralized database of information is obtained by simple telephoning the data base. However, computers utilize digital data—bits—for communication, but the telephone system has been designed for analog data—human speech. What this means is that the digital data must be converted to analog tones that can be conveyed on the telephone network. In essence, these tones carry the digital data over the telephone network.

The tone is a carrier signal, and certain of its parameters are changed in accordance with the digital data. The process of causing these changes is called *modulation*, while the process of extracting the digital data from the tones is called *demodulation*.

Data Transmission

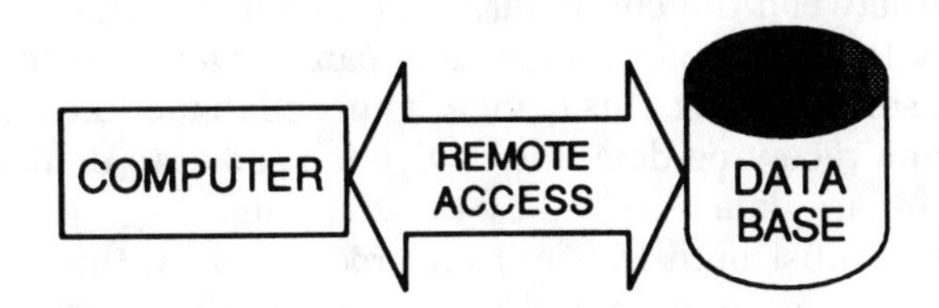

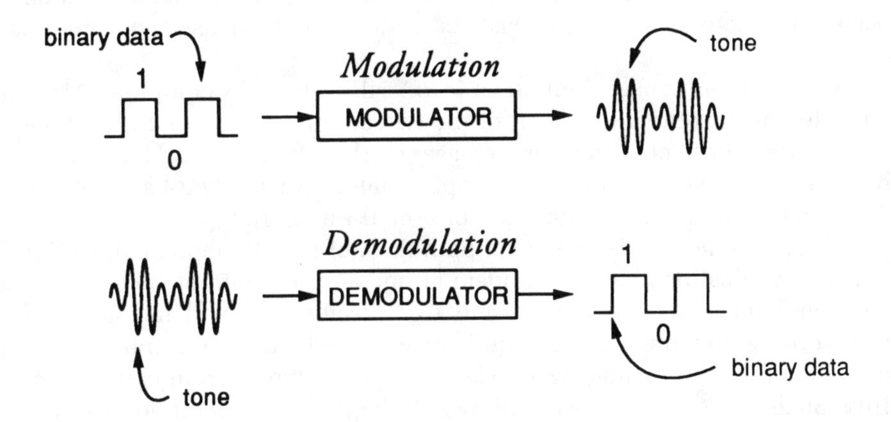

MODEM

A device is needed between the computer, which requires binary data in the form of bits, and the telephone network, which can convey only tones. When the computer is transmitting data, this device is needed to modulate tones with the binary data. When receiving data, the device is needed to demodulate the tones to obtain the binary data required by the computer.

Since computers must both transmit and receive data, the device must be able to perform both modulation and demodulation. The device is both a modulator and a demodulator, or *modem* for short.

By using a modem with a computer, people are able to access remote databases containing all kinds of information. People are also able to access special data networks that give fast access to many databases and other computers.

A very popular use of a modem is to "call" a friend's computer. Messages can be left at the other computer, or the two computers can be linked together to enable real-time "chatting," or messages can be left in a central large computer and accessed by specific or many people. These forms of text-based electronic messages are called electronic mail, or e-mail for short.

Clearly, a modem is needed at both ends of the telephone connection for data communications. So, the modem at one end needs to be able to "speak" to the modem at the other end, and this creates the need for standards to enable different modems to communicate with each other. A number of modem variables have been standardized to facilitate this intercommunication compatibility. Such items as the speed with which the data is transmitted, the modulation scheme that is used, the physical connection to the computer, the nature of the communications link, and the assignment of bits, all need to be specified in a standard fashion.

Modems are available as separate pieces of equipment that connect a computer to a telephone line or as cards that are installed internally within the computer.

Modem

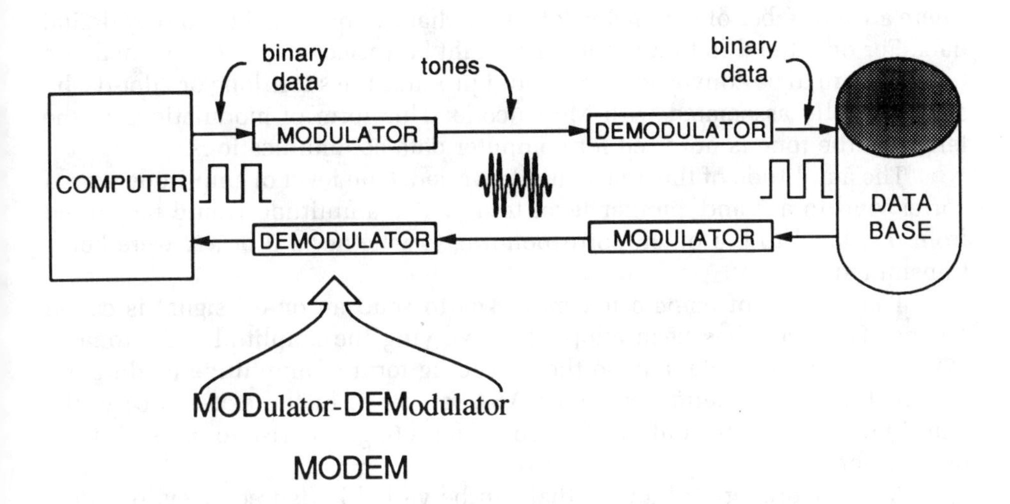
binary
data
tones
binary
data
COMPUTER
MODULATOR
DEMODULATOR
DEMODULATOR
MODULATOR
DATA
BASE
MODulator-DEModulator
MODEM

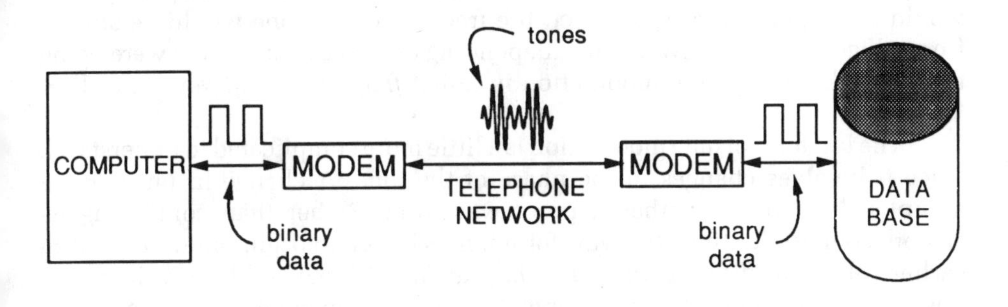
tones
COMPUTER
MODEM
MODEM
TELEPHONE
NETWORK
DATA
BASE
binary
data
binary
data

MODULATION

There are a number of parameters of a tone that can be varied to convey digital data. For one, the duration of the tone might be varied according to whether a 0 or a 1 were to be conveyed. This would produce the short-long or "dot-dash" signal usually associated with Morse code. This form of modulations of the length of the tone is not used for computer data communications.

The amplitude of the tone could be varied. One level of amplitude would correspond to a 0 and another level to a 1. The amplitude would be shifted from one level to the other, corresponding to whether a 0 or a 1 were being transmitted.

The process of using a telegraph key to send an on-off signal is called *keying*. This term has been adopted for varying the amplitude of a tone in response to digital data, and so the preceding form of amplitude modulation is called *amplitude shift keying*, or ASK for short. In the extreme case, the amplitude of the tone could be turned off for a 0, giving rise to *on-off keying*, or OOK for short.

Another property of a tone that can be varied is its frequency, or *pitch*. A high-frequency tone might correspond to a 1, while a low-frequency tone would correspond to a 0. In effect, the frequency of a tone would be shifted from a high value to a lower value, depending on whether a 1 or a 0 were being transmitted. This type of modulation is called *frequency shift keying*, or FSK for short.

The last form of data modulation is a little more complicated to understand, since it involves changes in the phase of the tone. A change in phase is an abrupt discontinuity in the shape of waveform. Rather than continuing its smooth course with time, the waveform jumps back and begins anew, repeating earlier values. In *differential phase shift keying*, abbreviated DPSK, the phase might be either changed or not changed, depending on whether a 0 or a 1 were being transmitted. Alternatively, the amount of phase change could be used to determine whether a 0 or a 1 were transmitted.

The tones used for data communications are pure, single-frequency tones, called sine waves.

Data Modulation Methods

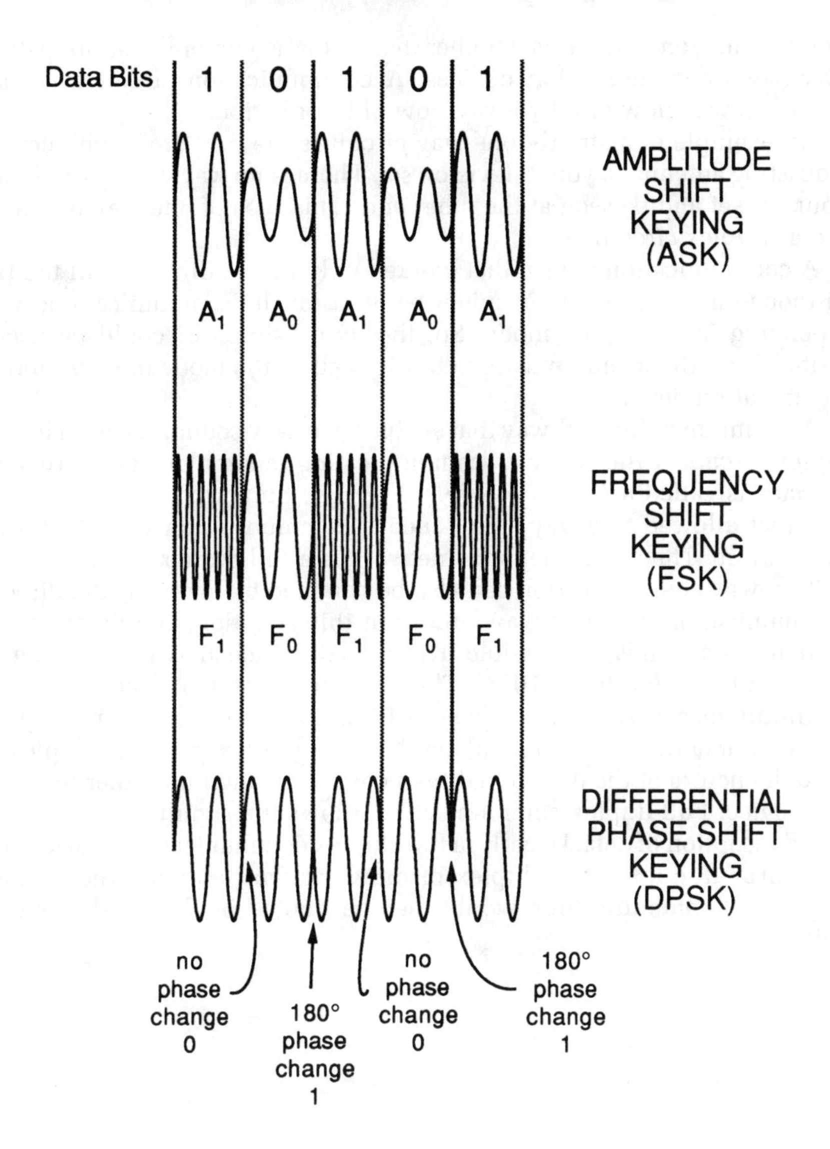

COMMUNICATION DIRECTIONALITY

A communication circuit is another name for a communication pathway between two communication devices. A communication circuit can enable either a one-way flow or a two-way flow of information.

An example of a strictly one-way circuit is the path from the television broadcasting antenna to your television set. There is no way you can sit in front of your TV set and be seen at the other end. This type of one-way pathway is called a *simplex circuit.*

A communication device that worked only in one direction all the time, even though a two-way circuit might be used for the communication, would be operating in a simplex mode. So, the term "simplex" could be used to describe either the communication circuit itself or the mode of operation of a communication device.

A communication pathway that enables two-way communication is called a *duplex circuit.* If the two-way communication is simultaneously two-way, the circuit is called a *full-duplex* (FDX) circuit. Similarly, a communication device that allowed two-way simultaneous communication would be a full-duplex device. The voice telephone network is a full-duplex system.

Two-way communication can also be obtained by reversing the direction of communication on a one-way circuit. In this situation, simultaneous two-way communication is not possible. This type of switched, two-way communication is called *half-duplex* (HDX). The push-to-talk system used by airplanes is a half-duplex system. Citizens band (CB) radio is also a half-duplex system.

As we mentioned, the terms simplex, full duplex, and half duplex can refer to the nature of the directionality of communication for either the circuit or the device. Full-duplex communication allows the simultaneous transmission and reception of data. Usually, a full-duplex communications device could be used on a full-duplex, half-duplex, or simplex circuit, although the capability of the overall communication would then be limited by the capability of the circuit.

Directionality

Simplex

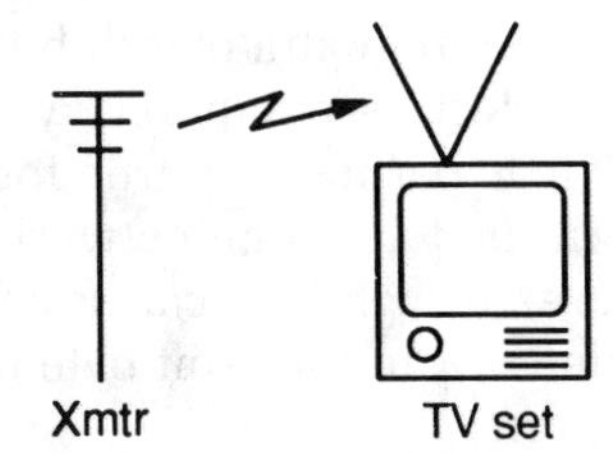

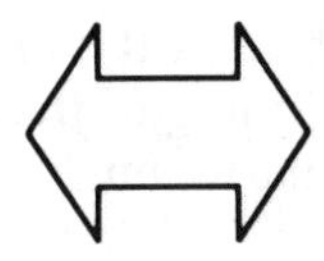

Full
Duplex
(FDX)

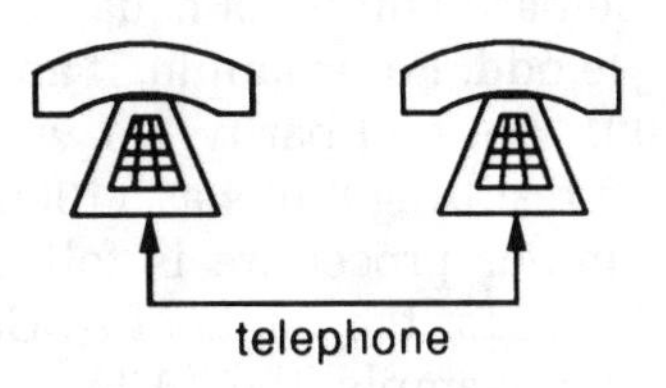

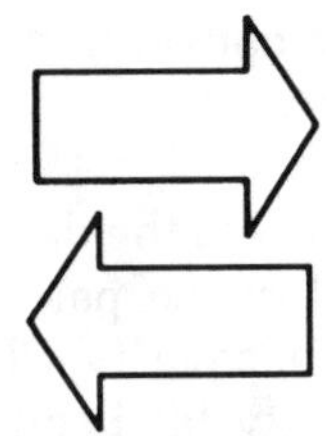

Half
Duplex
(HDX)

push
to
talk
microphone

PARITY

An ASCII character is 7-bits long, but a byte is 8-bits long. Of what use then is this apparently extra eighth bit?

As perfect as computer systems are claimed to be, errors nevertheless occur. This is particularly true for data communication in which a short burst of noise in the telephone network may obliterate a bit, causing a transmission error. The extra eighth bit can be encoded in such a fashion as to enable detection of a single error in the 8-bit byte of data. Used in this fashion, the eighth bit is called a *parity bit.*

There are two categories of parity: *odd-even* and *mark-space*. For odd parity, the parity bit is set equal to a 0 or a 1 to make the total number of ones in the byte odd. For example, if an "A" were transmitted, the ASCII 7-bit code is 1000001. For odd parity, a 1 would be placed at the beginning of the code, giving the resulting 8 bits as 11000001. The first bit is the parity bit.

A similar procedure is followed for even parity, except that the total number of 1s for the 7 bits of ASCII plus the parity bit itself must be an even number; for example, the "A" would be 01000001.

With odd or even parity, a single-bit error can be detected. This is because an error will cause 1 bit to change its value, and so the count of 1s will no longer be odd or even as required by the particular parity convention being used. If two single-bit errors were to occur, the parity convention would be satisfied, and so a 2-bit error could not be detected.

There are far more sophisticated techniques that enable both the detection and the correction of errors in data transmission. In mark-space parity, the parity bit is always set to a 1 for mark parity and to a 0 for space parity. Clearly, an error in transmission could be detected only if the parity bit itself were affected.

Many modems do not process the parity bit, but simply treat it as another bit of data to be passed on to the computer to which the modem is connected. So, these modems will work with odd, even, mark, or space parity.

Parity

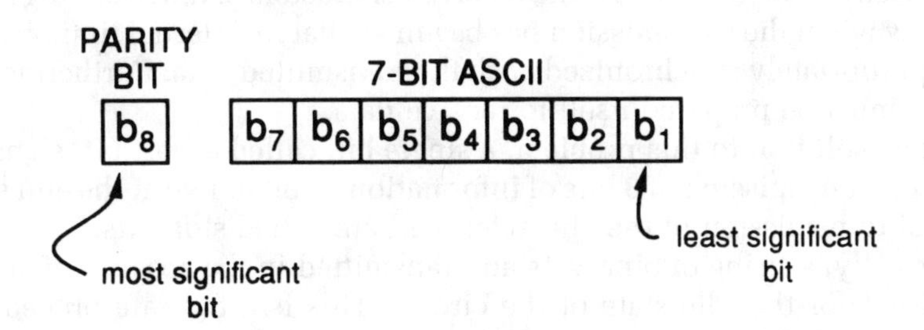

ODD/EVEN:

Odd Parity: parity bit set to 1 or 0 to make total number of 1s an odd number

Even Parity: parity bit set to 1 or 0 to make total number of 1s an even number

MARK/SPACE:

Mark Parity: parity bit always set to 1

Space Parity: parity bit always set to 0

TRANSMISSION TIMING

In order to be decoded correctly, the received characters of the receiving modem must know when the transmission has begun so that its internal timing circuits can be appropriately synchronized with the transmitted data. Furthermore, the computer must be prepared itself to receive data.

In one solution to this problem, a single bit, called a start bit, signals the initiation of transmission of 8 bits of information. A bit or two at the end signals the end of transmission of the character, and are called stop bits.

Normally, a string of binary 1s are transmitted in the absence of any data transmission for the idle state of the circuit. This is a fail-safe procedure so that if the circuit fails for any reason the modems will sense the absence of the signal. A binary 1 is called a *mark*, and a binary 0 is called a *space*, adopting telegraphy terminology.

The start bit must be a change from the idle transmission condition. So, the start bit must be a space (binary 0). The stop bit or bits must allow time for the circuit to return to its idle state, and so must be a mark (binary 1). The initial change in transmission "wakes up" the receiving modem and computer to be prepared to receive 8 bits of ASCII information (a 7-bit ASCII character plus 1 parity bit) or binary data.

The preceding form of transmission timing is used when a single character is transmitted at a time. The user of the computer types a character, and it is immediately transmitted. This type of transmission timing is called *asynchronous transmission*, since a single character can be sent at any time.

In a second transmission scheme, a long string of characters is sent in a single block with no start or stop bits between them. One 8-bit ASCII character is immediately followed by another 8-bit ASCII character. To signal the beginning of transmission and to give the receiving modem the information necessary to synchronize its internal timing circuits with the transmitted data, a special pattern of bits (usually 16 or 32 bits) is sent at the beginning of the string of characters. This is called *synchronous transmission*.

Another special pattern of bits signals the end of the block transmission. As the user types each character, nothing will be transmitted until the end of the line is reached. The characters will be saved in the computer, or in the modem, in a special temporary storage area called a buffer, and are finally transmitted in a burst at the appropriate time, when either the end of the line is reached or perhaps the buffer is full.

Synchronous transmission is somewhat more efficient than asynchronous transmission, since the overhead of a number of start and stop bits is avoided. However, since a long string of bits is being transmitted with synchronous transmission, the requirements on the accuracy of timing are quite stringent, and so more complicated and costly circuitry is required.

Synchronization

Asynchronous:

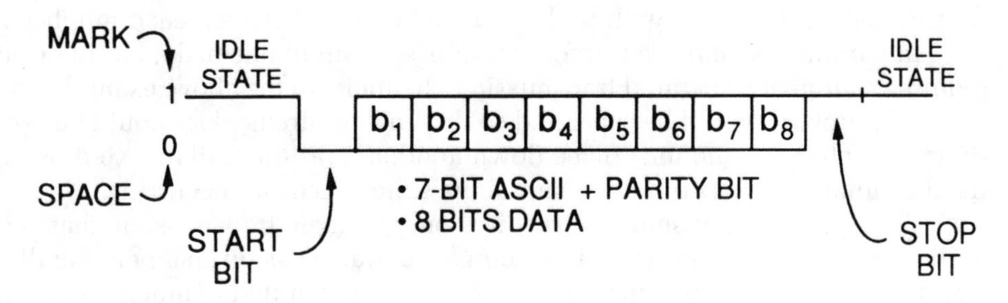

Synchronous:

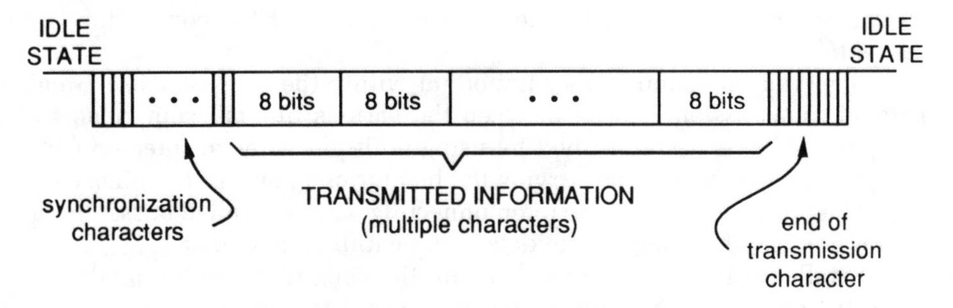

SERIAL-PARALLEL TRANSMISSION

Information can be transmitted in either *serial* fashion or *parallel* fashion. In serial transmission, each symbol is transmitted consecutively, one right after the other. For example, the decimal numbers 964, 278, and 531 would be transmitted in sequence, with perhaps a slight pause between each number.

In parallel transmission, information is sent simultaneously, but to do so requires a number of parallel transmission channels. In the above example, the decimal numbers would be separated so that the hundreds place could be sent down one channel, the tens place down another, and the units place down a third channel, with all three transmissions simultaneously occurring.

Clearly, serial transmission requires only a single transmission channel, while parallel transmission requires a number of transmission channels. Parallel transmission enables the same information to be transmitted in much less time, but at the expense of additional channels.

The telephone network offers a single communications path between two locations. So, it is particularly appropriate for serial transmission. A serial modem transmits binary data, one bit after the other, in a serial fashion. Serial transmission is used for both the telephone line and the connection to the computer.

Speed of operation is very important within the computer itself, and so parallel transmission is used between the various internal components of a computer. This is accomplished by using multiple wires to interconnect the components and each wire carrying the bits for one place in the binary word. As we mentioned earlier, these interconnecting wires are called busses. Parallel busses are used for many of the data paths within a computer.

Parallel transmission is used within the computer, but the modem uses serial transmission. So, conversation from serial to parallel representation of data, and vice versa, is required, and is performed by the I/O interfaces within the computer. The universal asynchronous receiver-transmitter (UART) is a circuit that performs serial-parallel conversion for asynchronous transmission.

Serial/Parallel

Serial:

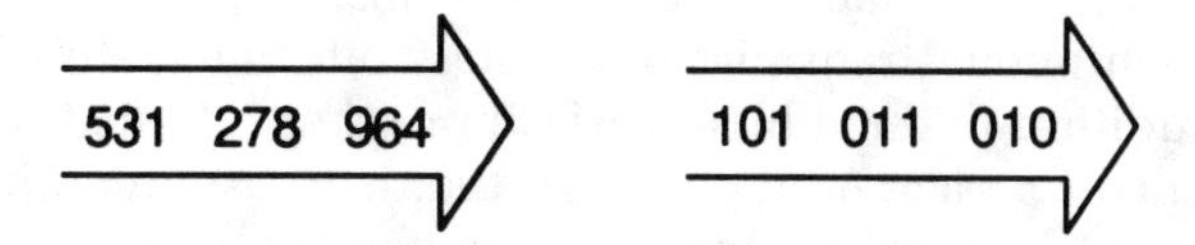

Parallel:

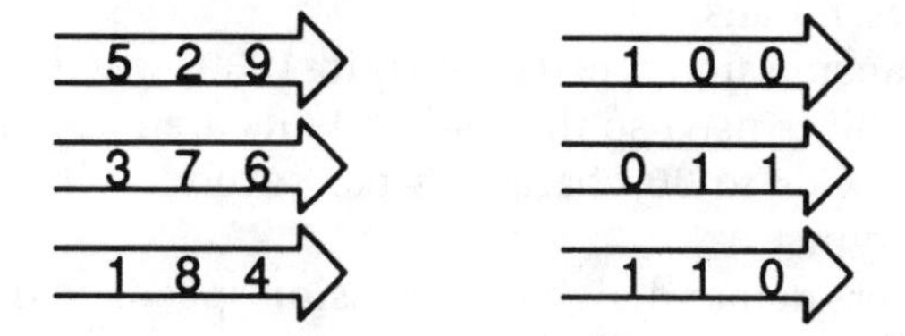

Serial/Parallel Coversion:

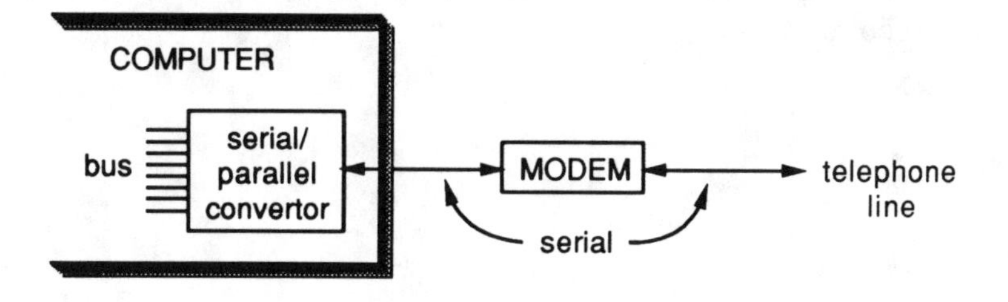

TRANSMISSION SPEED

Perhaps, the single most important specification for a modem is the speed at which it transmits and receives data. This speed is called the *bit rate*, and is expressed in bits per second, abbreviated as bps, or b/s.

For asynchronous transmission, a start bit and a stop bit are usually required in addition to the 8 bits of ASCII or other information. So, a total of 10 bits is needed for each 8-bit character that is transmitted. If 2 stop bits are used, a total of 11 bits are required for each character.

A good typist can type about 10 characters per second. If 11 bits are used for each character, this corresponds to a data transmission speed of 110 bps. It so happens that 110 bps is the transmission speed used by some older models of teletypewriter terminals.

A popular transmission speed of the mid-1980s was 300 bps. At this speed, only a single stop bit is used so that only 10 bits are needed per character. So, this speed corresponds to 30 characters per second, which is about twice as fast as the fastest typist.

Computer users demand high transmission speeds so that they can browse or scroll quickly through information. So, modems used with home computers migrated to the higher speeds of 4,800 and 9,600 bps which correspond to 480 and 960 characters per second in an asynchronous mode. Today's modems operate at speeds of 14,400 and 28,800 bits per second.

Standard bit rates are 300; 1,200; 2,400; 4,800; 9,600; 14,400 and 28,800 bits per second.

Bit Rate

Rates:

Bit Rate = bits transmitted per second

Character Rate = characters transmitted per second

Equivalency:

ASCII character	=	7 bits
parity	=	1 bit
start/stop bits	=	2 bits
		10 bits per character

120 chars / sec → 1200 bps
480 chars / sec → 4800 bps
960 chars / sec → 9600 bps

BAUD RATE

In addition to bit rate, another term used to describe transmission speed is the *baud rate*. Unfortunately, a fair amount of confusion clouds this terminology. Since the term hertz is synonymous with cycles per second, many people falsely believe that baud is synonymous with bits per second. Although it is true in many instances that the bit rate and the baud rate are identical, the two terms are conceptually quite different.

The bit rate refers to the actual rate at which bits are transmitted. The baud rate refers to the rate at which the signaling elements used to represent bits are transmitted. Since one signaling element usually encodes one bit, the two rates are then identical.

A signaling element is a short burst of a tone in which amplitude, frequency, or phase can represent one bit. For example, if the signaling element can take on one of two different amplitudes, then either a one or a zero can be encoded, according to the amplitude value. However, suppose each signaling element can take on one of four amplitudes. Then, each amplitude can correspond not to just one bit, but to a specific two-bit pair, since there are four amplitudes, and there are four possible combinations of two bits (00, 01, 10, 11). In this case, the bit rate would be twice the baud rate. This type of encoding is used for bit rates higher than 1200 b/s.

The bit rate is of concern to the computer user, while the baud rate concerns the telecommunication specialist. To be safe and to avoid contributing to confusion, we try to use only bit rates in bits per second. You will find that many people incorrectly use the terms synonymously, and that you may need to conform to their incorrect usage.

Baud Rate

BAUD RATE = number of signaling elements per second

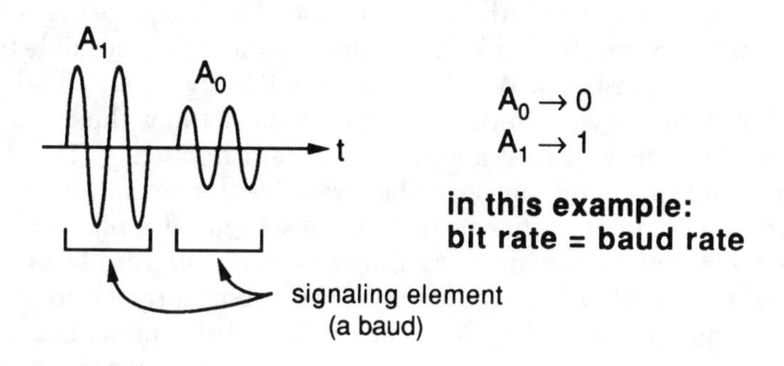

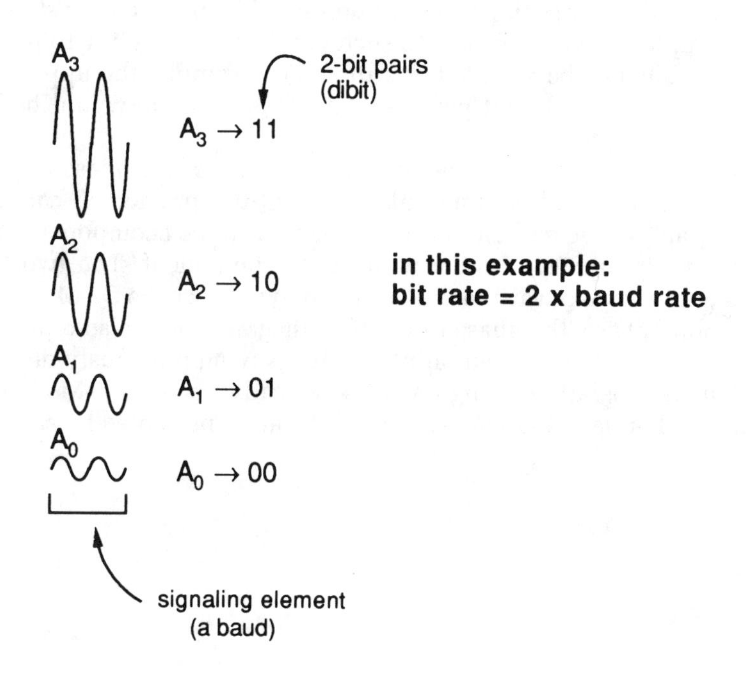

6

QUADRATURE AMPLITUDE MODULATION

We saw earlier that the bandwidth needed for a digital signal is half the bit rate. The bandwidth of a communication channel determines the maximum baud rate, but through appropriate representation, the bit rate can be much in excess of twice the bandwidth. We also saw that by choosing a number of possible maximum amplitudes for a signaling element, it is possible to represent more than one bit per baud. We shall now see that by varying both maximum amplitude and phase, we can carry a bit rate many times the baud rate.

A signaling element is at a known, constant frequency. This leaves only maximum amplitude and phase to be available for encoding bits. A given maximum amplitude and phase can be represented on a phase diagram. The radius to the point corresponds to the maximum amplitude of a signaling element, and the angle of the radius with the *x*-axis corresponds to the phase of the signaling element. The *x*-axis projection of the point is called the in-phase, or I, component. The *y*-axis projection is called the quadrature, or Q, component. Mathematically, $I = R \cos \varphi$ and $Q = R \sin \varphi$.

The possible amplitudes and phase can all be plotted on a state diagram. In the example shown, there are 16 such possible states. With 16 possibilities, a total of 4 bits can be encoded at a time. For example, the upper right state might correspond to the bit pattern 0011. This example would be called 16 QAM.

The process of modulating both the amplitude and phase of a carrier is called quadrature amplitude modulation, or QAM for short. It can be accomplished by amplitude modulating a sine wave with the I component and amplitude modulating a cosine wave with the Q component. The two separately amplitude-modulated waves are then added to produce the QAM signal. Rather than absolute phase, the phase in the state diagram could also represent phase change similar to the differential phase shift keying previously described.

Modems operating at high speeds typically use QAM. Modems at 14.4 kbps and higher use 64 QAM in which 6 bits are represented by each signaling element.

QAM

Phase Diagram:

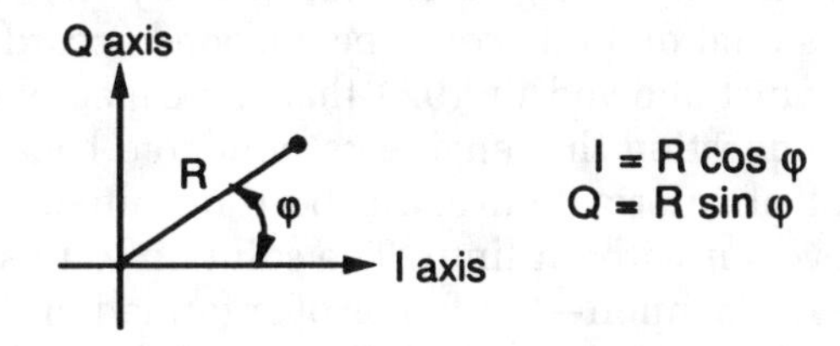

State Diagram:

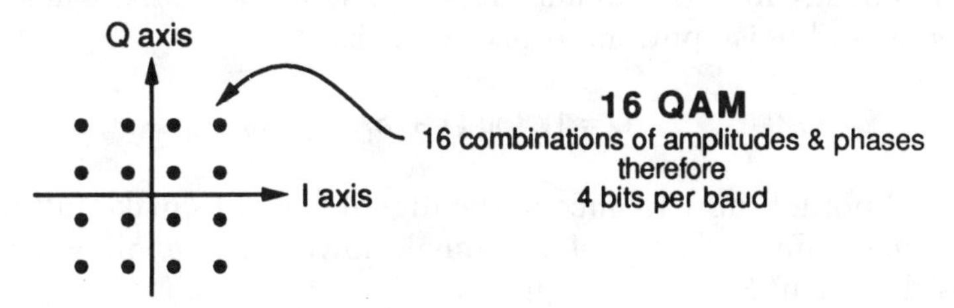

Quadrature Modulator:

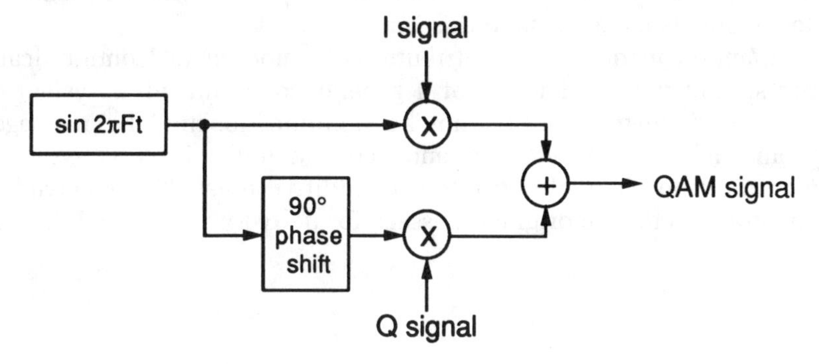

CHANNEL CAPACITY

In an earlier chapter, we saw that the capacity of a communication channel with a bandwidth of F_{max} is $2F_{max}$ bits per second for a digital signal varying between two levels. Suppose the digital signal can take on any one of n levels. For example, with 8 levels, a total of 3 bits could be encoded according to the transmitted level. Harry Nyquist showed in 1924 that the capacity of such a system was $2F_{max} \log_2 n$. The question that engineers pondered back then was whether an infinite amount of information could be transmitted simply by increasing the number of levels n without limit. The solution to this question was determined by Claude E. Shannon—the father of information theory—in his theoretical paper entitled "A Mathematical Theory of Communication" published in 1948 in the *Bell System Technical Journal*. Shannon proved that the maximum capacity C of a communication channel in the presence of white noise with a bandwidth W and a signal-to-noise ratio of S/N, where S and N are the signal and noise powers respectively, is:

$$C = W \log_2(1 + S/N)$$

Shannon did not tell how to encode the digital signal to make full use of the theoretical maximum capacity of a channel, however. The following example illustrates the use of Shannon's equation.

A telephone circuit has a signal-to-noise ratio of about 30 dB and a bandwidth of about 4 kHz. A signal-to-noise ratio of 30 dB is a ratio of 1,000 to 1, and 2^{10} equals 1,024. According to Shannon, the maximum capacity of such a circuit is $4{,}000 \log_2 (1 + 1{,}000) \approx 4000 \times 10 = 40{,}000$ bits per second. Today's fastest modems come close to this limit.

Shannon made many contributions to modern telecommunication. His paper started with a diagram of a general communication system and that diagram is known today as Shannon's communication model. A message created by some information source is encoded and transmitted as a signal over a communication channel corrupted by additive noise. The received signal is decoded to obtain the original message for delivery to the final destination.

Channel Capacity

Channel Capacity:

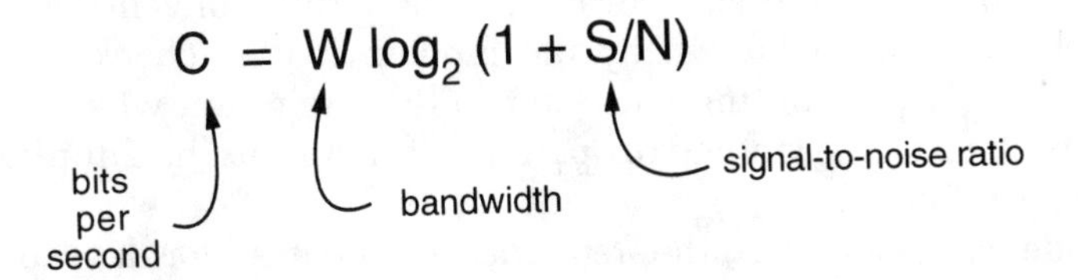

Shannon Model:

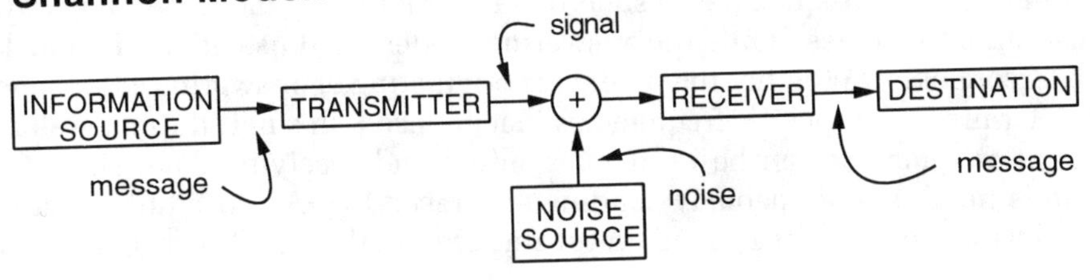

MODEM PARAMETERS

For a modem to be connected to a computer and used to access a remote database, a number of parameters must be specified. Perhaps the most essential is the speed of transmission in bits per second. The data format is nearly always serial binary and asynchronous. The parity convention being used by the database must be known by the modem if it is checking parity. Many modems simply pass on the parity bit to the computer, where it can perform any parity checks. The method of operation, full duplex, half duplex, or simplex, must be specified.

Modems must choose the frequencies to be used for data communication according to which modem originates the communication. This is because the same frequency cannot be used simultaneously both to transmit and to receive data in a full-duplex system. Different frequencies then must be used to send and to receive data in a full-duplex system. The procedure, or *protocol*, used to determine these frequencies is such that the originating modem will always use certain frequencies, while the answering modem will use other previously agreed frequencies. Whether the modem is originating or answering a transmission determines the specific frequencies that are used. In a half-duplex system, the same frequencies can be used for sending and receiving data, since the system is never simultaneously sending and receiving. A procedure must be agreed by the modems at each end for turning around the direction of communications on the circuit.

A modem connects to the telephone network and to the computer. Standards are needed to specify how these connections are made. Modems contain electrical circuits to protect the telephone network from any harm and are connected directly to the telephone line by using a standard modular plug. The connection to the computer is through a 25-pin connector, called a DB-25 connector. The electrical signals on the pins are specified as the RS232C interface standard. Pin 7 is ground; pin 2 carries a serial stream of binary data to be transmitted over the telephone line; and pin 3 carries the received data, also in a serial format. When the modem detects the carrier tone on the line, the modem sends a control signal to the computer on pin 8. The computer signals the modem that it is ready to receive or to send data on pin 20. Incoming ringing may be detected by the modem, and a signal is sent on pin 22 to the computer. Other functions are accomplished by signals on other pins.

Modem Parameters

Specifications:

- modulation method
- data rate
- synchronization
- parity
- operation (FDX, HDX)
- mode (originate, answer)
- interfaces

Mode:

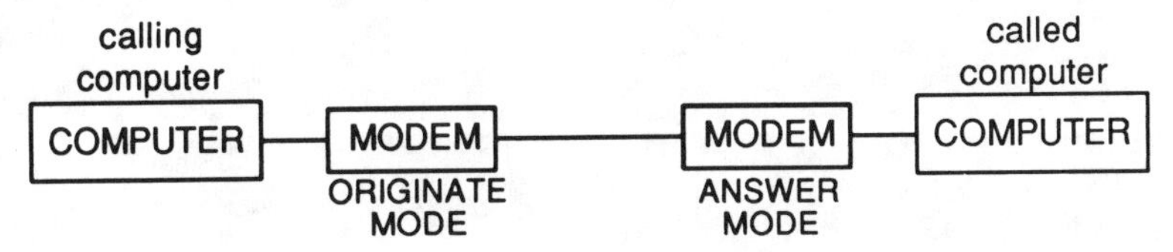

Interfaces:

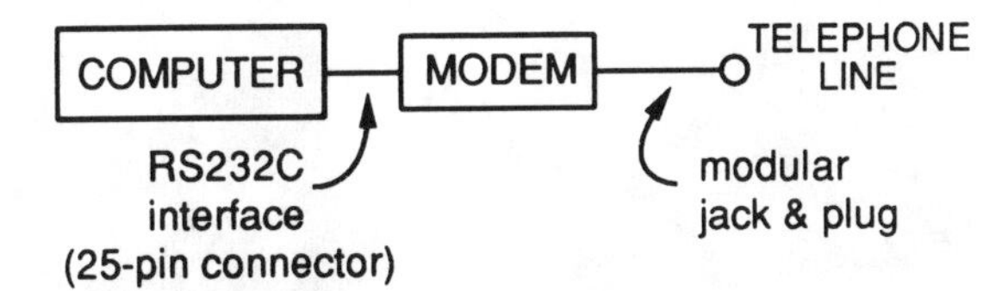

Pin	Function
1	protective ground
2	transmit data
3	receive data
7	signal ground
8	carrier detect
20	data terminal ready
22	ring detect

PROBLEMS

6.1. Convert the following numbers from binary to decimal or vice versa:
 (a) 101010 = ? decimal
 (b) 111111 = ? decimal
 (c) 45 = ? binary
 (d) 25 = ? binary

6.2. Write your name in 8-bit ASCII by using odd-parity for the parity bit. (Show all steps.)

Appendix A
Answers to Problems

SIGNALS AND WAVES

1.1. (a) 1 ms
(b) 0.5 μs
(c) 0.2 ms
(d) 0.2 ns
(e) 0.01 s
(f) 2 s

1.2. (a) 2 Hz
(b) 0.2 Hz
(c) 0.5 MHz
(d) 100 Hz
(e) 0.5 Hz
(f) 0.04 MHz

1.3. 10 Hz

1.4. 0.3 meter, or about 1 foot

ELECTRICITY

2.1. 5 mA

2.2. 100 V

2.3. 1 Ω

2.4. 2 mA

2.5. 6 mA in 2,000 Ω resistor
3 mA in 4,000 Ω resistor
1.33 kΩ equivalent resistance

2.6. 1 A

2.7. 12 A

2.8. 120 kW-hrs
$14.40

2.9. 155 V

2.10. 1,000 W
500 W

2.11. 100 V

SYSTEM ELEMENTS

3.1. 40 dB
3.2. 20 dB
3.3. (a) 20 dBV
(b) 40 dBV
(c) –20 dBV
(d) –40 dBV
(e) 10 dBW
(f) –10 dBW
(g) 10 dBW or 40 dBm

MODULATION

4.3. Frequency modulation is used for cellular radio.
Modulation index = 2.4
Bandwidth = 34 kHz

DIGITAL SIGNALS

5.1. 8,400,000 samples per second
67.2 Mb/s
5.2. 256 levels
5.3. 320 kHz

COMPUTERS AND DATA COMMUNICATIONS

6.1. (a) 42
(b) 63
(c) 101101
(d) 11001

BASIC MATHEMATICS

First set:
(a) 7.2×10^6
(b) 6.1×10^3
(c) 7.2×10^2
(d) 10^{-3}
(e) 5.0×10^{-7}
Second set:
(a) 10^{12}
(b) 10^2
(c) 10
(d) 10^7
(e) 10^{-7}

(f) 1

(g) 1

Third set:

(a) 10^{10}

(b) 7.8×10^{-1} (or 0.78)

(c) 6.6×10^{3}

(d) 9.0×10^{7}

(e) 2.0×10^{3}

(f) 2.0×10^{3}

(g) 6.6×10^{4}

(h) 4.3×10^{-2}

(i) 7.5×10^{-3}

Appendix B
Review of Basic Mathematics

INTRODUCTION

Nonengineering students sometimes have difficulties in understanding communication technology because numbers and mathematics are involved. To help remedy this situation, this appendix provides a short review of the basic mathematics and arithmetic used in electrical engineering and throughout our book.

For many students, this appendix will probably be sufficient to help them review and remedy their problems. For other students, more study may be required. The only way to remedy the situation is to build confidence in working with numbers by solving many problems.

SCIENTIFIC NOTATION AND POWERS OF 10

Very large and very small numbers are frequently encountered in communication technology, for example, a bandwidth of 4,500,000 Hz or a pulsewidth of 0.0000125 seconds. To simplify the performing of calculations with such numbers, scientific notation is used.

With scientific notation, numbers are expressed in terms of powers of 10. Powers of 10 are written as 10^n, where n is the exponent or power. For example, 10^2 (read as ten to the second power) means 10 multiplied by 10, which equals 100.

$$10^2 = 10 \times 10 = 100$$

As another example, 10^4 (read ten to the fourth power) means 10 times 10 times 10 times 10, which equals 10,000.

$$10^4 = 10 \times 10 \times 10 \times 10 = 10{,}000$$

A way of remembering this is that the exponent indicates the number of zeros after the one.

The exponent can also be a negative number. A negative exponent means that a fractional power of 10 is indicated. So, for example, 10^{-2} means one divided by 10^2 or 1/100, which equals 0.01.

$$10^{-2} = \frac{1}{10^2} = \frac{1}{100} = 0.01$$

As another example, 10^{-4} (read ten to the minus fourth power) means 1/10 times 1/10 times 1/10 times 1/10, which equals 1/10,000 or 0.0001.

$$10^{-4} = \frac{1}{10^4} = \frac{1}{10{,}000} = 0.0001$$

A way of remembering this is that a negative exponent indicates the number of decimal places—not zeros—before the one.

With scientific notation, the power of 10 is multiplied by the appropriate decimal number to give the desired number. Some examples will help to clarify this. In scientific notation, the earlier bandwidth of 4,500,000 Hz is written as 4.5×10^6 Hz, since 4,500,000 is $4.5 \times 1{,}000{,}000$ and 1,000,000 is 10^6. The pulsewidth of 0.0000125 seconds is then written as 1.25×10^{-5} seconds.

Ten can also be raised to the zeroth power. Since any number raised to the zeroth power equals 1, $10^0 = 1$. A table of power of 10 can be constructed as follows.

$$
\begin{aligned}
1{,}000{,}000 &= 10^6 \\
100{,}000 &= 10^5 \\
10{,}000 &= 10^4 \\
1{,}000 &= 10^3 \\
100 &= 10^2 \\
10 &= 10^1 \\
1 &= 10^0 \\
0.1 &= 10^{-1} \\
0.01 &= 10^{-2} \\
0.001 &= 10^{-3} \\
0.0001 &= 10^{-4} \\
0.00001 &= 10^{-5} \\
0.000001 &= 10^{-6}
\end{aligned}
$$

The following are problems to practice your ability to express numbers in scientific notation. Express the following in scientific notation.

(a) 7,200,000 =
(b) 6,100 =
(c) 720 =
(d) 0.001 =
(e) 0.0000005 =

RULES OF EXPONENTS

An advantage of expressing numbers in terms of powers of 10 is that arithmetic calculations are simplified. If two powers of 10 are multiplied, the exponents simply add. For example, $10^2 \times 10^3$ is equivalent to 100 times 1,000, which equals 100,000 or 10^5. So, adding together the exponents 2 and 3 gives the resulting exponent of 5.

If two powers of 10 are divided one into the other, the exponent of the divisor is subtracted from the exponent of the dividend. For example, $10^3/10^2$ is equivalent to 1,000 divided by 100, which equals 10 or 10^1. So, subtracting the exponent 2 from 3 gives the resulting exponent of 1.

If one or both of the exponents are negative, then the addition or subtraction has to be performed algebraically. Again, some examples should be helpful. Assume 10^2 is to be multiplied by 10^{-3}. The exponents 2 and −3 are added algebraically to give $2 + (-3) = 2-3 = -1$. So, $10^2 \times 10^{-3} = 10^{-1}$. This indeed checks, since $10^2 \times 10^{-3}$ is equivalent to $100 \times 1/1000$, which equals 1/10 or 0.1, which is 10^{-1}.

As another example, assume 10^2 is to be divided by 10^{-3}. The exponents are subtracted algebraically to give $2 - (-3) = 2 + 3 = 5$. Thus, $10^2/10^{-3} = 10^5$. This also checks, since $10^2/10^{-3}$ is equivalent to

$$\frac{100}{1/1000}$$

and multiplying the numerator and the denominator by 1000 to clear the fraction in the denominator, as follows.

$$\frac{100}{1/1000} \times \frac{1000}{1000}$$

gives 100×1000 or 10^5.

Equations and algebraic expressions are ways to express operations that are to be performed by using any numbers in general. So, the expression 10^a means ten raised to the ath power, where a stands for any number in general. The rules of exponents then are

- *Multiplication*

  $$10^a \times 10^b = 10^{a+b}$$

- *Division*

  $$\frac{10^a}{10^b} = 10^{a-b}$$

The following are some problems for practicing your ability to perform multiplication and division of powers of 10:

(a) $10^5 \times 10^7 =$
(b) $10^{-5} \times 10^7 =$
(c) $10^3/10^2 =$
(d) $10^3/10^{-4} =$
(e) $10^{-3}/10^4$

(f) $10^{-6} \times 10^{6} =$
(g) $10^{-8}/10^{-8} =$

SCIENTIFIC-NOTATION ARITHMETIC

The rules of exponents apply to the power-of-10 portion of numbers expressed in scientific notation, and conventional arithmetic applies to the decimal portion.

Again, some examples should help to clarify this. Assume 4×10^3 is multiplied by 2×10^2. Conventional arithmetic applies to 4×2, which equals 8, and the rules of exponents apply to $10^3 \times 10^2$, which equals 10^5. So, $(4 \times 10^3) \times (2 \times 10^2) = 8 \times 10^5$.

Assume that 7.2×10^6 is to be divided by 3.6×10^{-3}. First, 7.2 is divided by 3.6 to give 2. Then, 10^6 is divided by 10^{-3}, which gives $10^{6-(-3)}$ or 10^9. Then,

$$\frac{7.2 \times 10^6}{3.6 \times 10^{-3}} = 2 \times 10^9$$

Some words are necessary concerning addition and subtraction of numbers expressed in scientific notation. The addition or subtraction of two numbers expressed in scientific notation is much easier if both numbers are expressed as the same power of 10. The decimal portions can then simply be added or subtracted.

As an example, consider adding 5×10^3 to 3×10^4. The second number, 3×10^4, is expressed in terms of 10^3 as 30×10^3 by removing one power of 10 from the power-of-ten portion and multiplying it by the decimal portion. Since both numbers are now expressed in terms of the same power of 10, their decimal portions can be simply added:

$$\begin{array}{r} 5 \times 10^3 \\ +30 \times 10^3 \\ \hline 35 \times 10^3 \end{array}$$

The result 35×10^3 could be alternatively written as 3.5×10^4.

The following problems apply to arithmetic using scientific notation.

(a) $(5 \times 10^3) \times (2 \times 10^6) =$

(b) $(3.9 \times 10^2) \times (2 \times 10^{-3}) =$

(c) $(2.2 \times 10^{-3}) \times (3 \times 10^6) =$

(d) $\dfrac{9.9 \times 10^9}{1.1 \times 10^2} =$

(e) $\dfrac{6 \times 10^6}{3 \times 10^3} =$

(f) $\dfrac{80 \times 10^5}{40 \times 10^2} =$

(g) $(6 \times 10^3) + (6 \times 10^4) =$

(h) $(3 \times 10^{-3}) + (4 \times 10^{-2}) =$

(i) $(8 \times 10^{-3}) - (5 \times 10^{-4}) =$

EQUATIONS

Even though no one has ever been attacked or injured by an equation or algebraic expression, considerable fear is nevertheless caused by equations and algebra. The purpose of this section is to help dispel that fear.

Algebra and equations are a way of expressing mathematical operations that are to be performed in general, rather than with specific numbers. For example, 2 plus 3 equals 5 can be written in the form of a mathematical expression as $2 + 3 = 5$. In a general form, the addition of any two numbers can be written as

$$x + y = z$$

where x and y stand for the addends, and z stands for the result of the addition. If x were 2 and y were 3, then z would be $2 + 3$, or 5.

Another example is the relationship between the period in seconds of a waveform and corresponding fundamental frequency in cycles per second. This reciprocal relationship can be expressed algebraically as

$$F = \frac{1}{T}$$

where T is the period and F is the fundamental frequency. If an actual period is 0.001 or 10^{-3} seconds, this equation states that the fundamental frequency is 1/0.001 or 1,000 cycles per second.

Sometimes equations contain an unknown quantity symbolized by x. The rules of algebra explain the way to solve these equations for x. These rules are many, and go beyond what will be needed as a fundamental knowledge of basic arithmetic.

Annotated Bibliography

The Editors of *Consumer Guide, The Illustrated Computer Dictionary,* Bantam Books (New York), 1983. Superb dictionary of technical terms and concepts related to computers and data communications. Very clear and concise.

Curran, Susan, and Ray Curnow, *Overcoming Computer Illiteracy: A Friendly Introduction to Computers,* Penguin Books (Middlesex, UK), 1983. Very understandable, readable, and thorough treatment of all aspects of computers, including data communications.

Glossbrenner, Alfred, *The Complete Handbook of Personal Computer Communications.* St. Martin's Press (New York), 1983. Very readable explanation of data communications. Considerable information on the broad variety of services available through data communications.

Jacobowitz, Henry, *Electricity Made Simple,* Doubleday and Company (New York), 1959. Extremely understandable explanations of electricity and basic dc and ac circuits, including resistance, capacitance, and inductance.

Martin, James, *Telecommunications and the Computer,* Prentice-Hall (Englewood Cliffs, NJ), 1976. Tremendous wealth of material presented in varying depths and clarity. Well respected in field. Good explanations of time-division and frequency-division multiplexing.

Mileaf, Harry (Editor), *Electronics One,* Hayden Book Company (Rochelle Park, NJ), 1976. Superb explanations and diagrams of signals, amplitude modulation, and frequency modulation.

Mileaf, Harry (Editor), *Electricity One-Seven,* Hayden Book Company (Rochelle Park, NJ), 1978. Excellent explanations and diagrams of basic electricity, including, dc, ac, circuits, capacitance, inductance, and measurements.

Mims, Forrest M., III, *Getting Started in Electronics,* Radio Shack (Ft. Worth, TX), 1983. Very readable and understandable introduction to electricity and transistor circuitry.

Noll, A. Michael, *Introduction to Telephones and Telephone Systems,* Second Edition, Artech House (Norwood, MA), 1991. Demystifies modern telephone systems, including coverage of station apparatus, transmission, switching, and signaling.

Noll, A. Michael, *Television Technology: Fundamentals and Future Prospects,* Artech House (Norwood, MA), 1988. Demystifies the basic principles of modern monochrome and color television, including coverage of VCRs, CATV, HDTV, and VBI.

Pierce, John R., and A. Michael Noll, *Signals: The Science of Telecommunications,* Scientific American Library (New York), 1990. The excitement of modern telecommunication technology is well captured in a clear and understandable fashion with many color drawings and illustrations.

About the Author

A. Michael Noll is a professor at the Annenberg School for Communication at the University of Southern California. He has held industry positions with AT&T, both in the Consumer Products and Marketing Group, and at Bell Labs. He has published over 65 papers and is the coauthor of *Signals* (Scientific American Library, 1991) and three books published by Artech House. He holds a Ph.D. in electrical engineering from the Polytechnic Institute of Brooklyn and is a senior member of the IEEE.

Index

The Artech House Telecommunications Library

Vinton G. Cerf, Series Editor

Access Networks: Technology and V5 Interfacing, Alex Gillespie

Advanced High-Frequency Radio Communications, Eric E. Johnson, Robert I. Desourdis, Jr., *et al.*

Advanced Technology for Road Transport: IVHS and ATT, Ian Catling, editor

Advances in Computer Systems Security, Vol. 3, Rein Turn, editor

Advances in Telecommunications Networks, William S. Lee and Derrick C. Brown

Advances in Transport Network Technologies: Photonics Networks, ATM, and SDH, Ken-ichi Sato

An Introduction to International Telecommunications Law, Charles H. Kennedy and M. Veronica Pastor

Asynchronous Transfer Mode Networks: Performance Issues, Second Edition, Raif O. Onvural

ATM Switches, Edwin R. Coover

ATM Switching Systems, Thomas M. Chen and Stephen S. Liu

Broadband: Business Services, Technologies, and Strategic Impact, David Wright

Broadband Network Analysis and Design, Daniel Minoli

Broadband Telecommunications Technology, Byeong Lee, Minho Kang and Jonghee Lee

Cellular Mobile Systems Engineering, Saleh Faruque

Cellular Radio: Analog and Digital Systems, Asha Mehrotra

Cellular Radio: Performance Engineering, Asha Mehrotra

Cellular Radio Systems, D. M. Balston and R. C. V. Macario, editors

CDMA for Wireless Personal Communications, Ramjee Prasad

Client/Server Computing: Architecture, Applications, and Distributed Systems Management, Bruce Elbert and Bobby Martyna

Communication and Computing for Distributed Multimedia Systems, Guojun Lu

Communications Technology Guide for Business, Richard Downey, et al.

Community Networks: Lessons from Blacksburg, Virginia, Andrew Cohill and Andrea Kavanaugh, editors

Computer Networks: Architecture, Protocols, and Software, John Y. Hsu

Computer Mediated Communications: Multimedia Applications, Rob Walters

Computer Telephony Integration, Second Edition, Rob Walters

Convolutional Coding: Fundamentals and Applications, Charles Lee

Corporate Networks: The Strategic Use of Telecommunications, Thomas Valovic

The Definitive Guide to Business Resumption Planning, Leo A. Wrobel

Desktop Encyclopedia of the Internet, Nathan J. Muller

Digital Beamforming in Wireless Communications, John Litva and Titus Kwok-Yeung Lo

Digital Cellular Radio, George Calhoun

Digital Hardware Testing: Transistor-Level Fault Modeling and Testing, Rochit Rajsuman, editor

Digital Switching Control Architectures, Giuseppe Fantauzzi

Digital Video Communications, Martyn J. Riley and Iain E. G. Richardson

Distributed Multimedia Through Broadband Communications Services, Daniel Minoli and Robert Keinath

Distance Learning Technology and Applications, Daniel Minoli

EDI Security, Control, and Audit, Albert J. Marcella and Sally Chen

Electronic Mail, Jacob Palme

Enterprise Networking: Fractional T1 to SONET, Frame Relay to BISDN, Daniel Minoli

Expert Systems Applications in Integrated Network Management, E. C. Ericson, L. T. Ericson, and D. Minoli, editors

FAX: Digital Facsimile Technology and Applications, Second Edition, Dennis Bodson, Kenneth McConnell, and Richard Schaphorst

FDDI and FDDI-II: Architecture, Protocols, and Performance, Bernhard Albert and Anura P. Jayasumana

Fiber Network Service Survivability, Tsong-Ho Wu

Future Codes: Essays in Advanced Computer Technology and the Law, Curtis E. A. Karnow

Guide to Telecommunications Transmission Systems, Anton A. Huurdeman

A Guide to the TCP/IP Protocol Suite, Floyd Wilder

Implementing EDI, Mike Hendry

Implementing X.400 and X.500: The PP and QUIPU Systems, Steve Kille

Inbound Call Centers: Design, Implementation, and Management, Robert A. Gable

Information Superhighways Revisited: The Economics of Multimedia, Bruce Egan

Integrated Broadband Networks, Amit Bhargava

International Telecommunications Management, Bruce R. Elbert

International Telecommunication Standards Organizations, Andrew Macpherson

Internetworking LANs: Operation, Design, and Management, Robert Davidson and Nathan Muller

Introduction to Document Image Processing Techniques, Ronald G. Matteson

Introduction to Error-Correcting Codes, Michael Purser

An Introduction to GSM, Siegmund Redl, Matthias K. Weber and Malcom W. Oliphant

Introduction to Radio Propagation for Fixed and Mobile Communications, John Doble

Introduction to Satellite Communication, Second Edition, Bruce R. Elbert

Introduction to T1/T3 Networking, Regis J. (Bud) Bates

Introduction to Telecommunications Network Engineering, Tarmo Anttalainen

Introduction to Telephones and Telephone Systems, Second Edition, A. Michael Noll

Introduction to X.400, Cemil Betanov

LAN, ATM, and LAN Emulation Technologies, Daniel Minoli and Anthony Alles

Land-Mobile Radio System Engineering, Garry C. Hess

LAN/WAN Optimization Techniques, Harrell Van Norman

LANs to WANs: Network Management in the 1990s, Nathan J. Muller and Robert P. Davidson

The Law and Regulation of Telecommunications Carriers, Henk Brands and Evan T. Leo

Minimum Risk Strategy for Acquiring Communications Equipment and Services, Nathan J. Muller

Mobile Antenna Systems Handbook, Kyohei Fujimoto and J. R. James, editors

Mobile Communications in the U.S. and Europe: Regulation, Technology, and Markets, Michael Paetsch

Mobile Data Communications Systems, Peter Wong and David Britland

Mobile Information Systems, John Walker

Networking Strategies for Information Technology, Bruce Elbert

Packet Switching Evolution from Narrowband to Broadband ISDN, M. Smouts

Packet Video: Modeling and Signal Processing, Naohisa Ohta

Performance Evaluation of Communication Networks, Gary N. Higginbottom

Personal Communication Networks: Practical Implementation, Alan Hadden

Personal Communication Systems and Technologies, John Gardiner and Barry West, editors

Practical Computer Network Security, Mike Hendry

Principles of Secure Communication Systems, Second Edition, Don J. Torrieri

Principles of Signaling for Cell Relay and Frame Relay, Daniel Minoli and George Dobrowski

Principles of Signals and Systems: Deterministic Signals, B. Picinbono

Private Telecommunication Networks, Bruce Elbert

Pulse Code Modulation Systems Design, William N. Waggener

Radio-Relay Systems, Anton A. Huurdeman

RF and Microwave Circuit Design for Wireless Communications, Lawrence E. Larson

The Satellite Communication Applications Handbook, Bruce R. Elbert

Secure Data Networking, Michael Purser

Service Management in Computing and Telecommunications, Richard Hallows

Signaling in ATM Networks, Raif O. Onvural, Rao Cherukuri

Smart Cards, José Manuel Otón and José Luis Zoreda

Smart Card Security and Applications, Mike Hendry

Smart Highways, Smart Cars, Richard Whelan

SNMP-Based ATM Network Management, Heng Pan

Successful Business Strategies Using Telecommunications Services, Martin F. Bartholomew

Super-High-Definition Images: Beyond HDTV, Naohisa Ohta, et al.

Telecommunications Deregulation, James Shaw

Television Technology: Fundamentals and Future Prospects, A. Michael Noll

Telecommunications Technology Handbook, Daniel Minoli

Telecommuting, Osman Eldib and Daniel Minoli

Telemetry Systems Design, Frank Carden

Teletraffic Technologies in ATM Networks, Hiroshi Saito

Toll-Free Services: A Complete Guide to Design, Implementation, and Management, Robert A. Gable

Transmission Networking: SONET and the SDH, Mike Sexton and Andy Reid

Troposcatter Radio Links, G. Roda

Understanding Emerging Network Services, Pricing, and Regulation, Leo A. Wrobel and Eddie M. Pope

Understanding GPS: Principles and Applications, Elliot D. Kaplan, editor

Understanding Networking Technology: Concepts, Terms and Trends, Mark Norris

UNIX Internetworking, Second Edition, Uday O. Pabrai

Videoconferencing and Videotelephony: Technology and Standards, Richard Schaphorst

Voice Recognition, Richard L. Klevans and Robert D. Rodman

Wireless Access and the Local Telephone Network, George Calhoun

Wireless Communications in Developing Countries: Cellular and Satellite Systems, Rachael E. Schwartz

Wireless Communications for Intelligent Transportation Systems, Scott D. Elliot and Daniel J. Dailey

Wireless Data Networking, Nathan J. Muller

Wireless LAN Systems, A. Santamaría and F. J. López-Hernández

Wireless: The Revolution in Personal Telecommunications, Ira Brodsky

World-Class Telecommunications Service Development, Ellen P. Ward

Writing Disaster Recovery Plans for Telecommunications Networks and LANs, Leo A. Wrobel

X Window System User's Guide, Uday O. Pabrai

For further information on these and other Artech House titles, including previously considered out-of-print books now available through our In-Print-Forever™ (IPF™) program, contact:

Artech House
685 Canton Street
Norwood, MA 02062
781-769-9750
Fax: 781-769-6334
Telex: 951-659
email: artech@artech-house.com

Artech House
Portland House, Stag Place
London SW1E 5XA England
+44 (0) 171-973-8077
Fax: +44 (0) 171-630-0166
Telex: 951-659
email: artech-uk@artech-house.com

Find us on the World Wide Web at:
www.artech-house.com